D0404091

STATISTICS

IN A NUTSHELL

Other resources from O'Reilly

Related titles

Baseball Hacks™
Head First Statistics

Programming Collective Intelligence
Statistics Hacks™

oreilly.com

oreilly.com is more than a complete catalog of O'Reilly books. You'll also find links to news, events, articles, weblogs, sample chapters, and code examples.

oreillynet.com is the essential portal for developers interested in open and emerging technologies, including new platforms, programming languages, and operating systems.

Conferences

O'Reilly brings diverse innovators together to nurture the ideas that spark revolutionary industries. We specialize in documenting the latest tools and systems, translating the innovator's knowledge into useful skills for those in the trenches. Visit *conferences.oreilly.com* for our upcoming events.

Safari Bookshelf (*safari.oreilly.com*) is the premier online reference library for programmers and IT professionals. Conduct searches across more than 1,000 books. Subscribers can zero in on answers to time-critical questions in a matter of seconds. Read the books on your Bookshelf from cover to cover or simply flip to the page you need. Try it today for free.

STATISTICS

IN A NUTSHELL

Sarah Boslaugh and Paul Andrew Watters

O'REILLY®

Beijing • Cambridge • Farnham • Köln • Sebastopol • Tokyo

Statistics in a Nutshell

by Sarah Boslaugh and Paul Andrew Watters

Published by O'Reilly Media, Inc., 1005 Gravenstein Highway North, Sebastopol, CA 95472.

O'Reilly books may be purchased for educational, business, or sales promotional use. Online editions are also available for most titles (*safari.oreilly.com*). For more information, contact our corporate/institutional sales department: (800) 998-9938 or *corporate@oreilly.com*.

Editor: Mary Treseler	**Indexer:** John Bickelhaupt
Production Editor: Sumita Mukherji	**Cover Designer:** Karen Montgomery
Copyeditor: Colleen Gorman	**Interior Designer:** David Futato
Proofreader: Emily Quill	**Illustrator:** Robert Romano

Printing History:

July 2008:	First Edition.

ISBN: 978-0-596-51049-7

[LSI]   [2011-01-27]

Table of Contents

Preface

One thing I (Sarah) have learned over the last 20 or so years is that a sure way to derail a promising conversation at a party is to tell people what I do for a living. And rest assured that I'm neither a tax auditor nor captain of a sludge barge. No, I'm merely a biostatistician and statistics instructor, a revelation which invariably provokes a response such as "statistics was my worst class in school" or the sudden inspiration to quote that old chestnut popularized by Mark Twain that there are three kinds of lies: lies, damned lies, and statistics.

Personally, I find statistics fascinating and I love working in this field. I like teaching statistics as well, and I like to believe that I communicate some of this enthusiasm to my students, most of whom are physicians or other healthcare professionals required to take my classes as part of their fellowship studies. It's often an uphill battle, however: some of them arrive with a negative attitude toward everything statistical, possibly augmented by the belief that statistics is some kind of magical procedure that will do their thinking for them, or a set of tricks and manipulations whose purpose is to twist reality in order to mislead other people.

I'm not sure how statistics got such a bad reputation, or why so many people have a negative attitude toward it. I do know that most of them can't afford it: the need to be competent in statistics is fast becoming a necessity in many fields of work. It's also becoming a requirement to be a thoughtful participant in modern society, as we are bombarded daily by statistical information and arguments, many of questionable merit. I have long since ceased to hope that I can keep everyone from misusing statistics: instead I have placed my hopes in cultivating a statistics-educated populace who will be able to recognize when statistics are being misused and discount the speaker's credibility accordingly. We (Sarah and Paul) have tried to address both concerns in this book: statistics as a professional necessity, and statistics as part of the intellectual content required for informed citizenship.

What Is Statistics?

Before we jump into the technical details of learning and using statistics, let's step back for a minute and consider what can be meant by the word "statistics." Don't worry if you don't understand all the vocabulary immediately: it will become clear over the course of this book.

When people speak of statistics, they usually mean one or more of the following:

1. Numerical data such as the unemployment rate, the number of persons who die annually from bee stings, or the racial makeup of the population of New York City in 2006 as compared to 1906.

2. Numbers used to describe samples (subsets) of data, such as the mean (average), as opposed to numbers used to describe populations (entire sets of data); for instance, if we work for an advertising firm interested in the average age of people who subscribe to *Sports Illustrated*, we can draw a sample of subscribers and calculate the mean of that sample (a statistic), which is an estimate of the mean of the entire population of subscribers.

3. Particular procedures used to analyze data, and the results of those procedures, such as the t statistic or the chi-square statistic.

4. A field of study that develops and uses mathematical procedures to describe data and make decisions regarding it.

The type of statistics referred to in definition #1 is not the primary concern of this book: if you simply want to find the latest figures on unemployment, health, or any of the myriad other topics on which governments and private organizations regularly release statistical data, your best bet is to consult a reference librarian or subject expert. If, however, you want to know how to interpret those figures (to understand why the mean is often misleading as a statement of average value, for instance, or the difference between crude and standardized mortality rates), *Statistics in a Nutshell* can definitely help you out.

The concepts included in definition #2 will be discussed in Chapter 7, which introduces inferential statistics, but they also permeate this book. It is partly a question of vocabulary (*statistics* are numbers that describe *samples*, while *parameters* are numbers that describe *populations*), but also underscores a fundamental point about the practice of statistics. The concept of using information gained from studying a sample to make statements about a population is the basis of inferential statistics, and inferential statistics is the primary focus of this book (as it is of most books about statistics).

Definition #3 is also fundamental to most chapters of this book. The process of learning statistics is to some extent the process of learning particular statistical procedures, including how to calculate and interpret them, how to choose the appropriate statistic for a given situation, and so on. In fact, many new students of statistics subscribe to this definition: learning statistics to them means learning to execute a set of statistical procedures. This is not an invalid approach to statistics so much as it is incomplete: learning to execute statistical procedures is a necessary part of the practice of statistics, but it is far from being the entire story. What's more, since computer software has made it increasingly easy for anyone, regardless of mathematical background, to produce statistical analyses, the need to understand and interpret statistics has far outstripped the need to learn how to do the calculations themselves.

Definition #4 is nearest to my heart, since I chose statistics as my professional field. If you are a secondary or post-secondary student you are probably aware of this definition of statistics, as many universities and colleges today either have a separate department of statistics or include statistics as a field of specialization within mathematics. Statistics is increasingly taught in high school as well: in the U.S., enrollment in the A.P. (Advanced Placement) Statistics classes is increasing more rapidly than enrollment in any other A.P. area.

Statistics is too important to be left to the statisticians, however, and university study in many subjects requires one or more semesters of statistics classes. Many basic techniques in modern statistics have been developed by people who learned and used statistics as part of their studies in another field. For instance, Stephen Raudenbush, a pioneer in the development of hierarchical linear modeling, studied Policy Analysis and Evaluation Research at Harvard, and Edward Tufte, perhaps the world's leading expert on statistical graphics, began his career as a political scientist: his Ph.D. dissertation at Yale was on the American Civil Rights Movement.

With the increasing use of statistics in many professions, and at all levels from top to bottom, basic knowledge of statistics has become a necessity for many people who have been out of school for years. Such individuals are often ill-served by textbooks aimed at introductory college courses, which are too specialized, too focused on calculation, and too expensive.

Finally, statistics cannot be left to the statisticians because it's also a necessity to understand much of what you read in the newspaper or hear on television and the radio. A working knowledge of statistics is the best check against the proliferation of misleading or outright false claims (whether by politicians, advertisers, or social reformers), which seem to occupy an ever-increasing portion of our daily news diet. There's a reason that Darryl Huff's 1954 classic *How to Lie with Statistics* (W.W. Norton) remains in print: statistics are easy to misuse, the common techniques of statistical distortion have been around for decades, and the best defense against those who would lie with statistics is to educate yourself so you can spot the lies and stop the lying liars in their tracks.

The Focus of This Book

There are so many statistics books already on the market that you might well wonder why we feel the need to add another to the pile. The primary reason is that we haven't found any statistics books that answer the needs we have addressed in *Statistics in a Nutshell*. In fact, if I may wax poetic for a moment, the situation is, to paraphrase the plight of Coleridge's Ancient Mariner, "books, books everywhere, nor any with which to learn." The issues we have tried to address with this book are:

1. The need for a book that focuses on using and understanding statistics in a research or applications context, not as a discrete set of mathematical techniques but as part of the process of reasoning with numbers.

2. The need to integrate discussion of issues such as measurement and data management into an introductory statistics text.

3. The need for a book that isn't focused on a particular subject area. Elementary statistics is largely the same across subjects (a *t*-test is pretty much the same whether the data comes from medicine, finance, or criminal justice), so there's no need for a proliferation of texts presenting the same information with slightly different spin.

4. The need for an introductory statistics book that is compact, inexpensive, and easy for beginners to understand without being condescending or overly simplistic.

So who is the intended audience of *Statistics in a Nutshell?* We see three in particular:

1. Students taking introductory statistics classes in high schools, colleges, and universities.

2. Adults who need to learn statistics as part of their current jobs or in order to be eligible for promotion.

3. People who are interested in learning about statistics out of intellectual curiosity.

Our focus throughout *Statistics in a Nutshell* is not on particular techniques, although many are taught within this work, but on statistical reasoning. You might say that our focus is not on *doing statistics*, but on *thinking statistically*. What does that mean? Several things are necessary in order to be able to focus on the process of thinking with numbers. More particularly, we focus on thinking about data, and using statistics to aid in that process.

Statistics in the Age of Information

It's become fashionable to say that we're living in the Age of Information, where so many facts are collected and disseminated that no one could possibly keep up with them. Well, this is one of those clichés that is based on truth: we are drowning in data and the problem is only going to get worse. Wide access to computing technology and electronic means of data storage and dissemination have made information easier to access, which is great from the researcher's point of view, since you no longer have to travel to a particular library or archive to peruse printed copies of records.

Whether your interest is the U.S. population in 1790, annual oil production and consumption in different countries, or the worldwide burden of disease, an Internet search will point you to data sources that can be accessed electronically, often directly from your home computer. However, data has no meaning in and of itself: it has to be organized and interpreted by human beings. So part of participating fully in the Information Age requires becoming fluent in understanding data, including the ways it is collected, analyzed, and interpreted. And because the same data can often be interpreted in many ways, to support radically different conclusions, even people who don't engage in statistical work themselves need to understand how statistics work and how to spot valid versus invalid claims, however solidly they may seem to be backed by numbers.

Organization of This Book

Statistics in a Nutshell is organized into four parts: introductory material (Chapters 1–6) that lays the necessary foundation for the chapters that follow; elementary inferential statistical techniques (Chapters 7–11); more advanced techniques (Chapters 12-16); and specialized techniques (Chapters 17–19).

Here's a more detailed breakdown of the chapters:

Chapter 1, *Basic Concepts of Measurement*
Discusses foundational issues for statistics, including levels of measurement, operationalization, proxy measurement, random and systematic error, measures of agreement, and types of bias. Statistics demonstrated include percent agreement and kappa.

Chapter 2, *Probability*
Introduces the basic vocabulary and laws of probability, including trials, events, independence, mutual exclusivity, the addition and multiplication laws, and conditional probability. Procedures demonstrated include calculation of basic probabilities, permutations and combinations, and Bayes's theorem.

Chapter 3, *Data Management*
Discusses practical issues in data management, including procedures to troubleshoot an existing file, methods for storing data electronically, data types, and missing data.

Chapter 4, *Descriptive Statistics and Graphics*
Explains the differences between descriptive and inferential statistics and between populations and samples, and introduces common measures of central tendency and variability and frequently used graphs and charts. Statistics demonstrated include mean, median, mode, range, interquartile range, variance, and standard deviation. Graphical methods demonstrated include frequency tables, bar charts, pie charts, Pareto charts, stem and leaf plots, boxplots, histograms, scatterplots, and line graphs.

Chapter 5, *Research Design*
Discusses observational and experimental studies, common elements of good research designs, the steps involved in data collection, types of validity, and methods to limit or eliminate the influence of bias.

Chapter 6, *Critiquing Statistics Presented by Others*
Offers guidelines for reviewing the use of statistics, including a checklist of questions to ask of any statistical presentation and examples of when legitimate statistical procedures may be manipulated to appear to support questionable conclusions.

Chapter 7, *Inferential Statistics*
Introduces the basic concepts of inferential statistics, including probability distributions, independent and dependent variables and the different names under which they are known, common sampling designs, the central limit theorem, hypothesis testing, Type I and Type II error, confidence intervals and p-values, and data transformation. Procedures demonstrated include

converting raw scores to Z-scores, calculation of binomial probabilities, and the square-root and log data transformations.

Chapter 8, *The t-Test*

Discusses the *t*-distribution, the different types of *t*-tests, and the influence of effect size on power in *t*-tests. Statistics demonstrated include the one-sample *t*-test, the two independent samples *t*-test, the two repeated measures *t*-test, and the unequal variance *t*-test.

Chapter 9, *The Correlation Coefficient*

Introduces the concept of association with graphics displaying different strengths of association between two variables, and discusses common statistics used to measure association. Statistics demonstrated include Pearson's product-moment correlation, the *t*-test for statistical significance of Pearson's correlation, the coefficient of determination, Spearman's rank-order coefficient, the point-biserial coefficient, and phi.

Chapter 10, *Categorical Data*

Reviews the concepts of categorical and interval data, including the Likert scale, and introduces the R×C table. Statistics demonstrated include the chi-squared tests for independence, equality of proportions, and goodness of fit, Fisher's exact test, McNemar's test, gamma, Kendall's tau-a, tau-b, and tau-c, and Somers's d.

Chapter 11, *Nonparametric Statistics*

Discusses when to use nonparametric rather than parametric statistics, and presents nonparametric statistics for between-subjects and within-subjects designs. Statistics demonstrated include the Wilcoxon Rank Sum and Mann-Whitney U tests, the median test, the Kruskal-Wallis H test, the Wilcoxon matched pairs signed rank test, and the Friedman test.

Chapter 12, *Introduction to the General Linear Model*

Introduces linear regression and ANOVA through the concept of the General Linear Model, and discusses assumptions made when using these designs. Statistical procedures demonstrated include simple (bivariate) regression, one-way ANOVA, and post-hoc testing.

Chapter 13, *Extensions of Analysis of Variance*

Discusses more complex ANOVA designs. Statistical procedures demonstrated include two-way and three-way ANOVA, MANOVA, ANCOVA, repeated measures ANOVA, and mixed designs.

Chapter 14, *Multiple Linear Regression*

Extends the ideas introduced in Chapter 12 to models with multiple predictors. Topics covered include relationships among predictor variables, standardized coefficients, dummy variables, methods of model building, and violations of assumptions of linear regression, including nonlinearity, autocorrelation, and heteroscedasticity.

Chapter 15, *Other Types of Regression*

Extends the technique of regression to data with binary outcomes (logistic regression) and nonlinear models (polynomial regression), and discusses the problem of overfitting a model.

Chapter 16, *Other Statistical Techniques*

Demonstrates several advanced statistical procedures, including factor analysis, cluster analysis, discriminant function analysis, and multidimensional scaling, including discussion of the types of problems for which each technique may be useful.

Chapter 17, *Business and Quality Improvement Statistics*

Demonstrates statistical procedures commonly used in business and quality improvement contexts. Analytical and statistical procedures covered include construction and use of simple and composite indexes, time series, the minimax, maximax, and maximin decision criteria, decision making under risk, decision trees, and control charts.

Chapter 18, *Medical and Epidemiological Statistics*

Introduces concepts and demonstrates statistical procedures particularly relevant to medicine and epidemiology. Concepts and statistics covered include the definition and use of ratios, proportions, and rates, measures of prevalence and incidence, crude and standardized rates, direct and indirect standardization, measures of risk, confounding, the simple and Mantel-Haenszel odds ratio, and precision, power, and sample size calculations.

Chapter 19, *Educational and Psychological Statistics*

Introduces concepts and statistical procedures commonly used in the fields of education and psychology. Concepts and procedures demonstrated include percentiles, standardized scores, methods of test construction, the true score model of classical test theory, reliability of a composite test, measures of internal consistency including coefficient alpha, and procedures for item analysis. An overview of item response theory is also provided

Two appendixes cover topics that are a necessary background to the material covered in the main text, and a third provides references to supplemental reading:

Appendix A

Provides a self-test and review of basic arithmetic and algebra for people whose memory of their last math course is fast receding on the distant horizon. Topics covered include the laws of arithmetic, exponents, roots and logs, methods to solve equations and systems of equations, fractions, factorials, permutations, and combinations.

Appendix B

Provides an introduction to some of the most common computer programs used for statistical applications, demonstrates basic analyses in each program, and discusses their relative strengths and weaknesses. Programs covered include Minitab, SPSS, SAS, and R; the use of Microsoft Excel (not a statistical package) for statistical analysis is also discussed.

Appendix C

An annotated bibliography organized by chapter, which includes published works and websites cited in the text and others that are good starting points for people researching a particular topic.

You should think of these chapters as tools, whose best use depends on the individual reader's, background and needs. Even the introductory chapters may not be relevant immediately to everyone: for instance, many introductory statistics classes do not require students to master topics such as data management or measurement theory. In that case, these chapters can serve as references when the topics become necessary (expertise in data management is often an expectation of research assistants, for instance, although it is rarely directly taught).

Classification of what is "elementary" and what is "advanced" depends on an individual's background and purposes. We designed *Statistics in a Nutshell* to answer the needs of many different types of users. For this reason, there's no perfect way to organize the material to meet everyone's needs, which brings us to an important point: there's no reason you should feel the need to read the chapters in the order they are presented here. Statistics presents many chicken-and-egg dilemmas: for instance, you can't design experiments without knowing what statistics are available to you, but you can't understand how statistics are used without knowing something about research design. Similarly, it might seem that a chapter on data management would be most useful to individuals who have already done some statistical analysis, but I've advised many research assistants and project managers who are put in charge of large data sets before they've had a single course in statistics. So use the chapters in the way that best facilitates your specific purposes, and don't be shy about skipping around and focusing on whatever meets your particular needs.

Some of the later chapters are also specialized and not relevant to everyone, most obviously Chapters 17–19, which are written with particular subject areas in mind. Chapters 15 and 16 also cover topics that are not often included in introductory statistics texts, but that are the statistical procedure of choice in particular contexts. Because we have planned this book to be useful for consumers of statistics and working professionals who deal with statistics even if they don't compute them themselves, we have included these topics, although beginning students may not feel the need to tackle them in their first statistics course.

It's wise to keep an open mind regarding what statistics you need to know. You may currently believe that you will never have the need to conduct a nonparametric test or a logistic regression analysis. However, you never know what will come in handy in the future. It's also a mistake to compartmentalize too much by subject field: because statistical techniques are ultimately about numbers rather than content, techniques developed in one field often prove to be useful in another. For instance, control charts (covered in Chapter 17) were developed in a manufacturing context, but are now used in many fields from medicine to education.

We have included more advanced material in other chapters, when it serves to illustrate a principle or make an interesting point. These sections are clearly identified as digressions from the main thread of the book, and beginners can skip over them without feeling that they are missing any vital concepts of basic statistics.

Occasionally in the text we refer to tables of statistical probabilities (for instance, for the t-statistic and standard normal distribution) which are commonly available on the internet as well as in many reference books. Here are a few locations of collections of such tables: *http://www.itl.nist.gov/div898/handbook/eda/section3/eda367.htm; http://www.math.unb.ca/~knight/utility/; http://www.stat.lsu.edu/exstweb/statlab/TABLES98-contents.html.*

Symbols Used in This Book

Symbol	Meaning
Names of statistics	
μ	Mean of a population
σ	Standard deviation of a population
σ^2	Variance of a population
Π	Proportion of a population
$\bar{x}$	Mean of a sample
s	Standard deviation of a sample
s^2	Variance of a sample
n	Number of cases in a sample
p	Proportion of a sample
K	Kappa (measure of agreement)
χ^2	Chi-squared (statistic, distribution)
Statistical formulas	
Σ	Summation
$!$	Factorial
C	Combination
P	Permutation
E	Expected value
O	Observed value
x_{ij}	Value of variable x for case ij
Set theory, Bayes Theorem	
$\sim$	Not
$\vert$	Conditional probability
$\cup$	Union
$\cap$	Intersection
Other	
α	Alpha (significance level; probability of Type I error)
β	Beta (probability of Type II error)
R	Number of rows in a table
C	Number of columns in a table

Conventions Used in This Book

The following typographical conventions are used in this book:

Plain text
> Indicates menu titles, menu options, menu buttons, and keyboard accelerators (such as Alt and Ctrl).

Italic

> Indicates new terms, URLs, email addresses, filenames, file extensions, path-names, directories, and Unix utilities

`Constant width`

> Indicates examples

 This icon signifies a tip, suggestion, or general note.

We'd Like to Hear From You

Please address comments and questions concerning this book to the publisher:

O'Reilly Media, Inc.
1005 Gravenstein Highway North
Sebastopol, CA 95472
800-998-9938 (in the United States or Canada)
707-829-0515 (international or local)
707-829-0104 (fax)

We have a web page for this book, where we list errata, examples, and any additional information. You can access this page at:

http://www.oreilly.com/catalog/9780596510497

To comment or ask technical questions about this book, send email to:

bookquestions@oreilly.com

For more information about our books, conferences, Resource Centers, and the O'Reilly Network, see our website at:

http://www.oreilly.com

Safari® Books Online

 When you see a Safari® Books Online icon on the cover of your favorite technology book, that means the book is available online through the O'Reilly Network Safari Bookshelf.

Safari offers a solution that's better than e-books. It's a virtual library that lets you easily search thousands of top tech books, cut and paste code samples, download chapters, and find quick answers when you need the most accurate, current information. Try it for free at *http://safari.oreilly.com*.

Acknowledgments

Only two authors are listed on the cover of this book, but the contributions of many people played a role in its creation.

Sarah Boslaugh

I would like to thank my agent, Neil Salkind, for his continued guidance and support; my colleagues at Washington University and BJC HealthCare for their willingness to share their wisdom and experience; the crew at O'Reilly, including Mary Treseler, Isabel Kunkle, Rachel Monaghan, and Colleen Gorman; and the statisticians who assisted in the technical review process, especially Dave McArthur at UCLA who is never shy about sharing his suggestions. I would also like to thank all my friends who keep pestering me to explain statistical concepts to them, and thus encouraged me to write this book. On a personal note, I would like to thank my colleague Rand Ross at Washington University for helping me remain sane throughout the writing process, and my husband Dan Peck for being the very model of a modern supportive spouse.

Paul Watters

Firstly, I would like to thank the academics who managed to make learning statistics interesting: Professor Rachel Heath (University of Newcastle) and Mr. James Alexander (University of Tasmania). An inspirational teacher is a rare and wonderful thing, especially in statistics! Secondly, a big thank you to my colleagues at the School of ITMS at the University of Ballarat, and our partners at Westpac, IBM, and the Victorian government, for their ongoing research support. Finally, I would like to acknowledge the patience of my wife Maya, and daughters Arwen and Bounty, as writing a book invariably takes away time from family.

Basic Concepts of Measurement

Before you can use statistics to analyze a problem, you must convert the basic materials of the problem to data. That is, you must establish or adopt a system of assigning values, most often numbers, to the objects or concepts that are central to the problem under study. This is not an esoteric process, but something you do every day. For instance, when you buy something at the store, the price you pay is a measurement: it assigns a number to the amount of currency that you have exchanged for the goods received. Similarly, when you step on the bathroom scale in the morning, the number you see is a measurement of your body weight. Depending on where you live, this number may be expressed in either pounds or kilograms, but the principle of assigning a number to a physical quantity (weight) holds true in either case.

Not all data need be numeric. For instance, the categories *male* and *female* are commonly used in both science and in everyday life to classify people, and there is nothing inherently numeric in these categories. Similarly, we often speak of the colors of objects in broad classes such as "red" or "blue": these categories of which represent a great simplification from the infinite variety of colors that exist in the world. This is such a common practice that we hardly give it a second thought.

How specific we want to be with these categories (for instance, is "garnet" a separate color from "red"? Should transgendered individuals be assigned to a separate category?) depends on the purpose at hand: a graphic artist may use many more mental categories for color than the average person, for instance. Similarly, the level of detail used in classification for a study depends on the purpose of the study and the importance of capturing the nuances of each variable.

Measurement

Measurement is the process of systematically assigning numbers to objects and their properties, to facilitate the use of mathematics in studying and describing objects and their relationships. Some types of measurement are fairly concrete: for instance, measuring a person's weight in pounds or kilograms, or their height in feet and inches or in meters. Note that the particular system of measurement used is not as important as a consistent set of rules: we can easily convert measurement in kilograms to pounds, for instance. Although any system of units may seem arbitrary (try defending feet and inches to someone who grew up with the metric system!), as long as the system has a consistent relationship with the property being measured, we can use the results in calculations.

Measurement is not limited to physical qualities like height and weight. Tests to measure abstractions like intelligence and scholastic aptitude are commonly used in education and psychology, for instance: the field of psychometrics is largely concerned with the development and refinement of methods to test just such abstract qualities. Establishing that a particular measurement is meaningful is more difficult when it can't be observed directly: while you can test the accuracy of a scale by comparing the results with those obtained from another scale known to be accurate, there is no simple way to know if a test of intelligence is accurate because there is no commonly agreed-upon way to measure the abstraction "intelligence." To put it another way, we don't know what someone's actual intelligence is because there is no certain way to measure it, and in fact we may not even be sure what "intelligence" really is, a situation quite different from that of measuring a person's height or weight. These issues are particularly relevant to the social sciences and education, where a great deal of research focuses on just such abstract concepts.

Levels of Measurement

Statisticians commonly distinguish four types or levels of measurement; the same terms may also be used to refer to data measured at each level. The levels of measurement differ both in terms of the meaning of the numbers and in the types of statistics that are appropriate for their analysis.

Nominal Data

With *nominal* data, as the name implies, the numbers function as a *name* or label and do not have numeric meaning. For instance, you might create a variable for gender, which takes the value 1 if the person is male and 0 if the person is female. The 0 and 1 have no numeric meaning but function simply as labels in the same way that you might record the values as "M" or "F." There are two main reasons to choose numeric rather than text values to code nominal data: data is more easily processed by some computer systems as numbers, and using numbers bypasses some issues in data entry such as the conflict between upper- and lower-case letters (to a computer, "M" is a different value than "m," but a person doing data entry may treat the two characters as equivalent). Nominal data is not limited to two categories: for instance, if you were studying the relationship between years of experience and salary in baseball players, you might classify the players according to their primary position by using the traditional system whereby 1 is assigned to pitchers, 2 to catchers, 3 to first basemen, and so on.

If you can't decide whether data is nominal or some other level of measurement, ask yourself this question: do the numbers assigned to this data represent some quality such that a higher value indicates that the object has more of that quality than a lower value? For instance, is there some quality "gender" which men have more than women? Clearly not, and the coding scheme would work as well if women were coded as 1 and men as 0. The same principle applies in the baseball example: there is no quality of "baseballness" of which outfielders have more than pitchers. The numbers are merely a convenient way to label subjects in the study, and the most important point is that every position is assigned a distinct value. Another name for nominal data is *categorical* data, referring to the fact that the measurements place objects into categories (male or female; catcher or first baseman) rather than measuring some intrinsic quality in them. Chapter 10 discusses methods of analysis appropriate for this type of data, and many techniques covered in Chapter 11, on nonparametric statistics, are also appropriate for categorical data.

When data can take on only two values, as in the male/female example, it may also be called *binary* data. This type of data is so common that special techniques have been developed to study it, including logistic regression (discussed in Chapter 15), which has applications in many fields. Many medical statistics such as the odds ratio and the risk ratio (discussed in Chapter 18) were developed to describe the relationship between two binary variables, because binary variables occur so frequently in medical research.

Ordinal Data

Ordinal data refers to data that has some meaningful *order*, so that higher values represent more of some characteristic than lower values. For instance, in medical practice burns are commonly described by their degree, which describes the amount of tissue damage caused by the burn. A first-degree burn is characterized by redness of the skin, minor pain, and damage to the epidermis only, while a second-degree burn includes blistering and involves the dermis, and a third-degree burn is characterized by charring of the skin and possibly destroyed nerve endings. These categories may be ranked in a logical order: first-degree burns are the least serious in terms of tissue damage, third-degree burns the most serious. However, there is no metric analogous to a ruler or scale to quantify how great the distance between categories is, nor is it possible to determine if the difference between first- and second-degree burns is the same as the difference between second- and third-degree burns.

Many ordinal scales involve ranks: for instance, candidates applying for a job may be ranked by the personnel department in order of desirability as a new hire. We could also rank the U.S. states in order of their population, geographic area, or federal tax revenue. The numbers used for measurement with ordinal data carry more meaning than those used in nominal data, and many statistical techniques have been developed to make full use of the information carried in the ordering, while not assuming any further properties of the scales. For instance, it is appropriate to calculate the median (central value) of ordinal data, but not the mean (which assumes interval data). Some of these techniques are discussed later in this chapter, and others are covered in Chapter 11.

Interval Data

Interval data has a meaningful order and also has the quality that equal intervals between measurements represent equal changes in the quantity of whatever is being measured. The most common example of interval data is the Fahrenheit temperature scale. If we describe temperature using the Fahrenheit scale, the difference between 10 degrees and 25 degrees (a difference of 15 degrees) represents the same amount of temperature change as the difference between 60 and 75 degrees. Addition and subtraction are appropriate with interval scales: a difference of 10 degrees represents the same amount over the entire scale of temperature. However, the Fahrenheit scale, like all interval scales, has no natural zero point, because 0 on the Fahrenheit scale does not represent an absence of temperature but simply a location relative to other temperatures. Multiplication and division are not appropriate with interval data: there is no mathematical sense in the statement that 80 degrees is twice as hot as 40 degrees. Interval scales are a rarity: in fact it's difficult to think of another common example. For this reason, the term "interval data" is sometimes used to describe both interval and ratio data (discussed in the next section).

Ratio Data

Ratio data has all the qualities of interval data (natural order, equal intervals) plus a natural zero point. Many physical measurements are ratio data: for instance, height, weight, and age all qualify. So does income: you can certainly earn 0 dollars in a year, or have 0 dollars in your bank account. With ratio-level data, it is appropriate to multiply and divide as well as add and subtract: it makes sense to say that someone with $100 has twice as much money as someone with $50, or that a person who is 30 years old is 3 times as old as someone who is 10 years old.

It should be noted that very few psychological measurements (IQ, aptitude, etc.) are truly interval, and many are in fact ordinal (e.g., value placed on education, as indicated by a Likert scale). Nonetheless, you will sometimes see interval or ratio techniques applied to such data (for instance, the calculation of means, which involves division). While incorrect from a statistical point of view, sometimes you have to go with the conventions of your field, or at least be aware of them. To put it another way, part of learning statistics is learning what is commonly accepted in your chosen field of endeavor, which may be a separate issue from what is acceptable from a purely mathematical standpoint.

Continuous and Discrete Data

Another distinction often made is that between *continuous* and *discrete* data. Continuous data can take any value, or any value within a range. Most data measured by interval and ratio scales, other than that based on counting, is continuous: for instance, weight, height, distance, and income are all continuous.

In the course of data analysis and model building, researchers sometimes recode continuous data in categories or larger units. For instance, weight may be recorded in pounds but analyzed in 10-pound increments, or age recorded in years but analyzed in terms of the categories *0–17, 18–65,* and *over 65.* From a statistical point of view, there is no absolute point when data become continuous or discrete for the purposes of using particular analytic techniques: if we record age in years, we are still imposing discrete categories on a continuous variable. Various rules of thumb have been proposed: for instance, some researchers say that when a variable has 10 or more categories (or alternately, 16 or more categories), it can safely be analyzed as continuous. This is another decision to be made on a case-by-case basis, informed by the usual standards and practices of your particular discipline and the type of analysis proposed.

Discrete data can only take on particular values, and has clear boundaries. As the old joke goes, you can have 2 children or 3 children, but not 2.37 children, so "number of children" is a discrete variable. In fact, any variable based on counting is discrete, whether you are counting the number of books purchased in a year or the number of prenatal care visits made during a pregnancy. Nominal data is also discrete, as are binary and rank-ordered data.

Operationalization

Beginners to a field often think that the difficulties of research rest primarily in statistical analysis, and focus their efforts on learning mathematical formulas and computer programming techniques in order to carry out statistical calculations. However, one major problem in research has very little to do with either mathematics or statistics, and everything to do with knowing your field of study and thinking carefully through practical problems. This is the problem of *operationalization*, which means the process of specifying how a concept will be defined and measured. Operationalization is a particular concern in the social sciences and education, but applies to other fields as well.

Operationalization is always necessary when a quality of interest cannot be measured directly. An obvious example is intelligence: there is no way to measure intelligence directly, so in the place of such a direct measurement we accept something that we can measure, such as the score on an IQ test. Similarly, there is no direct way to measure "disaster preparedness" for a city, but we can operationalize the concept by creating a checklist of tasks that should be performed and giving each city a "disaster preparedness" score based on the number of tasks completed and the quality or thoroughness of completion. For a third example, we may wish to measure the amount of physical activity performed by subjects in a study: if we do not have the capacity to directly monitor their exercise behavior, we may operationalize "amount of physical activity" as the amount indicated on a self-reported questionnaire or recorded in a diary.

Because many of the qualities studied in the social sciences are abstract, operationalization is a common topic of discussion in those fields. However, it is applicable to many other fields as well. For instance, the ultimate goals of the medical profession include reducing mortality (death) and reducing the burden of disease and suffering. Mortality is easily verified and quantified but is frequently too blunt an instrument to be useful, since it is a thankfully rare outcome for most diseases. "Burden of disease" and "suffering," on the other hand, are concepts that could be used to define appropriate outcomes for many studies, but that have no direct means of measurement and must therefore be operationalized. Examples of operationalization of burden of disease include measurement of viral levels in the bloodstream for patients with AIDS and measurement of tumor size for people with cancer. Decreased levels of suffering or improved quality of life may be operationalized as higher self-reported health state, higher score on a survey instrument designed to measure quality of life, improved mood state as measured through a personal interview, or reduction in the amount of morphine requested.

Some argue that measurement of even physical quantities such as length require operationalization, because there are different ways to measure length (a ruler might be the appropriate instrument in some circumstances, a micrometer in others). However, the problem of operationalization is much greater in the human sciences, when the object or qualities of interest often cannot be measured directly.

Proxy Measurement

The term *proxy measurement* refers to the process of substituting one measurement for another. Although deciding on proxy measurements can be considered as a subclass of operationalization, we will consider it as a separate topic. The most common use of proxy measurement is that of substituting a measurement that is inexpensive and easily obtainable for a different measurement that would be more difficult or costly, if not impossible, to collect.

For a simple example of proxy measurement, consider some of the methods used by police officers to evaluate the sobriety of individuals while in the field. Lacking a portable medical lab, an officer can't directly measure blood alcohol content to determine if a subject is legally drunk or not. So the officer relies on observation of signs associated with drunkenness, as well as some simple field tests that are believed to correlate well with blood alcohol content. Signs of alcohol intoxication include breath smelling of alcohol, slurred speech, and flushed skin. Field tests used to quickly evaluate alcohol intoxication generally require the subjects to perform tasks such as standing on one leg or tracking a moving object with their eyes. Neither the observed signs nor the performance measures are direct measures of inebriation, but they are quick and easy to administer in the field. Individuals suspected of drunkenness as evaluated by these proxy measures may then be subjected to more accurate testing of their blood alcohol content.

Another common (and sometimes controversial) use of proxy measurement are the various methods commonly used to evaluate the quality of health care provided by hospitals or physicians. Theoretically, it would be possible to get a direct measure of quality of care, for instance by directly observing the care provided and evaluating it in relationship to accepted standards (although that process would still be an operationalization of the abstract concept "quality of care"). However, implementing such a process would be prohibitively expensive as well as an invasion of the patients' privacy. A solution commonly adopted is to measure processes that are assumed to reflect higher quality of care: for instance whether anti-tobacco counseling was offered in an office visit or whether appropriate medications were administered promptly after a patient was admitted to the hospital.

Proxy measurements are most useful if, in addition to being relatively easy to obtain, they are good indicators of the true focus of interest. For instance, if correct execution of prescribed processes of medical care for a particular treatment is closely related to good patient outcomes for that condition, and if poor or nonexistent execution of those processes is closely related to poor patient outcomes, then execution of these processes is a useful proxy for quality. If that close relationship does not exist, then the usefulness of measurements of those processes as a proxy for quality of care is less certain. There is no mathematical test that will tell you whether one measure is a good proxy for another, although computing statistics like correlations or chi-squares between the measures may help evaluate this issue. Like many measurement issues, choosing good proxy measurements is a matter of judgment informed by knowledge of the subject area, usual practices in the field, and common sense.

True and Error Scores

We can safely assume that no measurement is completely accurate. Because the process of measurement involves assigning discrete numbers to a continuous world, even measurements conducted by the best-trained staff using the finest available scientific instruments are not completely without error. One concern of measurement theory is conceptualizing and quantifying the degree of error present in a particular set of measurements, and evaluating the sources and consequences of that error.

Classical measurement theory conceives of any measurement or observed score as consisting of two parts: true score, and error. This is expressed in the following formula:

$$X = T + E$$

where X is the observed measurement, T is the true score, and E is the error. For instance, the bathroom scale might measure someone's weight as 120 pounds, when that person's true weight was 118 pounds and the error of 2 pounds was due to the inaccuracy of the scale. This would be expressed mathematically as:

$$120 = 118 + 2$$

which is simply a mathematical equality expressing the relationship between the three components. However, both T and E are hypothetical constructs: in the real world, we never know the precise value of the true score and therefore cannot know the value of the error score, either. Much of the process of measurement involves estimating both quantities and maximizing the true component while minimizing error. For instance, if we took a number of measurements of body weight in a short period of time (so that true weight could be assumed to have remained constant), using the most accurate scales available, we might accept the average of all the measurements as a good estimate of true weight. We would then consider the variance between this average and each individual measurement as the error due to the measurement process, such as slight inaccuracies in each scale.

Random and Systematic Error

Because we live in the real world rather than a Platonic universe, we assume that all measurements contain some error. But not all error is created equal. *Random error* is due to chance: it takes no particular pattern and is assumed to cancel itself out over repeated measurements. For instance, the error scores over a number of measurements of the same object are assumed to have a mean of zero. So if someone is weighed 10 times in succession on the same scale, we may observe slight differences in the number returned to us: some will be higher than the true value, and some will be lower. Assuming the true weight is 120 pounds, perhaps the first measurement will return an observed weight of 119 pounds (including an error of −1 pound), the second an observed weight of 122 pounds (for an error of +2 pounds), the third an observed weight of 118.5 pounds (an error of −1.5 pounds) and so on. If the scale is accurate and the only error is random, the average error over many trials will be zero, and the average observed weight will be 120 pounds. We can strive to reduce the amount of random error by using more accurate instruments, training our technicians to use them correctly, and so on, but we cannot expect to eliminate random error entirely.

Two other conditions are assumed to apply to random error: it must be unrelated to the true score, and the correlation between errors is assumed to be zero. The first condition means that the value of the error component is not related to the value of the true score. If we measured the weights of a number of different individuals whose true weights differed, we would not expect the error component to have any relationship to their true weights. For instance, the error component should not systematically be larger when the true weight is larger. The second condition means that the error for each score is independent and unrelated to the error for any other score: for instance, there should not be a pattern of the size of error increasing over time (which might indicate that the scale was drifting out of calibration).

In contrast, *systematic error* has an observable pattern, is not due to chance, and often has a cause or causes that can be identified and remedied. For instance, the scale might be incorrectly calibrated to show a result that is five pounds over the true weight, so the average of the above measurements would be 125 pounds, not 120. Systematic error can also be due to human factors: perhaps we are reading the scale's display at an angle so that we see the needle as registering five pounds higher than it is truly indicating. A scale drifting higher (so the error components are random at the beginning of the experiment, but later on are consistently high) is another example of systematic error. A great deal of effort has been expended to identify sources of systematic error and devise methods to identify and eliminate them: this is discussed further in the upcoming section on measurement bias.

Reliability and Validity

There are many ways to assign numbers or categories to data, and not all are equally useful. Two standards we use to evaluate measurements are *reliability* and *validity*. Ideally, every measure we use should be both reliable and valid. In reality, these qualities are not absolutes but are matters of degree and often specific to circumstance: a measure that is highly reliable when used with one group of people may be unreliable when used with a different group, for instance. For this reason it is more useful to evaluate how valid and reliable a measure is for a particular purpose and whether the levels of reliability and validity are acceptable in the context at hand. Reliability and validity are also discussed in Chapter 5, in the context of research design, and in Chapter 19, in the context of educational and psychological testing.

Reliability

Reliability refers to how consistent or repeatable measurements are. For instance, if we give the same person the same test on two different occasions, will the scores be similar on both occasions? If we train three people to use a rating scale designed to measure the quality of social interaction among individuals, then showed each of them the same film of a group of people interacting and asked them to evaluate the social interaction exhibited in the film, will their ratings be similar? If we have a technician measure the same part 10 times, using the same instrument, will the measurements be similar each time? In each case, if the answer is yes, we can say the test, scale, or instrument is reliable.

Much of the theory and practice of reliability was developed in the field of educational psychology, and for this reason, measures of reliability are often described in terms of evaluating the reliability of tests. But considerations of reliability are not limited to educational testing: the same concepts apply to many other types of measurements including opinion polling, satisfaction surveys, and behavioral ratings.

The discussion in this chapter will be kept at a fairly basic level: information about calculating specific measures of reliability are discussed in more detail in Chapter 19, in connection with test theory. In addition, many of the measures of reliability draw on the *correlation coefficient* (also called simply the *correlation*), which is discussed in detail in Chapter 9, so beginning statisticians may want to concentrate on the logic of reliability and validity and leave the details of evaluating them until after they have mastered the concept of the correlation coefficient.

There are three primary approaches to measuring reliability, each useful in particular contexts and each having particular advantages and disadvantages:

- Multiple-occasions reliability
- Multiple-forms reliability
- Internal consistency reliability

Multiple-occasions reliability, sometimes called *test-retest reliability*, refers to how similarly a test or scale performs over repeated testings. For this reason it is sometimes referred to as an index of *temporal stability*, meaning stability over time. For instance, we might have the same person do a psychological assessment of a patient based on a videotaped interview, with the assessments performed two weeks apart based on the same taped interview. For this type of reliability to make sense, you must assume that the quantity being measured has not changed: hence the use of the same videotaped interview, rather than separate live interviews with a patient whose state may have changed over the two-week period. Multiple-occasions reliability is not a suitable measure for volatile qualities, such as mood state. It is also unsuitable if the focus of measurement may have changed over the time period between tests (for instance, if the student learned more about a subject between the testing periods) or may be changed as a result of the first testing (for instance, if a student remembers what questions were asked on the first test administration). A common technique for assessing multiple-occasions reliability is to compute the correlation coefficient between the scores from each occasion of testing: this is called the *coefficient of stability*.

Multiple-forms reliability (also called *parallel-forms reliability*) refers to how similarly different versions of a test or questionnaire perform in measuring the same entity. A common type of multiple forms reliability is *split-half reliability*, in which a pool of items believed to be homogeneous is created and half the items are allocated to form A and half to form B. If the two (or more) forms of the test are administered to the same people on the same occasion, the correlation between the scores received on each form is an estimate of multiple-forms reliability. This correlation is sometimes called the *coefficient of equivalence*. Multiple-forms reliability is important for standardized tests that exist in multiple versions: for instance, different forms of the SAT (Scholastic Aptitude Test, used to measure academic ability among students applying to American colleges and universities) are calibrated so the scores achieved are equivalent no matter which form is used.

Internal consistency reliability refers to how well the items that make up a test reflect the same construct. To put it another way, internal consistency reliability measures how much the items on a test are measuring the same thing. This type of reliability may be assessed by administering a single test on a single occasion. Internal consistency reliability is a more complex quantity to measure than multiple-occasions or parallel-forms reliability, and several different methods have been developed to evaluate it: these are further discussed in Chapter 19. However, all depend primarily on the inter-item correlation, i.e., the correlation of each item on the scale with each other item. If such correlations are high, that is interpreted as evidence that the items are measuring the same thing and the various statistics used to measure internal consistency reliability will all be high. If the inter-item correlations are low or inconsistent, the internal consistency reliability statistics will be low and this is interpreted as evidence that the items are not measuring the same thing.

Two simple measures of internal consistency that are most useful for tests made up of multiple items covering the same topic, of similar difficulty, and that will be scored as a composite, are the *average inter-item correlation* and *average item-total correlation*. To calculate the average inter-item correlation, we find the correlation between each pair of items and take the average of all the correlations. To calculate the average item-total correlation, we create a total score by adding up scores on each individual item on the scale, then compute the correlation of each item with the total. The average item-total correlation is the average of those individual item-total correlations.

Split-half reliability, described above, is another method of determining internal consistency. This method has the disadvantage that, if the items are not truly homogeneous, different splits will create forms of disparate difficulty and the reliability coefficient will be different for each pair of forms. A method that overcomes this difficulty is Cronbach's alpha (coefficient alpha), which is equivalent to the average of all possible split-half estimates. For more about Cronbach's alpha, including a demonstration of how to compute it, see Chapter 19.

Measures of Agreement

The types of reliability described above are useful primarily for continuous measurements. When a measurement problem concerns categorical judgments, for instance classifying machine parts as acceptable or defective, measurements of agreement are more appropriate. For instance, we might want to evaluate the consistency of results from two different diagnostic tests for the presence or absence of disease. Or we might want to evaluate the consistency of results from three raters who are classifying classroom behavior as acceptable or unacceptable. In each case, each rater assigns a single score from a limited set of choices, and we are interested in how well these scores agree across the tests or raters.

Percent agreement is the simplest measure of agreement: it is calculated by dividing the number of cases in which the raters agreed by the total number of ratings. In the example below, percent agreement is (50 + 30)/100 or 0.80. A major disadvantage of simple percent agreement is that a high degree of agreement may be obtained simply by chance, and thus it is impossible to compare percent agreement across different situations where the distribution of data differs.

This shortcoming can be overcome by using another common measure of agreement called *Cohen's kappa*, or simply *kappa*, which was originally devised to compare two raters or tests and has been extended for larger numbers of raters. Kappa is preferable to percent agreement because it is corrected for agreement due to chance (although statisticians argue about how successful this correction really is: see the sidebar below for a brief introduction to the issues). Kappa is easily computed by sorting the responses into a symmetrical grid and performing calculations as indicated in Table 1-1. This hypothetical example concerns two tests for the presence (D+) or absence (D–) of disease.

Table 1-1. Agreement of two rates on a dichotomous outcome

		Test 2		
		+	–	
Test 1	+	50	10	60
	–	10	30	40
		60	40	100

The four cells containing data are commonly identified as follows:

	+	–
+	a	b
–	c	d

Cells *a* and *d* represent agreement (*a* contains the cases classified as having the disease by both tests, *d* contains the cases classified as not having the disease by both tests), while cells *b* and *c* represent disagreement.

The formula for kappa is:

$$\kappa = \frac{p_o - p_e}{1 - p_e}$$

where p_o = observed agreement and p_e = expected agreement.

$p_o = (a + d)/(a + b + c + d)$, i.e., the number of cases in agreement divided by the total number of cases.

p_e = the expected agreement, which can be calculated in two steps. First, for cells *a* and *d*, find the expected number of cases in each cell by multiplying the row and column totals and dividing by the total number of cases. For *a*, this is $(60 \times 60)/100$ or 36; for *d* it is $(40 \times 40)/100$ or 16. Second, find expected agreement by adding the expected number of cases in these two cells and dividing by the total number of cases. Expected agreement is therefore:

$$p_e = (36 + 16)/100 = 0.52$$

Kappa may therefore be calculated as:

$$\kappa = \frac{0.8 - 0.52}{1 - 0.52} = 0.583$$

Kappa has a range of -1–1: the value would be 0 if observed agreement were the same as chance agreement, and 1 if all cases were in agreement. There are no absolute standards by which to judge a particular kappa value as high or low; however, many researchers use the guidelines published by Landis and Koch (1977):

< 0   Poor

0–0.20   Slight

0.21–0.40   Fair

0.41–0.60   Moderate

0.61–0.81   Substantial

0.81–1.0   Almost perfect

Note that kappa is always less than or equal to the percent agreement because it is corrected for chance agreement.

For an alternative view of kappa (intended for more advanced statisticians), see the sidebar below.

Validity

Validity refers to how well a test or rating scale measures what is it supposed to measure. Some researchers define validation as the process of gathering evidence to support the types of inferences intended to be drawn from the measurements in

Controversies Over Kappa

Cohen's kappa is a commonly taught and widely used statistic, but its application is not without controversy. Kappa is usually defined as representing agreement beyond that expected by chance, or simply agreement corrected for chance. It has two uses: as a test statistic to determine if two sets of ratings agree more often than would be expected by chance (which is a yes/no decision), and as a measure of the level of agreement (which is expressed as a number between 0 and 1).

While most researchers have no problem with the first use of kappa, some object to the second. The problem is that calculating agreement expected by chance between any two entities, such as raters, is based on the assumption that the ratings are independent, a condition not usually met in practice. Because kappa is often used to quantify agreement for multiple individuals rating the same case, whether it is a child's classroom behavior or a chest X-ray from a person who may have tuberculosis, there is no reason to assume that ratings are independent. In fact quite the contrary—they are expected to agree.

Criticisms of kappa, including a lengthy bibliography of relevant articles, can be found on the website of John Uebersax, Ph.D., at *http://ourworld.compuserve. com/homepages/jsuebersax/kappa.htm.*

question. Researchers disagree about how many types of validity there are, and scholarly consensus has varied over the years as different types of validity are subsumed under a single heading one year, then later separated and treated as distinct. To keep things simple, we will adhere to a commonly accepted categorization of validity that recognizes four types: content validity, construct validity, concurrent validity, and predictive validity, with the addition of face validity, which is closely related to content validity. These types of validity are discussed further in the context of research design in Chapter 5.

Content validity refers to how well the process of measurement reflects the important content of the domain of interest. It is particularly important when the purpose of the measurement is to draw inferences about a larger domain of interest. For instance, potential employees seeking jobs as computer programmers may be asked to complete an examination that requires them to write and interpret programs in the languages they will be using. Only limited content and programming competencies may be included on such an examination, relative to what may actually be required to be a professional programmer. However, if the subset of content and competencies is well chosen, the score on such an exam may be a good indication of the individual's ability to contribute to the business as a programmer.

A closely related concept to content validity is known as *face validity*. A measure with good face validity appears, to a member of the general public or a typical person who may be evaluated, to be a fair assessment of the qualities under study. For instance, if students taking a classroom algebra test feel that the questions reflect what they have been studying in class, then the test has good face validity.

Face validity is important because if test subjects feel a measurement instrument is not fair or does not measure what it claims to measure, they may be disinclined to cooperate and put forth their best efforts, and their answers may not be a true reflection of their opinions or abilities.

Concurrent validity refers to how well inferences drawn from a measurement can be used to predict some other behavior or performance that is measured simultaneously. *Predictive validity* is similar but concerns the ability to draw inferences about some event in the future. For instance, if an achievement test score is highly related to contemporaneous school performance or to scores on other tests administered at the same time, it has high concurrent validity. If it is highly related to school performance or scores on other tests several years in the future, it has high predictive validity.

Triangulation

Because every system of measurement has its flaws, researchers often use several different methods to measure the same thing. For instance, colleges typically use multiple types of information to evaluate high school seniors' scholastic ability and the likelihood that they will do well in university studies. Measurements used for this purpose include scores on the SAT, high school grades, a personal statement or essay, and recommendations from teachers. In a similar vein, hiring decisions in a company are usually made after consideration of several types of information, including an evaluation of each applicant's work experience, education, the impression made during an interview, and possibly a work sample and one or more competency or personality tests.

This process of combining information from multiple sources in order to arrive at a "true" or at least more accurate value is called *triangulation,* a loose analogy to the process in geometry of finding the location of a point by measuring the angles and sides of the triangle formed by the unknown point and two other known locations. The operative concept in triangulation is that a single measurement of a concept may contain too much error (of either known or unknown types) to be either reliable or valid by itself, but by combining information from several types of measurements, at least some of whose characteristics are already known, we may arrive at an acceptable measurement of the unknown quantity. We expect that each measurement contains error, but we hope not the *same type* of error, so that through multiple measurements we can get a reasonable estimate of the quantity that is our focus.

Establishing a method for triangulation is not a simple matter. One historical attempt to do this is the multitrait, multimethod matrix (MTMM) developed by Campbell and Fiske (1959). Their particular concern was to separate the part of a measurement due to the quality of interest from that part due to the method of measurement used. Although their specific methodology is less used today, and full discussion of the MTMM technique is beyond the scope of a beginning text, the concept remains useful as an example of one way to think about measurement error and validity.

The MTMM is a matrix of correlations among measures of several concepts (the "traits") each measured in several ways (the "methods"); ideally, the same several methods will be used for each trait. Within this matrix, we expect different measures of the same trait to be highly related: for instance, scores measuring intelligence by different methods such as a pencil-and-paper test, practical problem solving, and a structured interview should all be highly correlated. By the same logic, scores reflecting different constructs that are measured in the same way should not be highly related: for instance, intelligence, deportment, and sociability as measured by a pencil-and-paper survey should not be highly correlated.

Measurement Bias

Consideration of *measurement bias* is important in every field, but is a particular concern in the human sciences. Many specific types of bias have been identified and defined: we won't try to name them all here, but will discuss a few common types. Most research design textbooks treat this topic in great detail and may be consulted for further discussion of this topic. The most important point is that the researcher must be alert to the possibility of bias in his study, because failure to consider and deal with issues related to bias may invalidate the results of an other-wise exemplary study.

Bias can enter studies in two primary ways: during the selection and retention of the objects of study, or in the way information is collected about the objects. In either case, the definitive feature of bias is that it is a source of systematic rather than random error. The result of bias is that the information analyzed in a study is incorrect in a systematic fashion, which can lead to false conclusions despite the application of correct statistical procedures and techniques. The next two sections discuss some of the more common types of bias, organized into two major catego-ries: bias in sample selection and retention, and bias resulting from information being collected or recorded differently for different subjects.

Bias in Sample Selection and Retention

Most studies take place on samples of subjects, whether patients with leukemia or widgets produced by a local factory, because it would be prohibitively expensive if not impossible to study the entire population of interest. The sample needs to be a good representation of the study population (the population to which the results are meant to apply), in order for the researcher to be comfortable using the results from the sample to describe the population. If the sample is biased, meaning that in some systematic way it is not representative of the study population, conclu-sions drawn from the study sample may not apply to the study population.

Selection bias exists if some potential subjects are more likely than others to be selected for the study sample. This term is usually reserved for bias that occurs due to the process of sampling. For instance, telephone surveys conducted using numbers from published directories unintentionally remove from the pool of potential respondents people with unpublished numbers or who have changed phone numbers since the directory was published. Random-digit-dialing (RDD) techniques overcome these problems but still fail to include people living in households without telephones, or who have only a cell phone. This is a problem for a research study if the people excluded differ systematically on a characteristic of interest, and because it is so likely that they do differ, this issue must be addressed by anyone conducting telephone surveys. For instances, people living in households with no telephone service tend to be poorer than those who have a telephone, and people who have only a cell phone (i.e., no "land line") tend to be younger than those who have conventional phone service.

Volunteer bias refers to the fact that people who volunteer to be in studies are usually not representative of the population as a whole. For this reason, results from entirely volunteer samples such as phone-in polls featured on some television programs are not useful for scientific purposes unless the population of interest is people who volunteer to participate in such polls (rather than the general public). Multiple layers of nonrandom selection may be at work: in order to respond, the person needs to be watching the television program in question, which probably means they are at home when responding (hence responses to polls conducted during the normal workday may draw an audience largely of retired people, housewives, and the unemployed), have ready access to a telephone, and have whatever personality traits would influence them to pick up their telephone and call a number they see on the television screen.

Nonresponse bias refers to the flip side of volunteer bias: just as people who volunteer to take part in a study are likely to differ systematically from those who do not volunteer, people who decline to participate in a study when invited to do so very likely differ from those who consent to participate. You probably know people who refuse to participate in any type of telephone survey (I'm such a person myself): do they seem to be a random selection from the general population? Probably not: the Joint Canada/U.S. Survey of Health found not only different response rates for Canadians versus Americans, but also found nonresponse bias for nearly all major health status and health care access measures (results summarized in *http://www.allacademic.com/meta/p_mla_apa_research_citation/0/1/6/8/4/p16845_index.html*).

Loss to follow-up can create bias in any longitudinal study (a study where data is collected over a period of time). Losing subjects during a long-term study is almost inevitable, but the real problem comes when subjects do not drop out at random but for reasons related to the study's purpose. Suppose we are comparing two medical treatments for a chronic disease by conducting a clinical trial in which subjects are randomly assigned to one of several treatment groups, and followed for five years to see how their disease progresses. Thanks to our use of a randomized design, we begin with a perfectly balanced pool of subjects. However, over time subjects for whom the assigned treatment is not proving effective will be more likely to drop out of the study, possibly to seek treatment elsewhere, leading to bias. The final sample of subjects we analyze will consist of those who remain in the trial until its conclusion, and if loss to follow-up was not random, the sample we analyze will no longer be the nicely randomized sample we began with. Instead, if dropping out was related to treatment ineffectiveness, the final subject pool will be biased in favor of those who responded effectively to their assigned treatment.

Information Bias

Even if the perfect sample is selected and retained, bias may enter the study through the methods used to collect and record data. This type of bias is often called *information bias* because it affects the validity of the information upon which the study is based, which may in turn invalidate the results of the study.

When data is collected using in-person or telephone interviews, a social relationship exists between the interviewer and subject for the course of the interview. This relationship can adversely affect the quality of the data collected. When bias is introduced into the data collected because of the attitudes or behavior of the interviewer, this is known as *interviewer bias*. This type of bias may be created unintentionally when the interviewer knows the purpose of the study or the status of the individuals being interviewed: for instance, interviewers might ask more probing questions to encourage the subject to recall toxic chemical exposures if they know the subject is suffering from a rare type of cancer related to chemical exposure. Interviewer bias may also be created if the interviewers display personal attitudes or opinions that signal to the subject that they disapprove of the behaviors being studied, such as promiscuity or drug use, making subjects less likely to report those behaviors.

Recall bias refers to the fact that people with life experiences such as serious disease or injury are more likely to remember events that they believe are related to the experience. For instance, women who suffered a miscarriage may have spent a great deal of time probing their memories for exposures or incidents that they believe could have caused the miscarriage. Women who had a normal birth may have had similar exposures but not given them further thought and thus will not recall them when asked on a survey.

Detection bias refers to the fact that certain characteristics may be more likely to be detected or reported in some people than in others. For instance, athletes in some sports are subject to regular testing for performance-enhancing drugs, and test results are publicly reported. World-class swimmers are regularly tested for anabolic steroids, for instance, and positive tests are officially recorded and often released to the news media as well. Athletes competing at a lower level or in other sports may be using the same drugs but because they are not tested as regularly, or because the test results are not publicly reported, there is no record of their drug use. It would be incorrect to assume, for instance, that because *reported* anabolic steroid use is higher in swimming than in baseball, that the *actual* rate of steroid use is higher in swimming than in baseball. The apparent difference in results could be due to more aggressive testing on the part of swimming officials, and more public disclosure of the test results.

Social desirability bias is caused by people's desire to present themselves in a favorable light. This often motivates them to give responses that they believe will please the person asking the question; this type of bias can operate even if the questioner is not actually present, for instance when subjects complete a pencil-and-paper survey. This is a particular problem in surveys that ask about behaviors or attitudes that are subject to societal disapproval, such as criminal behavior, or that are considered embarrassing, such as incontinence. Social desirability bias can also influence responses in surveys where questions are asked in such a way that they signal what the "right" answer is.

Exercises

Here's a review of the topics covered in this chapter.

Problem

Given the distribution of data in the table below, calculate percent agreement, expected values for cells a and d, and kappa for rater 1 and rater 2.

			Rater 2		
			+	−	
Rater 1		+	70	15	85
		−	30	25	55
			100	40	140

Solution

Percent agreement = (70 + 25)/140 = 0.679

Expected values:

60.7	
	15.7

a: $(85 \times 100)/140 = 60.7$
d: $(55 \times 40)/140 = 15.7$
p_o = observed agreement = (70 + 25)/140 = 0.679
p_e = expected agreement = (60.7 + 15.7)/140 = 0.546

$$\kappa = \frac{0.679 - 0.546}{1 - 0.546} = \frac{0.133}{0.454} = 0.293$$

The Likert Scale

The Likert scale may be the most common type of rating scale used in human subjects research. This type of scale was first described in 1932 by Rensis Likert (1903–1981), an organizational psychologist who served as director of the University of Michigan's Institute for Social Research from 1946 to 1970. Questions using the Likert scale typically present a statement and subjects are invited to choose their response to it from an ordered, odd-numbered set of choices (most often five, but sometimes seven or nine). Below is an example.

The United States should adopt a national system of health insurance.

1. Strongly agree
2. Agree
3. Neither agree nor disagree
4. Disagree
5. Strongly disagree

Sometimes an even number of responses are provided, so that there is no neutral middle choice: this is called the "forced choice" method because the respondent is forced to make the choice to agree or disagree with the statement. Often the order of responses is changed within a questionnaire so 1 = Strongly disagree and 5 = Strongly agree, to detect whether people are automatically selecting the first or last choices without reading the items.

Data gathered by Likert items is ordinal: although the choices are ordered, there is no reason to believe that there are equal intervals between them. For instance, we have no way of knowing if the distance between "Strongly agree" and "Agree" is the same as the distance between "Agree" and "Neither agree nor disagree."

Dewey Defeats Truman

Several United States presidential elections have featured inaccurate predictions based on biased samples. It's always humorous to see a respected publication or organization get it completely wrong, but these incidents also serve as a cautionary tale of what can happen when statistics conducted on a nonrepresentative sample are assumed to apply to the general population.

In 1936, the magazine *Literary Digest*, which had correctly predicted the winner of the presidential election in 1916, 1920, 1924, 1928, and 1932, predicted that Republican Alf Landon would defeat Democrat Franklin Roosevelt by a landslide. However, history shows that Roosevelt won the 1936 election in a landslide. The problem with the *Literary Digest* prediction was that although it was based on a large sample (over 2.3 million respondents out of 10 million invited to take part), the sample was biased because it consisted of people who owned automobiles or telephones, or who subscribed to the *Literary Digest*. In 1936, such individuals tended to be wealthier than the general population, and also more likely to be Republican. Because it was necessary to return a postcard to participate in the poll, the *Literary Digest* sample was subject to volunteer bias as well.

In 1948, every major poll predicted that the Republican Thomas Dewey would defeat the Democrat Harry S. Truman for president. The *Chicago Tribune* even printed papers with the front-page headline "Dewey Defeats Truman." Although polling techniques had improved since 1936, several sources of bias were still present in the polls, which led to this inaccurate prediction. One problem was that telephone surveys were used without statistical correction for the fact that telephone ownership was far more common among the affluent, who were also more likely to support Dewey. Another factor was that there were large numbers of undecided voters in the days leading up to the election, and none of the polls had a good method for predicting for whom these individuals would ultimately vote. A third problem, which related directly to the *Chicago Tribune* fiasco, was that Dewey's support was stronger in the East, and due to the differences in time zones, those election results were reported first. The *Tribune* decided to print papers based on those early results, which were based on a biased sample of results from eastern states. What the *Tribune* did not anticipate was that Truman would carry many western states, including California, and thus amass sufficient electoral votes to win the election.

2

Probability

In a conventional textbook, this chapter would be titled something like "A Brief Introduction to Probability Theory" or "Fundamentals of Probability for Beginners," in either case warning you that a Very Serious Topic was about to be broached, and that however forbidding the chapter might appear, it was only scratching the surface of a subject beyond the comprehension of most poor mortals.

Well, I don't buy any of it. Probability theory is the very basis of statistics, and it is a fascinating topic in its own right. But while it can become as complex as any other field of human endeavor, there's no reason why anyone willing to put in the time can't come to understand it. The basic principles of probability are simple to state, and intuitive and easy to comprehend. What's more, most people are already familiar with probabilistic statements, from the weather report that tells you there is a 30% chance of rain this afternoon to the warning on cigarette packages that smoking increases the risk for lung cancer.

If like most adults you hold one or more insurance policies, you are already engaged in a probabilistic enterprise. For instance, you probably have some kind of automobile insurance, which should really be called "automobile expenses insurance," since its primary purpose is to protect the policy-holder against extreme expenses that may be incurred due to an automobile accident. People don't purchase insurance policies because they are planning to get into a crash, or even because they expect such an occurrence: rather, it is an acknowledgment by the policy-holder that there is a nonzero probability of such events occurring in the future.

Governments often require automobile owners to have insurance policies for the same reason: it's not a judgment that you are a bad driver, just an acknowledgment that accidents do happen and few individuals would be able to cover the costs of a major accident out of their own pocket. The insurance industry employs a large cadre of statisticians to calculate how much you should be charged for a policy, taking into consideration (among other things) the probability that you will be in an accident or file a claim for any other reason, and the amount that each such claim would cost the company.

You need no more mathematical expertise than what you learned in junior high school to understand the basics of probability as presented in this chapter. That understanding will provide the basis to understanding the statistical techniques presented in subsequent chapters. It will also set you up in good stead to grasp a large proportion of the statistics you will ever encounter, unless you decide to take it up as a field of study. In addition, you will be able to understand probabilistic statements as used in everyday speech, and be able to recognize when they are used incorrectly. Finally, I hope you will come to enjoy, as I do, spending time working with the building blocks of statistical science: while the real world of statistical analysis can be chaotic and frustrating, the laws of probability are simple, and spending a few calm hours among them is an excellent preparation for more advanced analysis. Even experienced statisticians like to take a break from the complexity of real-world problems from time to time to review the basics of probability, just as a professional violinist warms up with simple scales or a golfer spends time grooving his swing on the practice range before a major tournament.

About Formulas

People who haven't done well in math or statistics in the past often dislike formulas, which many feel are an arcane system of communication invented by mathematicians to act as a barrier to keep the uninitiated from ever really learning math, thus reserving all the good jobs for themselves. (The part about the good jobs is not entirely a joke, considering that an understanding of mathematics and statistics is required for many lucrative professions today.) First-year calculus is often perceived as a barrier that prevents many otherwise able individuals from pursuing careers in science and engineering, and the specter of graduate-level statistics courses has probably discouraged an equal number from pursuing advanced work in the social sciences.

But the assumption that formulas are a barrier to understanding is wrong, and I hope I can reverse this attitude and convince you that formulas are your friends. They are simply a condensed and unambiguous way of communicating important information, and can be considered as a set of instructions written in the language of mathematics. As one of my calculus professors used to say, "Look at the formula, then do what the formula tells you to do."

Mathematical formulas have the advantage that they are not dependent on language, so mathematics can be communicated and understood among people regardless of their native language or national origin. This is reflected in the fact that mathematics is one of the most international fields of study today. It doesn't matter if you grew up speaking English or Russian or Farsi, you can communicate easily with your colleagues in mathematics without the barriers imposed by human languages.

Let's take the example of the formula for calculating the arithmetic mean, known in common language as the average, of a set of numbers. The formula to calculate the mean is:

$$\bar{x} = \frac{1}{n}\sum_{i=1}^{n} x_i$$

It may look like Greek to you (in fact, some of it is!), but it's really just a set of directions telling you how to do the necessary calculations. Let's break it down into parts:

- x is the number whose mean we are calculating.
- The symbol $\bar{x}$ means the mean of x, which is what we are calculating.
- The symbol x_i means a particular value of x.
- n means the number of values of x being used to compute the mean.
- The summation symbol, Σ, means to add together a number of cases, in this case all values of x. The notations above and below the summation symbol mean to add together all values of x, starting with the first value (x_1) and going to the last value (x_n).

The formula therefore tells you to calculate the mean by adding together all the values of x, then dividing by the number of cases that you just added together.

Suppose we want to calculate the mean of three numbers: 1, 3, and 5. In terms of variable notation, we would call them x_1, x_2, and x_3. In this example, $n = 3$ because we have three numbers. So to execute the formula, we add together the numbers from x_1 to x_3 and multiply by 1/3, giving us:

$$\bar{x} = \frac{1}{3}\sum_{i=1}^{3} x_i = \frac{1}{3}(1+3+5) = 3$$

You will encounter more complicated formulas as you progress in your statistical studies, but the process for using them is the same:

1. Identify the meaning of each symbol used and the operation required.
2. Identify the values to be substituted for each symbol.
3. Substitute the values into the equation, perform the specified operations, and you have your result.

Basic Definitions

Here are some basic concepts to know for a discussion of probability.

Trials

Probability is concerned with the outcome of *trials*, which are also called *experiments* or *observations*: the crucial fact, whichever term is used, is that they refer to an event whose outcome is unknown. If the outcome were known, after all, there would be no need to consider its probability. A trial can be as simple as flipping a coin or drawing a card from a deck, or as complex as observing whether a person diagnosed with breast cancer is still alive five years after diagnosis. We will reserve the term trial for a single observation, such as one coin flip, and the term experiment to refer to multiple trials, such as the results from flipping one coin five times.

Sample Space

The sample space, signified by S, is the set of all possible elementary outcomes of a trial. If the trial is flipping a coin once, then the sample space is S = {heads, tails} (often abbreviated S = {h,t}), because those two alternatives represent all the possible outcomes for the experiment. If the experiment is rolling a single die, the sample space is S = {1, 2, 3, 4, 5, 6}, representing the six faces of the die that may turn up in a single roll. These elementary outcomes are also referred to as sample points. If the experiment consists of multiple trials, all possible combinations of outcomes of the trials must be specified as part of the sample space. For instance, if the trial consists of flipping a coin twice, the sample space is S = {$(h, h), (h, t), (t, h), (t, t)$}.

Events

An *event*, usually signified by a capital letter other than S, is the specification of the outcome of a trial, and may consist of a single outcome or a set of outcomes. If the outcome or set of outcomes occurs, we say the outcome has "satisified the event" or "the event occurred." For instance, the event "heads in flipping a coin" could be specified as E = {*heads*} while the event "odd number in rolling a die" could be specified as E = {1, 3, 5}. A simple event is the outcome of a single experiment or observation, such as a single coin flip. Simple events may be combined into compound events, as in the union and intersection examples below. Events can be defined by listing the outcomes or by defining them logically, so that if the trial was rolling two dice, and we were interested in how often the sum would be less than 6, we could specify this as either E = {2, 3, 4, 5} or E = {*sum is less than 6*}.

A common way to graphically portray the probability of events and combinations of events is through Venn diagrams, in which a rectangle represents the sample space and circles represent particular events. Venn diagrams are used in Figures 2-1 through 2-4 below.

Union

The *union* of several simple events creates a compound event that occurs if one or more of the events occur. The union of E and F is written $E \cup F$ and means "either E or F, or both E and F." Note that the union symbol is similar to a capital letter U. The union of E and F is the shaded area in the Venn diagram in Figure 2-1. Note that this figure portrays two complete circles that partially overlap.

Venn Diagrams

Anyone who was brought up on "the new math" probably remembers Venn diagrams from their elementary school textbooks. While the wisdom of introducing set theory to fifth graders may be debatable, that is surely no fault of the British mathematician John Venn (1834–1923) or his diagrams, which are still widely used in mathematics and other fields to display the logical relationship between sets of objects, and have been adapted to other fields such as literature as well. Venn spent most of his adult life teaching at Caius College, Cambridge University, where his primary interest was in logic: he published three textbooks, including *Symbolic Logic* (1881), which introduced Venn diagrams. Caius students and faculty today have a daily reminder of Venn's accomplishments: he has been immortalized by stained glass windows in the college dining hall, which portray a Venn diagram with three overlapping sets signified by three circles of different colors.

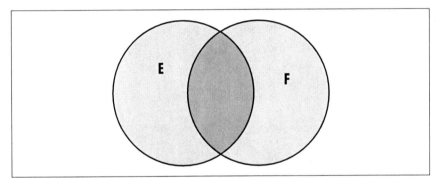

Figure 2-1. The union of E and F (shaded area)

Intersection

The intersection of two or more simple events creates a compound event that occurs only if all the simple events occur. The intersection of E and F is written $E \cap F$ and means "both E and F." The intersection of E and F is the shaded area in the Venn diagram in Figure 2-2.

Complement

The *complement* of an event means everything in the sample space that is not that event. The complement of event E is written variously as $\sim E$, E^c, or $\bar{E}$, and is read as "not E" or "E complement." For instance, if E = (*numbers* > 0), $\sim E$ = (*numbers* ≤ 0). If E = (probability breast cancer patient survives for at least five years), $\sim E$ = (probability breast cancer patient does not survive for at least five years). The complement of F is the shaded area in the Venn diagram in Figure 2-3.

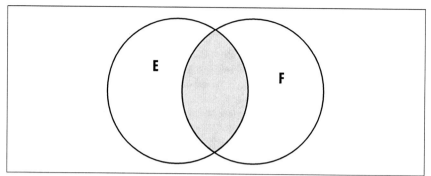

Figure 2-2. The intersection of E and F (shaded area)

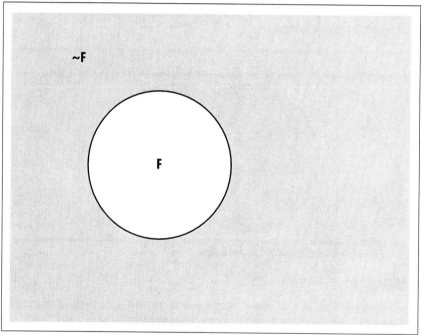

Figure 2-3. The complement of F (shaded area)

Mutual Exclusivity

If events cannot occur together, they are *mutually exclusive*. Following the same line of reasoning, if two sets have no events in common, they are mutually exclusive. For instance, the event A = (salary is greater than $100K) and event B = (salary is less than or equal to $100K) are mutually exclusive, as are the sets A = (even integers) and B = (odd integers). The mutually exclusive sets E and F are presented in the Venn diagram in Figure 2-4; note that they have no points in common.

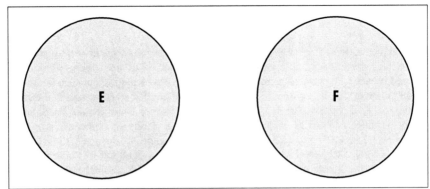

Figure 2-4. E and F are mutually exclusive: they have no points in common

Independence

If two trials are *independent*, that means that the outcome of one has no relationship to the outcome of another. To put it another way, knowing the outcome of one event gives you no information about the outcome of the second event. The classic example of independence is flipping an ordinary coin: if you flip the coin twice, the outcome of the first trial has no influence on the outcome of the second trial, and the probability of heads is the same on every flip.

Permutations

In probability theory, a *permutation* is all the possible ways elements in a set can be arranged. For instance, if a set consists of the elements (a, b, c), then the permutations of this set are (a, b, c), (a, c, b), (b, a, c), (b, c, a), (c, a, b), and (c, b, a). Note that the order of elements is important in permutations: (a, b, c) is a different permutation than (a, c, b).

You can calculate the number of permutations of any set of distinct elements (meaning that none of the elements repeat within the set) by using *factorials*, which are signified by an exclamation point. Many calculators have an $x!$ key to calculate factorials. 3! is read as "3 factorial" and means the product $3 \times 2 \times 1$ or 6, which agrees with the result we found by listing out the different permutations above. To find the factorial for any number, multiply every number from the starting number down to 1. This makes logical sense, because if you have three elements, you have three choices for the first element (a, b, c in our example), two choices for the second element (minus whatever was chosen for the first element), and one choice for the third element (whatever element remains after the first two are chosen). Therefore, you have $3 \times 2 \times 1 = 6$ different ways of arranging the elements. Permutations become large very quickly: for instance 5! = 120 and 10! = 3,628,800. 20! is so large it cannot be displayed on my calculator except through scientific notation: 20! = 2.432902008E18.

Scientific Notation

Scientific notation is used to display numbers that are very large or very small: it not only saves space but is more accurate because you do not have to write or read numbers with lots of zeros. The concept behind scientific notation is that any number can be written as a number greater than or equal to one and less than 10 (called the *coefficient*) multiplied by a power of 10 (called the *base*). So the number 1234 can be written as 1.234E3 (the E stands for *exponent*), which means 1.234×10^3, i.e., 1.234×1000. Similarly, 1.234E − 4 means 1.234×10^{-4} or 1.234×0.0001, which is 0.0001234. Another way to interpret E is how many places to the left or right to move the decimal point. So 1.234E3 tells you to move it three places to the right, producing 1,234, while 1.234E − 4 tells you to move it four places to the left, for 0.0001234.

Combinations

Combinations are similar to permutations, with the difference that the order of elements is not significant in combinations: (a, b, c) is the same combination as (b, a, c). For this reason there is only one combination of the set (a, b, c).

Combinations and permutations are used in statistics to calculate the number of ways a subset of specified size can be drawn from a set, which allows the calculation of the probability of drawing any particular subset. There are several different ways to denote permutations and combinations: these are demonstrated in Appendix A, along with a few problems. We'll stick to a simple system of notation for this section: the number of permutations possible when drawing two elements from a set of 3 is written 3P2, and the number of combinations possible as 3C2. Continuing with our example, 3P2 = 6 because there are 6 permutations of 2 elements drawn from a set of 3: (a, b), (a, c), (b, c), (b, a), (c, a) and (c, b). There are three combinations, so 3C2 = 3: (a, b), (a, c) and (b, c).

The number of permutations of subsets of size k drawn from a set of size n is calculated as:

$$\frac{n!}{(n-k)!}$$

Using this formula, the number of permutations of size 2 that can be drawn from a set of size 8 is:

$$8P2 = \frac{8!}{(8-2)!} = \frac{8!}{(6)!} = 56$$

Given the same values for n and k, there will always be fewer combinations than permutations, because a different order of the same elements counts as a different permutation, but not as a different combination. This is clear in the formula for a combination, which is the formula for the permutation divided by the factorial of the number of objects selected:

$$nCk = \frac{nPk}{k!}$$

Using this formula, we calculate the number of combinations of size 2 that can be drawn from a set of size 8 as:

$$8C2 = \frac{8P2}{2!} = \frac{56}{2} = 28$$

Defining Probability

There are several technical ways to define probability, but a definition useful for statistics is that probability tells us how often something is likely to occur when an experiment is repeated. For instance, the probability that a coin will come up heads can be estimated by executing a number of coin flips and observing how many times it is heads rather than tails. The most important single fact about probability is this:

The probability of an event is always between 0 and 1.

If the probability of an event is 0, that means there is no chance that it will occur, while if the probability of an event is 1, that means it is certain to occur. It is conventional in mathematics to specify probability using decimals, so we say that the probability of an event is between 0 and 1, but it is equally acceptable (and more common in everyday speech) to speak in terms of percentages, so it is equally correct to say that the probability of an event is always between 0% and 100%. To move from decimals to percent, multiply by 100 (per cent = per 100), so a probability of 0.4 is also a probability of 40% ($0.4 \times 100 = 40$) and a probability of 0.85 may also be stated as 85% probability.

Negative probability and probabilities greater than 100% exist only as figures of speech: they are logical impossibilities. The fact that probability is bounded by 0 and 1 has mathematical implications that we explore further when considering logistic regression in Chapter 15. This fact also provides a useful check on your calculations: if you come up with a probability below 0 or greater than 1, you have made a mistake somewhere along the way. Furthermore, if someone tells you there is a 200% chance that you will make a killing on the stock market if you follow their system, you should probably look for a new investment advisor.

Another useful fact about probability is that:

The probability of the sample space is always 1.

Because the sample space represents all possible outcomes of a trial, the total probability of the sample space must add up to 1. This is a useful fact because while we may know the probability of some events in a sample space, there may be others about which we have no information. However, since we know that the probability of the total sample space equals 1, we can assign a probability to those events about which we have no information, based on what remains after the known probabilities are considered.

A third useful fact that follows from the first two:

The probability of an event and its complement is always 1.

This fact follows from the definition of a complement: everything in the sample space that is not the event E is the complement of E, so E and $\sim E$ together must make up the entire sample space and the probability of E and $\sim E$ together must equal 1. This should be clear from Figure 2-3 above: the rectangular box represents the sample space, the circle the event F, and the shaded area within the box but outside the circle $\sim F$. Together, F and $\sim F$ comprise the entire sample space.

Expressing the Probability of an Event

It is typical to write probability statements as follows:

$P(E) = 0.5$

This is read as "the probability of event E is 0.5" or "the probability of event E is 50%". Our first fact about probability, that the probability of an event is always between 0 and 1, may be written:

$0 \leq P(E) \leq 1$

The second fact about probability, which follows from the definition of the sample space S as including all possible outcomes of a trial, may be stated as:

$P(S) = 1$

The third fact about probability, that the probability of an event and its complement is always equal to 1, can be written:

$P(F) + P(\sim F) = 1$

Which provides us with the important corollary:

$P(\sim F) = 1 - P(F)$

This will prove very handy in later calculations. If we know the probability of F, we automatically know the probability of $\sim F$, which is $1 - P(F)$.

Conditional Probabilities

Often we want to know the probability of some event, given that another event has occurred. This is expressed symbolically as $P(E|F)$ and read as "the probability of E given F." The second event is known as the "condition" and the process is sometimes referred to as "conditioning on F." Conditional probability is an important concept in statistics because often we are trying to establish that a factor has a relationship with an outcome, for instance that people who smoke cigarettes are more likely to develop lung cancer. This would be expressed symbolically as:

$P(\text{lung cancer} \mid \text{smoker}) > P(\text{lung cancer} \mid \text{nonsmoker})$

Conditional probabilities can also be used to define independence. Two variables are said to be independent if the following relationship holds:

$P(E|F) = P(E)$

To continue with the same example, the equation to state that the probability of having lung cancer is unrelated to smoking would be:

$P(\text{lung cancer} \mid \text{smoker}) = P(\text{lung cancer})$

Meaning that the probability of lung cancer for a person who smokes is the same as the probability for the population in general, smokers and nonsmokers alike. This is just an example, and I'm not implying that it is true: many studies have shown that the probability of lung cancer for a smoker is much higher than the rate in the general population.

Calculating the Probability of Multiple Events

To calculate the probability of any of several events occurring (the union of several events), add the probabilities of the individual events. The specific equation used will depend on whether the events are mutually exclusive (cannot both occur) or not.

Union of mutually exclusive events

If the events are mutually exclusive, as in Figure 2-4, the equation is simply:

$P(E \cup F) = P(E) + P(F)$

So imagine first a college that does not allow double majors. We will define the events E = (English major) as having a probability of 0.2, and F = (French major) as having a probability of 0.1. These events are mutually exclusive because students are allowed only one major, so we would calculate the probability the event (either English or French major) as:

$P(E \cup F) = 0.2 + 0.1 = 0.3$

Union of events that are not mutually exclusive

At a different college that does allow double majors, the events (English major) and (French major) are not mutually exclusive. The equation calculating P(English major or French major) must therefore include a term correcting for this overlap. Looking at Figure 2-2, the overlap is the area contained in both circles E and F (which is their intersection, represented by the shaded area). A college that allows students to elect more than one major could have people majoring in both English and French, and if we fail to take this into consideration we will be counting these people twice (people with double majors in French and English would be represented by the shaded area in Figure 2-2).

To correct for the probability of two events both occurring, we use the following equation to calculate the probability of either of two events that are not mutually exclusive:

$P(E \cup F) = P(E) + P(F) - P(E \cap F)$

Given P(English major) = 0.2, P(French major) = 0.1, and P(double major in French and English) = 0.05, the probability of a student being either an English or a French major is:

$P(E \cup F) = 0.2 + 0.1 - 0.05 = 0.25$

Intersection of independent events

To calculate the probability of all of several events occurring (the intersection of several events), multiply their individual probabilities. The specific formula used depends on whether the events are independent or not.

If they are independent, the probability of both E and F is calculated as simply:

$P(E \cap F) = P(E) \times P(F)$.

For instance, if we are flipping a coin twice, so E = (heads on first flip) and F = (heads on second flip), both trials are independent. If the probability of heads is 0.5, we can compute the probability (heads on both flips) as:

$P(E \cap F) = 0.5 * 0.5 = 0.25$

Intersection of nonindependent events

If two events are not independent, we have to know their conditional probability in order to be able to calculate the probability of both occurring. The formula to use is:

$P(E \cap F) = P(E) * P(F|E)$

For instance, if we are drawing two cards without replacement from a standard deck of 52, those events are not independent because the probability for the second draw depends on the result of the first draw. If we are interested in the probability of drawing two black cards, we can calculate this using the information that the probability E = (first card black) is 26/52 or 0.5, and the conditional probability $F|E$ = P(second card black|first card black) is 25/51 or 0.49. Therefore, the probability (both cards are black) is:

$P(E \cap F) = 0.5 \times 0.49 = 0.245$

Bayes's Theorem

Bayes's theorem, also known as Bayes's formula, is one of the most common applications of conditional probabilities. A typical use of Bayes's theorem in the medical field is to calculate the probability that a person who tests positive on a screening test for a particular disease actually has the disease. Bayes's formula also uses several of the basic concepts of probability introduced above and is therefore a good review for the entire chapter. Bayes's formula for any two events A and B is:

$$P(A|B) = \frac{P(A\&B)}{P(B)} = \frac{P(B|A)P(A)}{P(B|A)P(A) + P(B|\sim A)P(\sim A)}$$

The ampersand (&) is equivalent to the intersection symbol $\cap$ (i.e., it means "and"), so $P(A\&B)$ means the probability of both A and B. You would use this formula when you know $P(B|A)$ but want to know $P(A|B)$. The numerator of Bayes's theorem uses the fact that the probability of two events is the probability of the first event multiplied by the conditional probability of the second event given the first. In this example, the conditional probability of B given A is multiplied by the probability of A, giving us the probability of both A and B. The denominator uses this same fact plus the fact that any event plus its complement comprise the entire sample space and together have a probability of 1, so the sum of the conditional probabilities of (B given A) times the probability of A, and (B given ~A) times the probability of ~A, equals the probability of B.

Suppose we have a screening test that is 95% effective in detecting disease in those who have it, and 99% effective in not falsely diagnosing disease in those who are free of it. Clinicians would say that this test has 95% sensitivity and 99% specificity. Suppose also that the rate of disease in the population is 1%. Using the symbols D for disease, ~D for absence of disease, T for a positive test, and ~T for a negative test, these probabilities can be stated as:

Sensitivity: $P(T|D) = 0.95$

Specificity: $P(\sim T|\sim D) = 0.99$

$P(D) = 0.01$

These are very high values for sensitivity and specificity: many commonly used tests and procedures are less accurate. However, we know that some probability remains that a person who tests positive will not in fact have the disease (a false positive) and that a person who tests negative will in fact have the disease (a false negative). In the case of an individual who has tested positive, we want to know the probability that the person actually has the disease, written formally as $P(D|T)$. We can calculate this probability using Bayes's theorem plus the information about sensitivity, specificity, and disease rate in the population given above:

$$P(D|T) = \frac{P(D\&T)}{P(T)} = \frac{P(T|D)P(D)}{P(T|D)P(D) + P(T|\sim D)P(\sim D)}$$

Using the fact that an event plus its complement constitute the entire sample space and together have a probability of 1, we know that the false positive rate is 1-specificity, i.e., $P(T|\sim D) = 1 - 0.99$ or 0.01. For the same reason, we know that the probability in the population of not having the disease is $1 - P(D)$, so $P(\sim D) = 0.99$. Using this plus the information supplied above:

$$P(D|T) = \frac{0.95 \times 0.01}{(0.95 \times 0.01) + (0.01 \times 0.99)} = \frac{0.0095}{0.0095 + 0.0099} = 0.4897$$

This demonstrates that even with a highly specific and sensitive screening test, about half the people who test positive will be false positives, i.e., they won't have the disease. This is not necessarily a reason to not use the test, particularly if the disease has serious consequences and there is an accurate follow-up test to separate the true and false positives. However, any proposal to institute universal screening should always consider the false positive rate and the potential consequences of it.

It should be noted that this result is dependent on the rate of disease in the population as well as the sensitivity and specificity of the screening test. If the disease rate were 0.005 instead of 0.01, then fewer of the positives would be true positives, and more would be false positives:

$$P(D|T) = \frac{0.95 \times 0.005}{(0.95 \times 0.005) + (0.01 \times 0.995)} = \frac{0.00475}{0.00475 + 0.00995} = 0.3231$$

The Reverend Thomas Bayes

The ubiquitous Bayes's theorem was developed by the British Nonconformist minister the Reverend Thomas Bayes (1702–1761). Bayes studied logic and theology at the University of Edinburgh and earned his livelihood as a minister in Holborn and Tunbridge Wells, England. However, his fame today rests on his theory of probability, which was developed in an essay published after his death by the Royal Society of London. There is an entire field of study today known as Bayesian statistics, which have in common the notion of a probability as a statement of strength of belief, rather than a frequency of occurrence. However, it is uncertain whether Bayes himself would have embraced this definition, since he published relatively little on mathematics during his lifetime.

Enough Exposition, Let's Do Some Statistics!

Statistics is something you do, not something you read about, so the real purpose of the preceding theoretical presentation was to give you the information you need to perform simple calculations about the probability of events. It also introduced concepts, such as independence and mutual exclusivity, which you will need to understand in order to use more advanced statistical procedures.

The purpose of the problems that follow is to give you some experience in working with the concepts of basic probability. If you are a person who likes to work through a lot of problems in order to understand a topic, there are many excellent textbooks focusing on probability: several are suggested in Appendix C.

If you are new to solving problems in elementary probability, it may help to follow this procedure:

1. Define the trial and/or experiment.
2. Define the sample space.
3. Define the event.
4. Specify the relevant probabilities and do the calculations.

At some point you may not feel it is necessary to go through all these steps, but they may help you get started working with the exercises. In some cases, an alternative solution using a different approach to the problem is provided.

Dice, Coins, and Playing Cards

Because many of the examples in this chapter use dice, coins and playing cards, their characteristics are reviewed here:

Dice

The standard die (the singular of dice) used in the Western world is a cube with six sides, each displaying a different number of dots, from 1 to 6. A standard assumption in probability calculations is that all sides of the die are equally likely to land facing up when the die is rolled or thrown, so one roll of the die has six equally likely outcomes: 1, 2, 3, 4, 5, and 6. In technical terms, the set of outcomes from rolling one or more dice has a discrete uniform distribution because the possible outcomes can be enumerated and each outcome is equally likely. The results of two or more dice thrown at once (or multiple throws of the same die) are assumed to be independent of each other, so the probabilities of each combination of numbers are calculated by multiplying the probability of each separate result.

In the interests of precision, I should point out that the "equal probability for all sides" holds only for casino dice, in which the pips (circles used to mark the numbers on each side) are painted on. Cheaper dice, such as you may purchase at the dime store, do not have equal weight on all sides because the pips are drilled into the cube face rather than painted on. However, in theoretical discussions of probability, this nicety is usually ignored and we assume that all sides of the dice are equally probable.

Coins

The archetypal coin used in probability experiments has two sides, heads and tails. A fair coin is equally likely to come up heads or tails on any toss or flip. For any coin, fair or not, the probability of heads and tails is constant on each flip, so that the results of previous flips have no influence on later flips and the results of multiple flips are independent of each other. As with dice, the probability of an actual coin landing heads or tails is seldom exactly 50–50, for a number of physical reasons, including coin design and wear, and off-center technique on the part of the person performing the flip, but for the sake of probability exercises we assume it is unless otherwise specified. Sometimes experiments are conducted by spinning coins rather than flipping them (fewer projectiles flying through the air in a crowded classroom). However, the 50–50 assumption applies even less here, although for the purposes of doing calculations (as opposed to actually spinning coins and recording the results) we assume that it does. For more on these issues, see *http://www.sciencenews.org/articles/20040228/fob2.asp*.

Playing cards

The standard deck of playing cards today has 52 cards in four suits: spades, clubs, diamonds, and hearts. Spades and clubs are black cards, diamonds and hearts are red cards. There are 13 cards in each suit: an ace, numbered cards from 2 through 10, and 3 face cards—the jack, queen, and king.

Exercises

Problem

If I draw one card from an ordinary deck of 52 playing cards, what is the probability that it will be a red card?

Solution

1. The trial is a single draw of one card from a deck of 52.
2. The sample space is all the possible cards, each of which has an equal likelihood of being drawn.
3. The event is E = {red card}.
4. Since there are 52 cards in the deck and half (26) are red, the probability of drawing a red card is 26/52 or 0.5. The answer is that we have a 50% probability of drawing a red card.

Problem

If I roll a die once, what is the probability of getting a number lower than 5?

Solution

1. The trial is a single roll of a six-sided die.
2. The sample space is the numbers (1, 2, 3, 4, 5, 6), all of which are equally likely.
3. The event is E = (any of 1, 2, 3, 4), which can also be considered the union of four simple events, i.e., E = (E = 1) $\cup$ (E = 2) $\cup$ (E = 3) $\cup$ (E = 4).
4. Four of the six simple events or possible outcomes that constitute the sample space satisfy the event E, so the probability of E is 4/6 or 0.67 (rounded).

Alternative solution

Another way to look at this is to calculate the probability of each simple event that satisfies the event E and add them together, since the events are mutually exclusive. Using this approach, the probability of each simple event in E is 1/6, i.e., there is a 1 in 6 chance that the number will be 1, 1 in 6 that the number will be 2, and so on, so the probability of E is 1/6 + 1/6 + 1/6 + 1/6 or 4/6, which is the same answer as above.

Problem

If I flip a fair coin twice, what is the probability that I will get at least one head?

Solution

1. The experiment is two flips of a fair (P = 0.5 for either heads or tails) coin, i.e., two independent trials.
2. The sample space is {(h, h,), (h, t), (t, h), (t, t)}, all of which are equally likely.

3. The event is E = (at least one head); three of the events in the sample space satisfy this condition: (h, h), (h, t), and (t, h).

4. Each of the outcomes is equally likely, and three of the four satisfy the event E, so the probability of E is 3/4 or 0.75.

Alternative solution

We can also find this result mathematically by calculating the probability of the complement of this event, then subtracting it from 1 to get the probability of the event. If the event E is (at least one head) then its complement is $\sim E$ = (no heads, i.e., two tails). We know that the probability of getting a tail on any flip of a fair coin is 0.5, and the flips are independent, so the probability of (t,t) is 0.5×0.5 or 0.25. Using the definition of complement above, $1 - P(\sim E) = P(E)$, so $1 - 0.25 = 0.75$ or $P(E)$ and the probability of at least one head from two flips is 0.75.

Problem

If I draw one card from a standard 52-card deck, what is the probability that it will be a black (clubs or spades) face card (king, queen, or jack)?

Solution

1. The trial is drawing one card from a 52-card deck.

2. The sample space is all 52 cards, each of which has equal probability of being drawn.

3. The event is E = (black face card); six cards satisfy this condition, the jack, queen, or king of either spades or clubs.

4. The probability is 6/52 or 0.115.

Mathematical solution

P(face card) = 12/52 or 0.231

P(black card) = 26/52 or 0.5

P(black face card) = P(face card) * P(black card) = 0.231 * 0.5 = 0.116

 This solution is possible because the probability of drawing a black card, and the probability of drawing a face card are independent.

Problem

If I draw one card from a standard 52-card deck, what is the probability that it will be either black (clubs or spades) or a face card (king, queen, or jack)?

Solution

1. The trial is drawing one card from a 52-card deck.

2. The sample space is all 52 cards, each of which has an equal probability of being drawn.

3. The event is E = (either black card or face card), meaning any of the 26 black cards or any of the 12 face cards will satisfy the event.

4. The two types of cards that will satisfy the condition are not mutually exclusive: some black cards are also face cards, and vice versa. There are 26 black cards: ace through king of spades (13) and ace through king of clubs (13). There are 12 face cards: jack, king, and queen for each of hearts, diamonds, clubs, and spades. There are six cards in both categories: jack, king, and queen of spades and clubs. So there are 26 + 12 − 6 = 32 cards that satisfy this event, and the probability is 32/52 or 0.615.

Mathematical solution

P(black card) = 26/52 or 0.500

P(face card) = 12/52 or 0.2301

P(black face card) = 6/52 or 0.115

P(black card or face card) = 0.500 + 0.231 − 0.115 = 0.616

Problem:

If I draw a single card from a 52-card deck and it is black, what is the probability that its suit is clubs?

Solution

1. The trial is drawing one card from a 52-card deck.

2. The sample space is all black cards, since we are interested in the conditional probability of a card being a club, given that it is a black card. Our sample space is therefore the 26 black cards.

3. The event is E = (club | black card).

4. The probability of the card being a club, given that it is a black card, is 13/26 or 0.5.

 Note that this is a *conditional* probability (conditioned on the fact that the card is black); the *unconditional* probability of the card being a club, if we had no information about its color, is 13/52 or 0.25.

Mathematical solution

P(clubs | black card) = P(clubs)/P(black card) = 0.25/0.5 = 0.5

Problem

If order is not significant, how many ways are there to select a subset of 5 students from a classroom of 20?

Solution

This is a combinatorial problem that is too lengthy to solve by listing all possible subsets. Instead, we will use the combination formula. In this case, $n = 20$ and $k = 5$:

$$^nC_k = \frac{^nP_k}{k!}$$

$$^nP_k = \frac{n!}{(n-k)!} = \frac{20!}{(20-5)!} = 1860480$$

$$^nC_k = \frac{1860480}{5!} = 15504$$

Problem

There are 80 students attending a conference: 40 boys and 40 girls. 30 of the boys are majoring in math, as are 20 of the girls. We know that if you pick a boy at random, there is a 75% chance that he is a math major; however, if you pick a math major at random, what is the probability that the student is male?

Solution

P(male) = 40/80 = 0.5

P(~male) = 40/80 = 0.5

P(math|male) = 30/40 = 0.75

P(math|~male) = 20/40 = 0.5

$$P(male|math) = \frac{0.75 \times 0.5}{(0.75 \times 0.5) + (0.5 \times 0.5)} = \frac{0.375}{0.625} = 0.6$$

Closing Note: The Connection Between Statistics and Gambling

Statisticians like to illustrate probability using dice, coin flips, and playing cards as examples, objects that are also used in gambling (or gaming, in the industry's preferred terminology). One reason is that these objects are familiar to most people, the probabilities of the different outcomes are known and unchanging, and they supply examples that can be used to illustrate the basic concepts of probability, including independence and mutual exclusivity. They also have the advantage that problems can be solved using the concrete objects in question (for instance, by selecting from a standard deck of cards) as well as through mathematical equations.

Another reason is that many of the laws of probability were discovered in connection with games of chance and skill involving dice and playing cards. In fact, gamblers have been a regular source of inquiry into the probabilities of different combinations of events, in large part because their ability to win rather than lose money depends in large part on their understanding the probability of different events within their chosen game.

Many historians trace the beginning of modern probability theory to the Chevalier de Mere, a gentleman gambler in 17th century France. He was fond of betting that he would roll at least one six in four rolls of a single die: the wisdom of this bet will be demonstrated below. However, he also believed that it was a good bet to propose that he would roll one or more double sixes in 24 rolls of a pair of dice: this turned out to be a losing proposition. Fortunately for the future of probability, the Chevalier took this problem to his friend the philosopher Blaise Pascal, who discussed it with his friend the mathematician Pierre de Fermat and in the process developed, among other things, Pascal's triangle, the binomial distribution, and the modern concept of probability.

In an even bet among friends, when there is no "house" taking a percentage of the proceeds, a good bet is one you are likely to win more than 50% of the time, i.e., a bet where the likelihood of winning is 0.5 or greater. The Chevalier's first bet met this standard: the probability of rolling at least one six in four rolls of a die is 0.518. This is easily calculated by considering the probability of rolling no sixes in four trials, which is $(5/6)^4$. Rolling at least one six is the converse of rolling no sixes, so the P(at least one six in four trials) is $1 - (5/6)^4$ or $1 - 0.482$, which is 0.518. This means that about 52% of the time, the Chevalier won this bet.

However, rolling at least one double six in 24 rolls of a pair of dice is not a wise bet. There are 36 combinations of numbers in each of two rolls of a pair of dice, and only one combination is double sixes: therefore on each roll the probability is 35/36 that double sixes will not come up. Because each roll of the dice is independent, we can multiply the probabilities for each roll together: because the probabilities do not change, this means multiplying (35/36) by itself 24 times, which is the same as raising it to the power of 24. The probability of rolling at least one double six is $1 - P$(no double sixes) or $1 - 0.509$, which is 0.491. Since this probability is less than 0.5, this is a losing bet.

If you are interested in learning more about how probability theory applies to games of chance and skill such as roulette, craps, blackjack, horse racing, and poker, take a look at Edward Packel's *The Mathematics of Games and Gambling*, published by the Mathematical Association of America and listed in Appendix C.

Data Management

You may wonder what a chapter on data management is doing in a book about statistics. It's really very simple: statistics is about analyzing data, and the validity of the statistical result depends in large part on the validity of the data analyzed. So if you will be working with statistics, you need to know something about data management, whether you will be performing the necessary tasks yourself or delegating them to someone else. Oddly enough, data management is often ignored in conventional statistics classes, as well as in many offices and labs: professors and project managers alike sometimes seem to believe that data will magically organize itself without the need for human intervention. However, people who work with data on a daily basis are more likely to subscribe to the 80/20 rule, which says that you spend 80% of your time preparing the data for analysis, and only 20% of your time actually analyzing it. Additionally, even people who understand the need for data management often act as if everyone was born knowing how to do it, unlike matters such as doing linear algebra or riding a bicycle, which actually need to be learned. This is nonsense: data management is a skill that can be learned like any other, and while it is certainly possible to learn it on the job, a.k.a. The School of Hard Knocks, there's no reason not to take advantage of the collective wisdom of those who have gone before you.

The quality of analysis depends on the quality of the data, a fact enshrined in a phrase that originated in the world of computer programming: "Garbage In, Garbage Out," or GIGO. The same concept applies to statistics: the finest statistician cannot produce valid results if the data is a mess. However, recognition of this truth may be obscured by a cultural gulf between the practice of statistics and the practice of data management. If the discipline of statistics dwells on the Olympian heights of abstraction, in the ethereal world of formulas and idealized populations, data management must get down in the trenches and grapple with the reality of the data actually collected. Data by its nature is messy, and seldom does a data file arrive in perfect shape and ready for analysis. So between collecting the data and analyzing it, someone has to get their hands dirty working directly with the data file, cleaning, organizing, and otherwise getting it ready for analysis. There's no mystery about what needs to be done during this process, but it does require a systematic approach guided by knowledge of the data and the uses to which it will be put, and a skeptical attitude informed by common sense.

GIGO has another meaning that applies equally well to statistical analysis: "Garbage In, Gospel Out." This refers to the distressing tendency of some people to believe that anything produced by a computer must be correct, or by extension that any analytic results are correct because they were produced using statistical procedures. This problem may be intensified if the statistical results were produced using a computer package such as SPSS or SAS, adding the mystique of computing to the intimidation of statistics. Unfortunately, statistical processes don't know whether the data is good or bad: the fact that you can calculate the mean and variance of any set of numbers does not mean that those numbers have any meaning, let alone that they represent a reasonable summary of the data. The ease with which complex analyses can be produced using modern statistical software has increased the prevalence of this problem because results can be produced faster and with less effort on the analyst's part. The basic problem remains that the software assumes you have provided it with correct data and produces results according to the algorithms programmed within it, without regard to whether the data supplied is either accurate or meaningful.

If your interest is entirely in learning statistical procedures, for instance, if you are taking a class where the professor supplies you with data sets that have already been cleaned, or if your job responsibilities are only to analyze the data as it is supplied to you, you may want to skip this chapter. Similarly, if you have no practical experience working with data, this chapter may seem entirely abstract and you may want to skim or pass over it until you've actually handled some data. On the other hand, you may find it useful to consider the process of data management and become aware of what may happen when it isn't done correctly. In addition, job change is a regular feature of the modern world, so in the future you may find yourself applying for a position that includes data management, or supervision of others who perform those duties. If you can speak convincingly on the topics covered in this chapter, you'll have a head start on a good proportion of the candidates. In addition, if you get the job, you will start off with a good understanding of why data management is important and how it is done.

An Approach, Not a Set of Recipes

Because many different methods and computer programs are used to collect, store, and analyze data, it's impossible to write a chapter spelling out how to carry out particular procedures that will work in all of them. So I'm focusing in this chapter on an *approach* to data management, including issues common to many situations and a general process of moving from raw data to a data set ready for analysis.

If I had to give one piece of advice concerning data management, it would be this: *assume nothing*. Don't assume that the data file supplied to you is the file you are actually supposed to analyze. Don't assume that all the variables transferred correctly when you moved the file from Excel to SAS (volumes could be written on this subject alone, and every version of any software seems to include a new set of problems). Don't assume that quality control was exercised over the data entry process or that anyone else has examined the data for out-of-range or otherwise impossible values. Don't assume that the person who gave you the project is aware that a key variable is missing for 50% of the cases... you get the idea. Data collection and data entry are activities performed by human beings, who don't always know their jobs perfectly, and make mistakes now and then. A large part of the data management process is discovering where those mistakes were made and either correcting them or thinking of ways to work around them so the data may be analyzed as intended.

The Chain of Command

Without carrying the military metaphor too far, efficient data management for a large project requires establishing a structure or hierarchy of people who are responsible for different aspects of the process. Equally important, everyone involved in the project should know who is authorized to make what decisions, so that when a problem arises it can be resolved quickly and reasonably. This is common sense, but not always exercised in practice. If the data entry clerk notices that data is coming in with lots of variables missing, for instance, he should know exactly who to report the problem to so it can be corrected while the project is still in the data collection phase. If an analyst finds out-of-range values during initial inspection of the data file, she should know who can make the decision about what to do with those values, so they can be corrected or recoded before the main analysis takes place. Make it difficult for such issues to be resolved, and the staff is likely to impose their own ad hoc solutions or give up trying to deal with them, leaving you with a data set of uncertain quality.

Codebooks

The codebook is a classic tool of social science research, but the principle of the codebook applies to any project that involves collecting and analyzing data. Sometimes the codebook is an actual book, generally either a spiral notebook or a three-ring binder, which is used to collect and organize important information about a project. I have also worked on many projects where there was no actual code "book" in the sense of a physical object of paper and ink; instead, all information was stored electronically, in the data and syntax files themselves and ancillary electronic documents. Some projects use a hybrid system, in which most of the codebook information is stored electronically, but also printed and kept in a binder. The bottom line is that it doesn't matter what method you choose, as long as the vital information about the project and the data set is reliably recorded in some location for future reference.

On the whole, I would say that companies whose data consists of the records of their day-to-day business operations do a better job of documentation than academics and people working on small projects. That is probably a combination of two factors. When data reflects the main business of a company, the information technology department has a real incentive to get it right, and when the data collection and storage processes are ongoing and standardized, it is easier to establish a set of procedures and follow them. In addition, companies generally assign people to carry out the procedures of data management, and ensure that they are appropriately trained. The polar opposite is often found in academia, where numerous small projects, each with their own quirks, may be conducted simultaneously. In such circumstances, data management may be relegated to undergraduates with minimal experience or training, or to Ph.D.s or M.D.s who are subject matter experts but unfamiliar with (and possibly uninterested in) the day-to-day issues of data management.

The main reason you need a codebook or its equivalent is to create a repository of information about the project and its data, so that people who join the project later or analyze the data long after the collection process has ceased know what it is and how to interpret it. It's also helpful for people who have been involved from the start, because no one's memory is perfect and it's easy to forget what decisions were made six months or two years ago. Having codebook information easily accessible is also a great timesaver when it's time to write up your results or when you need to explain the project to a new analyst.

At a minimum, the codebook needs to include information in the following categories:

- The project itself and data collection procedures used
- Data entry procedures
- Decisions made about the data
- Coding procedures

Details about the project that should be recorded include the original purpose, timeline, funding, original personnel and any changes, and who is in charge of what. Data collection procedures should include when the data was collected, what procedures were used, and who actually did the data collection. If a form like a questionnaire was used, a copy should be included in the codebook, as should any instructions given to the data collectors.

Information about data entry procedures is particularly important when data is collected in one medium, for instance, on paper questionnaires, and analyzed in another, usually as an electronic file. However, even if a CATI (computer assisted telephone interviewing) system or other method of electronic data collection was used, the codebook should explain how the individual files were collected and transferred. Usually electronic file transfer works smoothly, but not always. Every time a file is transferred there is an opportunity for a data file to become corrupted, in which case it may be necessary to trace back the process in order to correct the error. Information about the training of data entry personnel and any quality control methods (such as double entry of a percentage of the data) should also be recorded.

Seldom is data ready to be analyzed exactly as it has been collected: someone needs to examine it and make decisions about such things as out-of-range values and missing data before the file is ready for analysis. All these decisions need to be recorded, as well as the location of each version of the file. An archived version of the original data file should be stored somewhere it can't be changed, in case you want to reverse a coding decision later, or the edited file becomes corrupt and has to be re-created. It's also sensible to store versions of the file after each major round of editing, in case you decide that decisions made in rounds 1, 2, 3, and 5 were valid but not those of round 4: being able to go back to version 3 of the file saves having to process the original version from scratch. The number of variables and cases in each version of the file, and the file layout, should also be recorded. Every time a file is transferred you need to confirm that the right number of cases and variables appear in the new version, and the file layout is useful when you need to refer to variables by position rather than name (for instance, if the last variable in the file didn't survive the transfer).

Coding procedures will probably occupy the largest part of your codebook. Information that should be recorded here includes variable names, labels added to variables and data values, definitions of missing value codes and how they were applied, and a list of any new variables and the process by which they were created (by using a function, categorizing continuous data, etc.).

The Rectangular Data File

There are many ways to store data electronically, but the most typical remains the rectangular data file. This format should be familiar to anyone who has used a spreadsheet program such as Excel, and although statistical packages such as SAS and SPSS can read data stored in many different packages, the rectangular data file remains a common currency in which files can be exchanged among many different programs.

The most important aspect of the layout of a rectangular data file is that each row represents a record, and each column represents a variable. In addition, data is often arranged so that each row represents one case as defined by the anticipated unit of analysis (see the upcoming "Unit of Analysis" sidebar for more about this), but this requirement is not strictly necessary.

Figure 3-1 displays an excerpt of data from the General Social Survey of 1993, a nationally representative survey that has been conducted by the National Opinion Research Center at the University of Chicago almost every year since 1972. Each line holds data collected from one individual, identified by the variable "id" in the first column. Each column represents data on a particular variable: for instance, the second column holds values for the variable "wrkstat", which is the individual's response to a question about their work status, and the third column holds values for the variable "marital", which is the individual's response to a question about their marital status.

Figure 3-2 shows the same excerpt from the same data set when opened in SPSS: the chief difference is that in Excel the first row is used to store variable names ("id", "wrkstat", etc.) while in SPSS variable names are stored separately. This means that when moving a data file from Excel to SPSS, there will appear to be one fewer case because the variable names occupy a data row in Excel, while in SPSS they do not.

Although other data arrangements are possible in spreadsheets, such as placing variables in rows and cases in columns, these methods should not be used for data that will be imported into a statistical program. In addition, while spreadsheets

Figure 3-1. Rectangular data file in Excel

	A	B	C	D	E	F	G	H	I	J	K
1	id	wrkstat	marital	agewed	sibs	childs	age	birthmo	zodiac	educ	degree
2	1	1	3	20	3	1	43	5	2	11	
3	2	1	5	0	2	0	44	8	6	16	
4	3	1	3	25	2	0	43	2	11	16	
5	4	2	5	0	4	0	45	99	99	15	
6	5	5	5	0	1	0	78	10	7	17	
7	6	5	1	25	2	2	83	3	12	11	
8	7	1	1	22	2	2	55	10	7	12	
9	8	5	1	24	3	2	75	11	9	12	
10	9	1	3	22	1	2	31	7	4	18	
11	10	2	5	0	1	0	54	3	12	18	
12	11	1	5	0	1	0	29	4	2	18	
13	12	1	5	0	0	0	23	10	8	15	
14	13	1	1	31	0	1	61	99	99	12	
15	14	5	4	24	3	4	63	3	1	4	
16	15	4	5	0	4	3	33	3	12	10	
17	16	1	5	0	0	1	36	11	8	14	
18	17	7	5	0	98	4	39	3	12	8	
19	18	1	1	22	9	0	55	1	10	15	
20	19	1	1	32	1	1	55	9	7	16	
21	20	1	1	24	2	2	34	4	2	16	
22	21	3	1	24	5	2	36	6	3	14	
23	22	2	1	23	0	3	44	8	5	18	
24	23	5	2	25	2	2	80	5	2	18	

	id	wrkstat	marital	agewed	sibs	childs	age	birthmo	zodiac	educ	degree
1	1	1	3	20	3	1	43	5	2	11	
2	2	1	5	0	2	0	44	8	6	16	
3	3	1	3	25	2	0	43	2	11	16	
4	4	2	5	0	4	0	45	99	99	15	
5	5	5	5	0	1	0	78	10	7	17	
6	6	5	1	25	2	2	83	3	12	11	
7	7	1	1	22	2	2	55	10	7	12	
8	8	5	1	24	3	2	75	11	9	12	
9	9	1	3	22	1	2	31	7	4	18	
10	10	2	5	0	1	0	54	3	12	18	
11	11	1	5	0	1	0	29	4	2	18	
12	12	1	5	0	0	0	23	10	8	15	
13	13	1	1	31	0	1	61	99	99	12	
14	14	5	4	24	3	4	63	3	1	4	
15	15	4	5	0	4	3	33	3	12	10	
16	16	1	5	0	0	1	36	11	8	14	
17	17	7	5	0	98	4	39	3	12	8	
18	18	1	1	22	9	0	55	1	10	15	
19	19	1	1	32	1	1	55	9	7	16	
20	20	1	1	24	2	2	34	4	2	16	
21	21	3	1	24	5	2	36	6	3	14	
22	22	2	1	23	0	3	44	8	5	18	

Figure 3-2. Rectangular data file in SPSS

allow for the inclusion of other types of information beyond data and variable names, such as titles and calculated fields, that information should be removed before importing into a statistical program.

The main point to keep in mind when considering how to store your data in an electronic file is that it must be formatted for use with the program you intend to use to analyze it, whether that is SPSS, SAS, R, or some other program. What all these programs have in common is that they assume the data is arranged in a particular way and will apply the algorithms for the procedure requested assuming that that expectation has been met. Most programs provide multiple ways of transforming data files and the burden is on the statistical analyst to find out the correct way to format the data for the intended analysis, and to get the data set into that format.

Unit of Analysis

The unit of analysis in a research project is the major entity the study focuses on: examples include individual students, classrooms, schools, arrests, visits to the emergency room, neighborhoods, and countries. We refer to the unit of *analysis* because the same data could be analyzed using different units: for instance, one analysis might look at the academic achievement of individual schoolchildren, while another analysis of the same data could look at achievement levels among different schools, and a third could look at academic achievement in different cities.

Data that is specific to one unit of analysis is often referred to as belonging to a particular level, so in the example above, the variables collected on individual schoolchildren would be called individual-level data and the variables collected concerning schools (such as enrollment or type of funding) would be called school-level data. Although in some fields it is still acceptable to mix data from different levels in a conventional statistical analysis, this can produce misleading results. Instead, it is becoming more and more the expectation that specialized techniques such as multilevel modeling will be used if data from different levels is included in a single analysis.

Spreadsheets and Relational Databases

Even if a project's data will ultimately be analyzed using a specialized statistical analysis program, it is common to collect and/or enter the data using a different program, such as Excel, Access, or Filemaker. These programs can be simpler to use for data entry than a statistical program and many people have them installed on their computer anyway (particularly Excel), limiting the number of licenses that must be purchased. Excel is a spreadsheet, while Access and Filemaker are relational databases. All three can open electronic files from other programs and write files that can be opened by other programs, making them good choices if data will be transferred among programs. In addition, all three can also be used to inspect the data and compute elementary statistics.

For small data projects with simple data, a spreadsheet may be completely adequate. The advantage of spreadsheets is their simplicity: you can create a new data file simply by opening a new spreadsheet and typing the data into the window, and the entire data set can be contained in a single document. Beginners find spreadsheets easy to use, and the spreadsheet format encourages entering data in the rectangular data file form, facilitating data sharing among programs.

Relational databases are a better choice for some larger or more complex projects. They consist of a number of separate tables, each of which looks similar to a spreadsheet page. In a well-designed database, each table holds only one particular type of data, and the tables are linked by key variables. This means that within the database, data for one case (for instance, for one person) would be contained in many separate, specialized tables. A student database might have one table for home addresses, one for birth dates, one for enrollment dates, and so on. If data needs to be transferred to a different program for analysis, the relational database program can write a rectangular data file that contains all the desired information in a single table. The chief advantage of a relational database is efficiency: data is never entered more than once, and multiple records can draw on the same data. In the school example, this would mean that several siblings could draw on the same home address record, while in a spreadsheet that information would have to be entered separately for each child, raising the possibility of typing or transcription errors.

Inspecting a New Data File

Let's assume you have just been sent a new data file to analyze. You have read the background information on the project and know what type of analysis you need to perform, but you need to confirm that the file is in good shape before you begin the analysis. In most cases, you need to answer the questions below (at least) before you begin to analyze the data. Procedures to determine the answer to these questions are available in many types of software, from Excel and Access through specialized statistical packages. Several books that cover data cleaning techniques in particular programs are listed in Appendix C.

1. How many cases are in the file?
2. How many variables are in the file?
3. Are there any (unintended) duplicate cases?
4. Did the variable values, names, and labels transfer correctly?
5. Is all the data within reasonable range?
6. How much data is missing and in what patterns?

You should know how many cases are expected to be in the data file you received: if that does not match up with the number actually in the file, either you were sent the wrong file (which happens all the time) or the file got corrupted during the transfer process (which also happens all the time). At this point you need to go back to the source and get the correct, uncorrupted file before continuing further in your investigation.

Assuming the number of cases is correct, you also need to confirm that the correct number of variables are included in the file. Aside from being sent the wrong data file, a common typical reason why a received file has the wrong number of variables is that the program you are using to open the file has a restriction on the file size or number of variables it can accommodate.

Assuming you have a file with the correct number of cases and variables, you next want to see if it contains any unintended duplicate cases. This requires communication with whoever is in charge of data collection on the project to find out what constitutes a duplicate case, and if a key variable (see the upcoming "Unique Identifiers" sidebar if this term is unfamiliar) is used to identify cases. The definition of a duplicate case depends on the unit of analysis: if the unit of analysis is hospital visits, it would be appropriate for the same person to have multiple records in the file (because they could have made multiple hospital visits). In a file of death records, on the other hand, you would expect only one record per individual. Methods for identifying duplicate records depend on the software being used as well as the specifics of the data set: sometimes it is as simple as confirming that no ID number appears more than once, while sometimes it requires searching for records that have the same values on all or more variables.

Unique Identifiers

The concept of the unique identifier is vital to data management and is familiar to people who work with databases, but may not be known to people who have never worked in data processing. An identifier is a code, usually a number, which is used to identify cases in a data set, and a unique identifier is one that is unique for each case.

Most data sets need at least one unique identifier for each potential unit of analysis. For instance, if data from a medical clinic could be analyzed at either the patient or visit level, one identifier is required that is unique for each patient but common to all the records for one patient, and a second identifier that identifies all the records belonging to a specific patient visit. The unique identifier is useful to confirm that there are no duplicate records, to identify common records belonging to one unit (for instance, all the clinic visits for an individual), and to avoid confusing records for different individuals. There may be multiple Bill Smiths in a large hospital file, and you wouldn't want to get them confused. By the same principle, a particular Bill Smith might come to the clinic five times in a year: when looking at his health history, you want to be able to pull out all the records relating to him.

Checking that variable values, names, and labels are correct is the next step in inspecting a data file. Correct transfer of data values is the most important issue: many unexpected things can happen to data in the file transfer process. Things to check include correct variable type (sometimes numeric variables are unexpectedly translated to string variables, or vice versa: more on string and numeric variables below), length of string variables (which are often truncated or padded), and correct values for dates. Date variables are a frequent source of trouble, because of the different ways they are stored in different systems. Generally data is stored as a number reflecting the number of units of time (days or seconds) from a particular reference date. Unfortunately, each program seems to use a different reference date, and some use different units as well, with the consequence that date values often do not transfer correctly from one program to another. If date values cannot be made to transfer correctly, they can be translated to string variables, which can be used to re-create the date values in the new program.

Variable names can also change unexpectedly during the file transfer process, due to different programs having different rules about what is allowable in a variable name. For instance, Excel allows variable names to begin with a number, while SAS and SPSS do not. Some programs allow names up to 64 characters in length, while others truncate names at 8 characters, a process that may result in duplicate variable names or the substitution of generic names such as "var1". Although data can be analyzed no matter how the individual variables are named, odd and nonmeaningful names impose an extra burden on the analyst and may make the analytical process less efficient. Some advance planning is in order if data will be shared among several programs: in particular, someone needs to look up the naming conventions for each program whose use is anticipated, and create variable names that will be compatible with all the programs that will be used.

Variable and value labels are a great convenience when working with a data file, but often create problems when files are moved from one program or platform to another. Variable labels are text phrases attached to a variable, often used to work around name length restrictions: for instance, the variable "wrkstat" in the GSS example could be assigned the label "Work status in the previous six months." Value labels are similar but are assigned to the values of individual variables. Continuing with the previous example, for the variable "wrkstat" we might assign the label "Full-time employment" to the value 1, "Part-time employment" to the value 2, and so on. However, due to differences in how variable and value labels are stored in different platforms and programs, often they don't transfer correctly. One solution, if you know that the data will be shared across several platforms and/or programs, is to use simple variable names such as "v1" and write a piece of code to be run on each platform or program that takes care of the assignment of variable and value labels.

The next step is to examine the actual values in the data set and see whether they seem reasonable. If you find errors at this point, they may be due to data recording or data entry problems. In either case, someone who is involved with the original project will have to make a judgment call about how to deal with apparently incorrect data values. You can inspect the range of values on a variable by running a frequency procedure if you are using a statistical package, by sorting the cases and inspecting them visually if you are using a spreadsheet package, or with software-specific procedures such as the Filter option in Microsoft Excel. The question you need to answer is whether the values in the data file make sense, because once you start analyzing the data, the program will assume that all the values entered are valid. Typical problems to be on the alert for include out-of-range data (someone with an age of 150 years), invalid values ("3" entered for a question that has only two valid values, "0" and "1") and incongruous patterns (newborn infants reported as college graduates).

The final step before beginning an analysis is to examine the amount of missing data and its patterns. Missing data is a complex problem that can only be touched on in this chapter: if you need to get more deeply into the subject, I recommend consulting the classic Little and Rubin text listed in Appendix C. Your first goal is to find out how much missing data there is, a task that can be accomplished using frequency procedures. The second is to examine the patterns of missing data across multiple variables. For instance, is data frequently missing on particular sets of variables? Are there cases with lots of missing data, while others are entirely or primarily complete? What are the different reasons why data is missing (for instance, because a person declined to provide information, versus because a question did not apply to them) and how are the different types of missing data coded? See the section "Missing Data" later in this chapter for further discussion.

String and Numeric Data

One distinction observed in most electronic data processing and statistical analysis systems (although they may use different names for the concept) is the difference between string and numeric variables. The values stored in string variables, which are also called character or alphanumeric variables, can include letters, numbers, blanks, and symbols such as "#" (the specific characters allowed vary across different systems). String variables are stored as a series of coded values: the coding systems most commonly used are EBCDIC (Extended Binary Coded Decimal Interchange Code) and ASCII (American Standard Code for Information Interchange). Because string variables are coded as a system of values, certain procedures are possible that refer to the position of the characters: for instance, selecting the first three characters in a variable and storing it in a new variable.

Numeric variables are stored as numeric values rather than as the characters that are used to write those values, and may be used in mathematical and statistical procedures (such as addition and subtraction), while string variables may not. In some systems, certain symbols such as the decimal point, comma, and dollar sign are also allowed within numeric variables.

The specific method used to store the values of numeric variables differs across platforms and systems, as does the precision with which values are stored. Each electronic data system (Excel, SAS, SPSS, etc.) has a different set of rules for what characters are allowed in string and numeric variables, and rules about which types of variables may be used in which procedures. You need to be aware that when transferring electronic files from one system to another, the variable type may change or certain values may be dropped. This is a problem that needs to be handled on a file-by-file basis: the specific problems that occur when transferring files from Excel to SPSS, for instance, may be different from those that occur when transferring files from Access to SAS.

Missing Data

Nearly every data set ever collected has included missing data at some point in its history. Despite the ubiquity of missing data, however, it is not a simple problem to deal with and analysts have to decide to what lengths they are willing and able to go in order to handle missing data. This discussion can only introduce the main concepts concerning missing data, and suggest some practical fixes. For a more in-depth and academic discussion, see the classic text *Statistical Analysis with Missing Data* by Little and Rubin (Wiley) listed in Appendix C.

Data can be missing for many reasons, and it is useful if the reasons are recorded within the data set. The method to accomplish this varies with different systems but is nearly always possible, often by using specific data codes to differentiate among them, using values (such as –8 and –9) that cannot appear as true values for the variable in question. For instance, on a survey, an individual might refuse a particular question, or might not have the information requested: those two types of responses could be assigned different codes and their meaning noted in the codebook. Many types of software also allow this information to be included in the data file through value labels. A third possibility is that the question was not asked because it was not applicable (such as a question about having a driving license when interviewing a 10-year-old), so that type of missing data would be assigned yet another code. The reason for differentiating among these cases is that you might want to include the reasons data is missing in an analysis. For instance, you might want to know if those who declined to answer differed demographically from those who did not know the answer to the question.

Missing data poses two problems: it reduces the number of cases available for analysis, thereby reducing statistical power (your ability to find true differences in the data, a topic discussed further in Chapter 18), and may also introduce bias into the data. The first point is based on the fact that, all things being equal, statistical power is increased as the number of cases increase, so any loss of cases may result in a loss of power. The second point requires an excursion into missing data theory.

Missing data is often divided into three types: missing completely at random (MCAR), missing at random (MAR), and nonignorable. MCAR means that the fact of a piece of data being missing is not related to either its own value or the value of other variables in the data set. This is the easiest type of missing data to deal with, since the complete cases may be considered to be a random sample of the entire data set. Unfortunately, MCAR data rarely occurs in practice. MAR data means that the fact that the data is missing is not related to its own value, but is related to the values of other variables in the analysis. For instance, failure to complete a survey item about household income may be related to an individual's level of education. Nonignorable missing data is unfortunately the most common type, and also the type most likely to introduce bias into a statistical analysis. Nonignorable refers to data whose missingness is related to its own value. For instance, overweight people may be less willing to supply information about how much they weigh, and people with low-prestige jobs may be less likely to fill out an occupational survey.

This discussion may seem a bit theoretical: how can you tell what type of missing data you have, when you by definition don't know the values of the data that is missing? The answer is that you have to make a judgment based on knowledge of the population surveyed and your experience in the field. Because the most common methods of statistical analysis assume you have complete, unbiased data, if a data set has a large quantity of missing data, you (or whoever is empowered to make such decisions) will have to decide how to deal with it. Implementing any of the solutions suggested below, other than the first or fourth, may require calling in a statistical consultant or using software designed specifically for dealing with missing data, so the departmental budget and availability of such experts and software will also play a role in the decision. Some potential solutions are included below: the most preferable is 1. Solutions 5 through 7 are seldom justified from a statistical point of view, although they are often used in practice.

1. Make an extra effort to collect the missing data by following up with the source, which solves the problem by making the missing data no longer missing.

2. Consider a different analytical design, such as a multilevel model rather than a classic repeated-measures model.

3. Impute values for the missing data using maximum likelihood methods such as those available in the SPSS MVA module, or use multiple imputation in SAS PROC MI to generate a distribution for the missing value and select a value for each missing case.

4. Include a dummy (0, 1) variable in your analysis that indicates that data was missing, along with an imputed value replacing the missing data.

5. Drop the cases or variables with large amounts of missing data from the analysis (only feasible if the problem is confined to a small percentage of cases and/or variables that are not central to your analysis, and may introduce bias if the data is not MCAR).

6. Conditional imputation: use available values to impute missing values (not recommended, as it may result in an underestimate of variance).

7. Simple imputation: substitute a value such as the population mean for the missing value (not recommended, as it nearly always results in an extreme underestimate of variance).

4

Descriptive Statistics and Graphics

Most of this book, as in most statistics books, is concerned with *statistical inference*, which is the practice of drawing conclusions about a population using statistics calculated on a sample considered to be representative of that population. However, this particular chapter is concerned with *descriptive statistics*, meaning the use of statistical and graphic techniques to present information about the data set being studied. Computing descriptive statistics and examining graphic displays of data is an advisable preliminary step in data analysis. You can never be too familiar with your data, and the time you spend examining the actual distribution of the data collected (as opposed to the distribution you expected it to assume) is always time well spent. Descriptive statistics and graphic displays are also the final product in some contexts: for instance, a business may want to monitor total volume of sales for its different locations without any desire to use that information to make inferences about other businesses.

Populations and Samples

The same data set may be considered as either a population or a sample, depending on the reason for its collection and analysis. For instance, the final exam grades of the students in a class are a population if the purpose of the analysis is to describe the distribution of scores in that class. They are a sample if the purpose of the analysis is to make some inference from those scores to the scores of students in other classes. Analyzing a population means you are performing your calculations on all members of the group in question, while analyzing a sample means you are working with a subset drawn from a larger population. Samples rather than populations are often analyzed for practical reasons, since it may be impossible or prohibitively expensive to study a large population directly.

Notational conventions and terminology differ from one author to the next, but as a general rule numbers that describe a population are referred to as *parameters* and are signified by Greek letters such as μ and σ, while numbers that describe a sample are referred to as *statistics* and are signified by Latin letters such as $\bar{x}$ and *s*. Sometimes computation formulas for a parameter and the corresponding statistic are the same, as in the population and sample mean. However, sometimes they differ: the most famous example is that of the population and sample variance and standard deviation. Somewhat confusingly, because most statistical practice is concerned with inferential statistics, sometimes statistical formulas properly meant for samples are applied to populations (when the parameter formula should be used instead). When the formulas differ, both will be provided in this chapter.

Measures of Central Tendency

Measures of central tendency, also known as measures of location, are typically among the first statistics computed for the continuous variables in a new data set. The main purpose of computing measures of central tendency is to give you an idea of what is a typical or common value for a given variable. The three most common measures of central tendency are the arithmetic mean, median, and mode.

The Mean

The arithmetic *mean*, or simply the mean, is more commonly known as the *average* of a set of values. It is appropriate for interval and ratio data, and can also be used for dichotomous variables that are coded as 0 or 1. For continuous data, for instance measures of height or scores on an IQ test, the mean is simply calculated by adding up all the values and dividing by the number of values. The mean of a population is denoted by the Greek letter *mu* (μ) while the mean of a sample is typically denoted by a bar over the variable symbol: for instance, the mean of *x* would be designated $\bar{x}$ and pronounced "x-bar." The bar notation is sometimes adapted for the names of variables also: for instance, some authors denote "the mean of the variable age" by $\overline{age}$, which would be pronounced "age-bar".

For instance, if we have the following values of the variable *x*:

  100, 115, 93, 102, 97

We calculate the mean by adding them up and dividing by 5 (the number of values):

  $\bar{x} = (100 + 115 + 93 + 102 + 97)/5 = 507/5 = 101.4$

Statisticians often use a convention called *summation notation*, introduced in Chapter 1, which defines a statistic by expressing how it is calculated. The computation of the mean is the same whether the numbers are considered to represent a population or a sample: the only difference is the symbol for the mean itself. The mean of a data set, as expressed in summation notation, is:

$$\bar{x} = \frac{1}{n}\sum_{i=1}^{n} x_i$$

Where $\bar{x}$ is the mean of x, n is the number of cases, and x_i is a particular value of x. The Greek letter sigma (Σ) means summation (adding together), and the figures above and below the sigma define the range over which the operation should be performed. In this case the notation says to sum all the values of x from 1 to n. The symbol i designates the position in the data set, so x_1 is the first value in the data set, x_2 the second value, and x_n the last value in the data set. The summation symbol means to add together or sum the values of x from the first (x_1) to x_n. The mean is therefore calculated by summing all the data in the data set, then dividing by the number of cases in the data set, which is the same thing as multiplying by $1/n$.

The mean is an intuitively easy measure of central tendency to understand. If the numbers represented weights on a beam, the mean would be the point where the beam would balance perfectly. However the mean is not an appropriate summary measure for every data set because it is sensitive to extreme values, also known as *outliers* (discussed further below), and may also be misleading for skewed (nonsymmetrical) data. For instance, if the last value in the data set were 297 instead of 97, the mean would be:

$$\bar{x} = (100 + 115 + 93 + 102 + 297)/5 = 707/5 = 141.4$$

This is not a typical value for this data: 80% of the data (the first four values) are below the mean, which is distorted by the presence of one extremely high value. A good practical example of when the mean is misleading as a measure of central tendency is household income data in the United States. A few very rich households make the mean household income a larger value than is truly representative of the average or typical household.

The mean can also be calculated using data from a *frequency table*, i.e., a table displaying data values and how often each occurs. Consider the following simple example in Table 4-1.

Table 4-1. Simple frequency table

Value	Frequency
1	7
2	5
3	12
4	2

To find the mean of these numbers, treat the frequency column as a weighting variable, i.e., multiply each value by its frequency. The mean is then calculated as:

$$\bar{x} = \frac{(1 \times 7) + (2 \times 5) + (3 \times 12) + (4 \times 2)}{7 + 5 + 12 + 2} = 2.35$$

This is the same result you would reach by adding together each individual score (1+1+1+1+...) and dividing by 26.

The mean for *grouped data*, in which data has been tabulated by range, is calculated in a similar manner. One additional step is necessary: the midpoint of each range must be calculated, and for the purposes of the calculation it is assumed that all data points in that range have the midpoint as their value. A mean calculated in this way is called a *grouped mean*. A grouped mean is not as precise as the mean calculated from the original data points, but it is often your only option if the original values are not available. Consider the following tiny grouped data set in Table 4-2.

Table 4-2. Grouped data

Range	Frequency	Midpoint
1–20	5	10.5
21–40	25	30.5
41–60	37	50.5
61–80	23	70.5
81–100	8	90.5

The mean is calculated by multiplying the midpoint of each interval by its frequency, and dividing by the total frequency:

$$\bar{x} = \frac{(10.5 \times 5) + (30.5 \times 25) + (50.5 \times 37) + (70.5 \times 23) + (90.5 \times 8)}{5 + 25 + 37 + 23 + 8}$$

$$= 51.32$$

One way to lessen the influence of outliers is by calculating a *trimmed mean*. As the name implies, a trimmed mean is calculated by trimming or discarding a certain percentage of the extreme values in a distribution, and calculating the mean of the remaining values. In the second distribution above, the trimmed mean (defined by discarding the highest and lowest values) would be:

$$\bar{x} = (100 + 115 + 102)/3 = 317/3 = 105.7$$

This is much closer to the typical values in the distribution than 141.4, the value of the mean of all the values. In a data set with many values, a percentage such as 10 percent or 20 percent of the highest and lowest values may be eliminated before calculating the trimmed mean.

The mean can also be calculated for dichotomous data using 0–1 coding, in which case the mean is equivalent to the percent of values with the number 1. For instance, if we have 10 subjects, 6 males and 4 females, coded 1 for male and 0 for female, computing the mean will give us the percentage of males in the population:

$$\bar{x} = (1+1+1+1+1+1+0+0+0+0)/10 = 6/10 = 0.6 \text{ or } 60\% \text{ males}$$

The Median

The *median* of a data set is the middle value when the values are ranked in ascending or descending order. If there are *n* values, the median is formally defined as the $(n+1)/2$th value. If $n = 7$, the middle value is the $(7+1)/2$th or fourth value. If there is an even number of values, the median is the average of the two middle values. This is formally defined as the average of the $(n/2)$th and $((n/2)+1)$th value. If there are six values, the median is the average of the $(6/2)$th and $((6/2)+1)$th value, or the third and fourth values. Both techniques are demonstrated below:

Odd number of values: 1, 2, 3, 4, 5, 6, 7 median = 4
Even number of values: 1, 2, 3, 4, 5, 6 median = $(3+4)/2 = 3.5$

The median is a better measure of central tendency than the mean for data that is asymmetrical or contains outliers. This is because the median is based on the ranks of data points rather than their actual values: 50 percent of the data values in a distribution lie below the median, and 50 percent above the median, without regard to the actual values in question. Therefore it does not matter if the data set contains some extremely large or small values, because they will not affect the median more than less extreme values. For instance, the median of all three distributions below is 4:

Distribution A: 1, 1, 3, 4, 5, 6, 7
Distribution B: 0.01, 3, 3, 4, 5, 5, 5
Distribution C: 1, 1, 2, 4, 5, 100, 2000

The Mode

A third measure of central tendency is the *mode*, which refers to the most frequently occurring value. The mode is most useful in describing ordinal or categorical data. For instance, imagine that the numbers below reflect the favored news sources of a group of college students, where 1 = newspapers, 2 = television, and 3 = Internet:

1, 1, 2, 2, 2, 2, 3, 3, 3, 3, 3, 3, 3

We can see that the Internet is the most popular source because 3 is the modal (most common) value in this data set.

In a symmetrical distribution (such as the normal distribution, discussed in Chapter 7), the mean, median, and mode are identical. In an asymmetrical or skewed distribution they differ, and the amount by which they differ is one way to evaluate the skewness of a distribution.

Measures of Dispersion

Dispersion refers to how variable or "spread out" data values are: for this reason measures of dispersions are sometimes called "measures of variability" or "measures of spread." Knowing the dispersion of data can be as important as knowing its central tendency: for instance, two populations of children may both have mean IQs of 100, but one could have a range of 70 to 130 (from mild retardation to very superior intelligence) while the other has a range of 90 to 110 (all within the normal range). Despite having the same average intelligence, the range of IQ scores for these two groups suggests that they will have different educational and social needs.

The Range and Interquartile Range

The simplest measure of dispersion is the *range*, which is simply the difference between the highest and lowest values. Often the minimum (smallest) and maximum (largest) values are reported as well as the range. For the data set (95, 98, 101, 105), the minimum is 95, the maximum is 105, and the range is 10 (105 − 95). If there are one or a few outliers in the data set, the range may not be a useful summary measure: for instance, in the data set (95, 98, 101, 105, 210), the range is 115 but most of the numbers lie within a range of 10 (95 to 105). Inspection of the range for any variable is a good data screening technique: an unusually wide range, or extreme minimum or maximum values, warrants further investigation. It may be due to a data entry error or to inclusion of a case that does not belong to the population under study (for instance, information from an adult that got mixed in with a data set concerned with children).

The *interquartile range* is an alternative measure of dispersion that is less influenced by extreme values than the range. The interquartile range is the range of the middle 50% of the values in a data set, which is calculated as the difference between the 75th and 25th percentile values. The interquartile range is easily obtained from most statistical computer programs but may also be calculated by hand using the following rules (n = the number of observations, k the percentile you wish to find):

1. Rank the observations from smallest to largest.

2. If $(nk)/100$ is an integer (a round number with no decimal or fractional part), the kth percentile of the observations is the average of the $((nk)/100$th)and $((nk)/100+1)$th largest observations.

3. If $(nk)/100$ is not an integer, the kth percentile of the observation is the $(j+1)$th largest measurement, where j is the largest integer less than $(nk)/100$.

Consider the following data set, with 13 observations:

- (1, 2, 3, 5, 7, 8, 11, 12, 15, 15, 18, 18, 20).
- First we want to find the 25th percentile, so $k = 25$.
- We have 13 observations, so $n = 13$.
- $(nk)/100 = (25 * 13)/100 = 3.25$, which is not an integer, so we will use the second method (#3 in the list above).
- $j = 3$ (the largest integer less than $(nk)/100$, i.e., less than 3.25).
- So the 25th percentile is the ($j + 1$)th or 4th observation, which has the value 5.

We can follow the same steps to find the 75th percentile:

- $(nk)/100 = (75*13)/100 = 9.75$, not an integer.
- $j = 9$, the smallest integer less than 9.75.
- So the 75th percentile is the 9 + 1 or 10th observation, which has the value 15.
- Therefore, the interquartile range is (5 to 15) or 10.

The resistance of the interquartile range to outliers should be clear. This data set has a range of 19 (20 − 1) and an interquartile range of 10; however, if the last value was 200 instead of 20, the range would be 199 (200 − 1) but the interquartile range would still be 10, and that number would better represent most of the values in the data set.

The Variance and Standard Deviation

The most common measures of dispersion for continuous data are the *variance* and *standard deviation*. Both describe how much the individual values in a data set vary from the mean or average value. The variance and standard deviation are calculated slightly differently depending on whether a population or a sample is being studied, but basically the variance is the average of the squared deviations from the mean, and the standard deviation is the square root of the variance. The variance of a population is signified by σ^2 (pronounced "sigma-squared": σ is the Greek letter sigma) and the standard deviation as σ, while the sample variance and standard deviation are signified by s^2 and s, respectively.

The deviation from the mean for one value in a data set is calculated as $(x_i - \bar{x})$ where x_i is value i from the data set and $\bar{x}$ is the mean of the data set. Written in summation notation, the formula to calculate the sum of all deviations from the mean for a data set with n observations is:

$$\sum_{i=1}^{n} (x_i - \bar{x})$$

Unfortunately this quantity is not useful because it will always equal zero. This is not surprising if you consider that the mean is computed as the average of all the values in the data set. This may be demonstrated with the tiny data set (1, 2, 3, 4, 5):

$$\bar{x} = (1 + 2 + 3 + 4 + 5)/5 = 3$$

$$\sum_{i=1}^{n} (x_i - \bar{x}) = (1-3) + (2-3) + (3-3) + (4-3) + (5-3)$$

$$= -2 + (-1) + 0 + 1 + 2 = 0$$

So we work with squared deviations (which are always positive) and divide their sum by n, the number of cases, to get the average deviation or variance for a population:

$$\sigma^2 = \frac{1}{n} \sum_{i=1}^{n} (x_i - \bar{x})^2$$

The sample formula for the variance requires dividing by $n - 1$ rather than n because we lose one degree of freedom when we calculate the mean. The formula for the variance of a sample, notated as s^2, is therefore:

$$s^2 = \frac{1}{(n-1)} \sum_{i=1}^{n} (x_i - \bar{x})^2$$

Continuing with our tiny data set, we can calculate the variance for this population as:

$$\sigma^2 = \frac{1}{5} \times ((-2)^2 + (-1)^2 + (0)^2 + (1)^2 + (2)^2)$$

$$= \frac{4 + 1 + 0 + 1 + 4}{5} = \frac{10}{5} = 2$$

If we consider these numbers to be a sample, the variance would be computed as:

$$s^2 = \frac{1}{(5-1)} \times ((-2)^2 + (-1)^2 + (0)^2 + (1)^2 + (2)^2)$$

$$= \frac{4 + 1 + 0 + 1 + 4}{4} = \frac{10}{4} = 2.5$$

Note that because of the different divisor, the sample formula for the variance will always return a larger result than the population formula, although if the sample size is close to the population size, this difference will be slight. The divisor $(n-1)$ is used so that the sample variance will be an unbiased estimator of the population variance.

Because squared numbers are always positive (outside the realm of imaginary numbers), the variance will always be equal to or greater than 0. The variance would be zero only if all values of a variable were the same (in which case the variable would really be a constant). However, in calculating the variance, we have changed from our original units to squared units, which may not be convenient to interpret. For instance, if we were measuring weight in pounds, we would probably want measures of central tendency and dispersion expressed in the same units, rather than having the mean expressed in pounds and variance in squared pounds. To get back to the original units, we take the square root of the variance: this is called the standard deviation and is signified by σ for a population and s for a sample.

For a population, the formula for the standard deviation is:

$$\sigma = \sqrt{\sigma^2} = \sqrt{\frac{1}{n} \sum_{i=1}^{n} (x_i - \bar{x})^2}$$

In the example above:

$$\sigma = \sqrt{2} = 1.41$$

The formula for the sample standard deviation is:

$$s = \sqrt{s^2} = \sqrt{\frac{1}{(n-1)} \sum_{i=1}^{n} (x_i - \bar{x})^2}$$

In the above example:

$$s = \sqrt{2.5} = 1.58$$

In general, for two variables measured with the same units (e.g., two groups of people both weighed in pounds), the group with the larger variance and standard deviation has more variability among their scores. However, the unit of measure affects the size of the variance: the same population weights, expressed in ounces rather than pounds, would have a larger variance and standard deviation. The *coefficient of variation* (CV), a measure of relative variability, gets around this difficulty and makes it possible to compare variability across variables measured in different units. The CV is shown here using sample notation, but could be calculated for a population by substituting σ for s. The CV is calculated by dividing the standard deviation by the mean, then multiplying by 100:

$$CV = \frac{s}{\bar{x}} \times 100\%$$

For the previous example, this would be:

$$CV = \frac{1.58}{3} \times 100\% = 52.7\%$$

Outliers

There is no absolute agreement among statisticians about how to define *outliers*, but nearly everyone agrees that it is important that they be identified and that appropriate analytical techniques be used for data sets that contain outliers. Basically, an outlier is a data point or observation whose value is quite different from the others in the data set being analyzed. This is sometimes described as a data point that seems to come from a different population, or is outside the typical pattern of the other data points. For instance, if the variable of interest was years of education and most of your subjects had 10–16 years of school (first year of high school through university graduation) but one subject had 0 years and another had 26, those two values might be defined as outliers. Identification and analysis of outliers is an important preliminary step in many types of data analysis, because the presence of just one or two outliers can completely distort the value of some common statistics, such as the mean.

It's also important to identify outliers because sometimes they represent data entry errors. In the above example, the first thing to do would be to check if the data was entered correctly: perhaps the correct values were 10 and 16, respectively. The second thing to do is to investigate whether the cases in question actually belong to the same population as the other cases: for instance, does the 0 refer to the years of education of a child when the data set was supposed to contain only information about adults?

If neither of these simple fixes solves the problem, the statistician is left to his own judgment as to what to do with them. It is possible to delete cases with outliers from the data set before analysis, but the acceptability of this practice varies from field to field. Sometimes a standard statistical fix already exists, such as the trimmed mean described above, although the acceptability of such fixes also varies from one field to the next. Other possibilities are to transform the data (discussed in Chapter 7) or use nonparametric statistical techniques (discussed in Chapter 11), which are less influenced by outliers.

Various rules of thumb have been developed to make the identification of outliers more consistent. One common definition of an outlier, which uses the concept of the interquartile range (IQR), is that mild outliers are those lower than the 25th quartile minus $1.5 \times IQR$, or greater than the 75th quartile plus $1.5 \times IQR$. Cases this extreme are expected in about 1 in 150 observations in normally distributed data. Extreme outliers are similarly defined with the substitution of $3 \times IQR$ for $1.5 \times IQR$; values this extreme are expected about once per 425,000 observations in a normal distribution.

Graphic Methods

There are innumerable graphic methods to present data, from the basic techniques included with spreadsheet software such as Microsoft Excel to the extremely specific and complex methods developed in the computer language R. Entire books have been written on the use and misuse of graphics in presenting data: the leading (if also controversial) expert in this field is Edward Tufte, a Yale professor (with a Master's degree in statistics and a PhD in political science). His most famous work is *The Visual Display of Quantitative Information* (listed in Appendix C), but all of Tufte's books are worthwhile for anyone seriously interested in the graphic display of data. This section concentrates on the most commonly used graphic methods for presenting data, and discusses issues concerning each. It is assumed throughout this section that graphics are a tool used in the service of communicating information about data rather than an end in themselves, and that the simplest presentation is often the best.

Frequency Tables

The first question to ask when considering a graphic method of presentation is whether one is needed at all. It's true that in some circumstances a picture may be worth a thousand words, but at other times *frequency tables* do a better job than graphs at presenting information. This is particularly true when the actual values of the numbers in different categories, rather than the general pattern among the categories, are of primary interest. Frequency tables are often an efficient way to present large quantities of data and represent a middle ground between text (paragraphs describing the data values) and pure graphics (such as a histogram).

Suppose a university is interested in collecting data on the general health of their entering classes of freshmen. Because obesity is a matter of growing concern in the United States, one of the statistics they collect is the Body Mass Index (BMI), calculated as weight in kilograms divided by squared height in meters. Although not without controversy, the ranges for the BMI shown in Table 4-3, established by the Centers for Disease Control and Prevention (CDC) and the World Health Organization (WHO), are generally accepted as useful and valid.

Table 4-3. WHO/CDC categories for BMI

BMI range	Categories
< 18.5	Underweight
18.5–24.9	Normal weight
25.0–29.9	Overweight
30.0 and above	Obese

So consider Table 4-4, an entirely fictitious list of BMI classifications for entering freshmen.

Table 4-4. Distribution of BMI in the freshman class of 2005

BMI range	Number
< 18.5	25
18.5–24.9	500
25.0–29.9	175
30.0 and above	50

This is a useful table: it tells us that most of the freshman are normal body weight or are moderately overweight, with a few who are underweight or obese. The BMI is not an infallible measure: for instance athletes often measure as either underweight (distance runners, gymnasts) or overweight or obese (football players, weight throwers). But it's an easily calculated measurement that is a reliable indicator of a healthy or unhealthy body weight for many people. This table presents raw numbers or counts for each category, which are sometimes referred to as *absolute frequencies*: they tell you how often each value appears, not in relation to any other value. This table could be made more useful by adding a column for *relative frequency*, which displays the percent of the total represented by each category. The relative frequency is calculated by dividing the number of cases in each category by the total number of cases (750), and multiplying by 100. Table 4-5 shows the column for relative frequency.

Table 4-5. Relative frequency of BMI in the freshmen class of 2005

BMI range	Number	Relative frequency
< 18.5	25	3.3%
18.5–24.9	500	66.7%
25.0–29.9	175	23.3%
30.0 and above	50	6.7%

Note that relative frequency should add up to approximately 100%, although it may be slightly off due to rounding error.

We can also add a column for *cumulative frequency*, which adds together the relative frequency for each category and those above it in the table, reading down

Table 4-6. The cumulative frequency for the final category should always be 100% except for rounding error.

Table 4-6. Cumulative frequency of BMI in the freshman class of 2005

BMI range	Number	Relative frequency	Cumulative frequency
< 18.5	25	3.3%	3.3%
18.5–24.9	500	66.7%	70.0%
25.0–29.9	175	23.3%	93.3%
30.0 and above	50	6.7%	100%

Cumulative frequency allows us to tell at a glance, for instance, that 70% of the entering class is normal weight or underweight. This is particularly useful in tables with many categories, as it allows the reader to quickly ascertain specific points in the distribution such as the lowest 10%, the median (50% cumulative frequency), or the top 5%.

You can also construct frequency tables to make comparisons between groups, for instance, the distribution of BMI in male and female freshmen, or for the class that entered in 2005 versus the entering classes of 2000 and 1995. When making comparisons of this type, raw numbers are less useful (because the size of the classes may differ) and relative and cumulative frequencies more useful. Another possibility is to create graphic presentations such as the charts described in the next section, which make such comparisons possible at a glance.

Bar Charts

The *bar chart* is particularly appropriate for displaying discrete data with only a few categories, as in our example of BMI among the freshman class. The bars in a bar chart are customarily separated from each other so they do not suggest continuity: although in this case our categories are based on categorizing a continuous variable, they could equally well be completely nominal categories, such as favorite sport or major field of study. Figure 4-1 shows the freshman BMI information presented in a bar chart (unless otherwise noted, the charts presented in this chapter were created using Microsoft Excel).

Absolute frequencies are useful when you need to know the number of people in a particular category: for instance, the number of students who are likely to need obesity counseling and services each year. Relative frequencies are more useful when you need to know the relationship of the numbers in each category, particular when comparing multiple groups: for instance, whether the proportion of obese students is rising or falling. The student BMI data is presented as relative frequencies in the chart in Figure 4-2. Note that the two charts are identical, except for the *y*-axis (vertical axis) labels, which are frequencies in Figure 4-1 and percentages in Figure 4-2.

The concept of relative frequencies becomes even more useful if we compare the distribution of BMI categories over several years. Consider the entirely fictitious frequency information in Table 4-7.

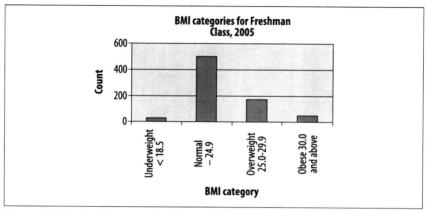

Figure 4-1. Absolute frequency of BMI categories in freshman class

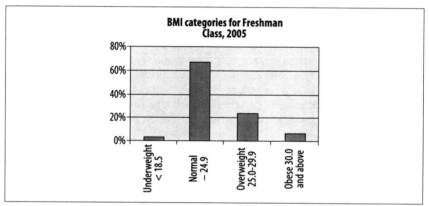

Figure 4-2. Relative frequency of BMI categories in freshman class

Table 4-7. Absolute and relative frequencies of BMI for three entering classes

BMI range	1995		2000		2005	
Underweight < 18.5	50	8.9%	45	6.8%	25	3.3%
Normal 18.5–24.9	400	71.4%	450	67.7%	500	66.7%
Overweight 25.0–29.9	100	17.9%	130	19.5%	175	23.3%
Obese 30.0 and above	10	1.8%	40	6.0%	50	6.7%
Total	560	100.0%	665	100.0%	750	100.0%

Because the class size is different in each year, the relative frequencies (%) are most useful in observing trends in weight category distribution. In this case, there has been a clear decrease in the proportion of underweight students and an increase in the number of overweight and obese students. This information can also be displayed using a bar chart, as in Figure 4-3.

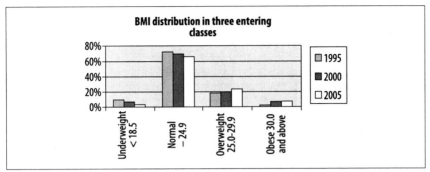

Figure 4-3. Bar chart of BMI distribution in three entering classes

This is a *grouped bar chart,* which shows that there is a small but definite trend over 10 years toward fewer underweight and normal weight students and more overweight and obese students (reflecting changes in the American population at large). Bear in mind that creating a chart is not the same thing as conducting a statistical test, so we can't tell from this chart alone whether these differences are statistically significant.

Another type of bar chart, which emphasizes the relative distribution of values within each group (in this case, the relative distribution of BMI categories in three entering classes), is the stacked bar chart, illustrated in Figure 4-4.

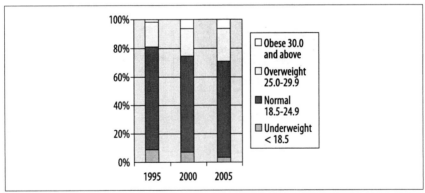

Figure 4-4. Stacked bar chart of BMI distribution in three entering classes

In this type of chart, the bar for each year totals 100 percent, and the relative percent in each category may be compared by the area within the bar allocated to each category. There are many more types of bar charts, some with quite fancy graphics, and some people hold strong opinions about their usefulness. Edward Tufte's term for graphic material that does not convey information is "chart-junk," which concisely conveys his opinion. Of course the standards for what is considered "junk" vary from one field of endeavor to another: Tufte also wrote a famous essay denouncing Microsoft PowerPoint, which is the presentation software of choice in my field of medicine and biostatistics. My advice is to use the simplest type of chart that clearly presents your information, while remaining aware of the expectations and standards within your profession or field of study.

Pie Charts

The familiar *pie chart* presents data in a manner similar to the stacked bar chart: it shows graphically what proportion each part occupies of the whole. Pie charts, like stacked bar charts, are most useful when there are only a few categories of information, and when the differences among those categories are fairly large. Many people have particularly strong opinions about pie charts: while they are still commonly used in some contexts (business presentations come to mind), they have also been aggressively denounced in other contexts as uninformative at best and potentially misleading at worst. So you can make your own decision based on context and convention; I will present the same BMI information in pie chart form and you may be the judge of whether it is useful (Figure 4-5). Note that this is a single pie chart showing one year of data, but other options are available including side-by-side charts (similar to Figure 4-4, to allow comparison of the proportions of different groups) and exploded sections (to show a more detailed breakdown of categories within a segment).

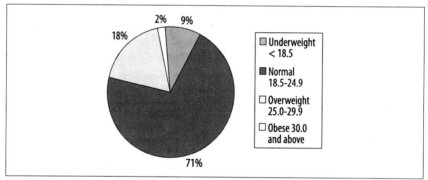

Figure 4-5. Pie chart showing BMI distribution for freshmen entering in 2005

Pareto Charts

The *Pareto chart* or *Pareto diagram* combines the properties of a bar chart, displaying frequency and relative frequency, with a line displaying cumulative frequency. The bar chart portion displays the number and percentage of cases, ordered in descending frequency from left to right (so the most common cause is the furthest to the left and the least common the furthest to the right). A cumulative frequency line is superimposed over the bars. Consider the hypothetical data set shown in Table 4-8, which displays the number of defects traceable to different aspects of the manufacturing process in an automobile factory.

Table 4-8. Manufacturing defects by department

Department	Number of defects
Accessory	350
Body	500
Electrical	120
Engine	150
Transmission	80

Figure 4-6 shows the same information presented in a Pareto chart, produced using SPSS.

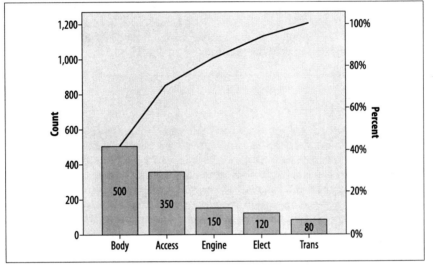

Figure 4-6. Major causes of manufacturing defects

This chart tells us immediately that the most common causes of defects are in the Body and Accessory manufacturing processes, which together account for about 75% of defects. We can see this by drawing a straight line from the "bend" in the cumulative frequency line (which represents the cumulative number of defects from the two largest sources, Body and Accessories, to the right-hand y-axis. This is a simplified example and violates the 80:20 rule because only a few major causes of defects are shown: typically there might be 30 or more competing causes and the Pareto chart is a simple way to sort them out and decide which processes to focus improvement efforts on. This simple example does serve to display the typical characteristics of a Pareto chart: the bars are sorted from highest to lowest, the frequency is displayed on the left y-axis and the percent on the right, and the actual number of cases for each cause are displayed within each bar.

Vilfredo Pareto

Vilfredo Pareto (1843–1923) was an Italian economist who discovered what is now called the Pareto principle, also known as the principle of "the vital few and the trivial many" or "the 80–20 rule." The Pareto principle states that in many circumstances, 80% of the activity or outcomes stem from 20% of the causes. For instance, in many countries, approximately 80% of the wealth is owned by approximately 20% of the people; it is often the case in industrial production that 20% of the errors in production are responsible for 80% of the defects; and in health services utilization, 20% of the patients typically make 80% of the visits. The "vital few" in the Pareto principle are the 20% of people, errors, etc., that account for most of the activity, and the "trivial many" are the other 80% that collectively account for only 20% of the activity. Pareto is best-known today for the Pareto chart, which is commonly used in quality control to help identify which processes are causing most of the difficulties, be they customer complaints or defective products.

The Stem-and-Leaf Plot

The types of charts discussed so far are most appropriate for displaying categorical data. Continuous data has its own set of graphic display methods. One of the simplest is the *stem-and-leaf plot*, which can easily be created by hand and presents a quick snapshot of the distribution of the data. To make a stem-and-leaf plot, divide your data into intervals (using your common sense and the level of detail appropriate to your purpose) and display each case using two columns. The "stem" is the leftmost column and contains one value per row, while the "leaf" is the rightmost column and contains one digit for each case belonging to that row. This creates a plot that displays the actual values of the data set but also assumes a shape that indicates which ranges of values are most common. The numbers can represent multiples of other numbers (for instance, units of 10,000 or of 0.01) as appropriate to the values in the distribution.

Here's a simple example. Suppose we have the final exam grades for 26 students and want to present them graphically. These are the grades:

> 61, 64, 68, 70, 70, 71, 73, 74, 74, 76, 79, 80, 80, 83, 84, 84, 87, 89, 89, 89, 90, 92, 95, 95, 98, 100

The logical division is units of 10 points, e.g., 60–69, 70–79, etc. So we construct the "stem" of the digits 6, 7, 8, 9 (the "tens place" for those of you who remember your grade school math) and create the "leaf" for each number with the digit in the "ones place," ordered left to right from smallest to largest. Figure 4-7 shows the final plot.

This display not only tells us the actual values of the scores and their range (61 to 100) but the basic shape of their distribution as well. In this case, most scores are in the 70s and 80s, with a few in the 60s and 90s, and one is 100. The shape of the "leaf" side is in fact a crude sort of histogram, rotated 90 degrees, with the bars

Stem	Leaf
6	148
7	00134469
8	003447999
9	02558
10	0

Figure 4-7. Stem-and-leaf plot of final exam grades

being units of 10; the shape in this case is approaching normality (given that there are only five bars to work with).

The Boxplot

The *boxplot*, also known as the "hinge plot" or the "box and whiskers plot," was devised by the statistician John Tukey as a compact way to summarize and display the distribution of a set of continuous data. Although boxplots can be drawn by hand (as can many other graphics, including bar charts and histograms), in practice they are nearly always created using software. Interestingly, the exact methods used to construct a boxplot vary from one software program to another, but they are always constructed to highlight five important characteristics of a data set: the median, the first and third quartiles (and hence the interquartile range as well), and the minimum and maximum. The central tendency, range, symmetry, and presence of outliers in a data set can be seen at a glance in a boxplot, and side-by-side boxplots make it easy to make comparisons among different distributions of data. Figure 4-8 is a boxplot of the final exam grades used in the stem-and-leaf plot above.

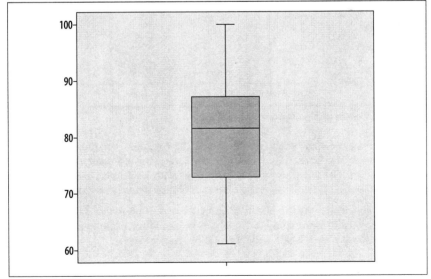

Figure 4-8. Boxplot of exam data (created in SPSS)

The dark line represents the median value, in this case 81.5. The shaded box encloses the interquartile range, so the lower boundary is the first quartile (25th percentile) of 72.5 and the upper boundary is the third quartile or 75th percentile of 87.75. Tukey called these quartiles "hinges," hence the name "hinge plot." The short horizontal lines at 61 and 100 represent the minimum and maximum values, and together with the lines connecting them to the interquartile range "box" are called "whiskers," hence the name "box and whiskers plot." We can see at a glance that this data set is basically symmetrical, because the median is approximately centered within the interquartile range, and the interquartile range is located approximately centrally within the complete range of the data.

This data set contains no outliers, i.e., no numbers that are far outside the range of the other data points. In order to demonstrate a boxplot that contains outliers, I have changed the score of 100 in this data set to 10 and renamed the data set "error." Figure 4-9 shows the boxplots of the two datasets side by side (the boxplot for the correct data is labeled "final").

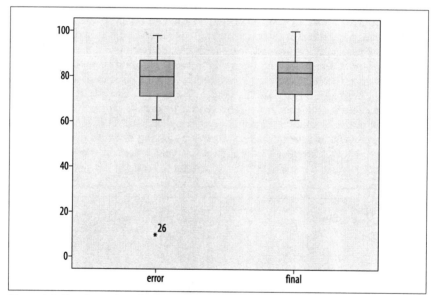

Figure 4-9. Boxplot with outlier (created in SPSS)

Note that except for the single outlier value, the two data sets look very similar: this is because the median and interquartile range are resistant to influence by extreme values. The outlying value is designated with an asterisk and labeled with its case number (26): the latter feature is not included in every statistical package.

A more typical use of the boxplot is to compare two or more real data sets side by side. Figure 4-10 shows a comparison of two years of final exam grades from 2007 and 2008, labeled "final2007" and "final2008", respectively.

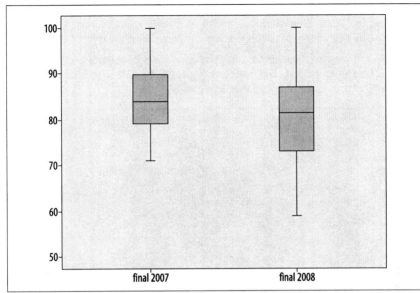

Figure 4-10. Boxplot comparing final exam scores from 2007 and 2008 (created in SPSS)

Without looking at any of the actual grades, I can see several differences between the two years:

- The highest scores are the same in both years.
- The lowest score is much lower in 2008.
- There is a greater range of scores in 2008, both in the interquartile range (middle 50% of the scores) and overall.
- The median is slightly lower in 2008.

The fact that the highest score was the same in both years is not surprising: the exam had a range of 0–100 and the highest score was achieved in both years. This is an example of a *ceiling effect*, which exists when scores by design can be no higher than a particular number, and people actually achieve that score. The analogous condition, if a score can be no lower than a specified number, is called a *floor effect*: in this case, the exam had a floor of 0 (the lowest possible score) but because no one achieved that score, no floor effect is present in the data.

The Histogram

The *histogram* is another popular choice for displaying continuous data. A histogram looks similar to a bar chart, but generally has many more individual bars, which represent ranges of a continuous variable. To emphasize the continuous nature of the variable displayed, the bars (also known as "bins," because you can think of them as bins into which values from a continuous distribution are sorted) in a histogram touch each other, unlike the bars in a bar chart. Bins do not have to be the same width, although frequently they are. The x-axis (vertical axis) represents a scale rather than simply a series of labels, and the area of each bar represents the percentage of values that are contained in that range.

Figure 4-11 shows the final exam data, presented as a histogram created in SPSS with four bars of width ten, and with a normal distribution superimposed, which looks quite similar to the shape of the stem-and-leaf plot.

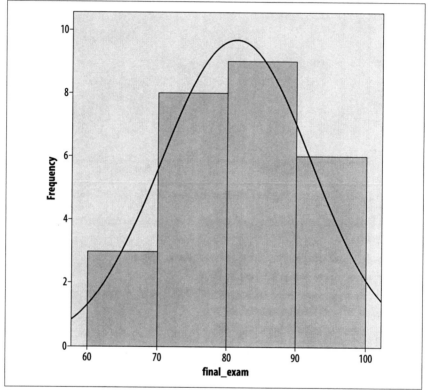

Figure 4-11. Histogram with a bin width of 10

The normal distribution is discussed in detail in Chapter 7; for now, suffice it to say that it is a commonly used theoretical distribution that assumes the familiar bell shape shown here. The normal distribution is often superimposed on histograms as a visual reference so we may judge how closely a data set fits a normal distribution.

For better or for worse, the choice of the number and width of bars can drastically affect the appearance of the histogram. Usually histograms have more than four bars; Figure 4-12 shows the same data with eight bars of width five.

It's the same data, but it doesn't look nearly as normal, does it? Figure 4-13 shows the same data with a bin width of two.

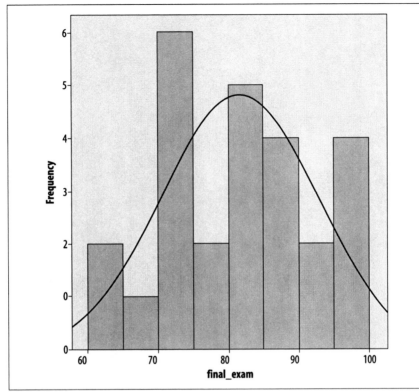

Figure 4-12. Histogram with a bin width of 5

So how do you decide how many bins to use? There are no absolute answers, but there are some rules of thumb. The bins need to encompass the full range of data values. Beyond that, a common rule of thumb is that the number of bins should equal the square root of the number of points in the data set. Another is that the number of bins should never be less than about six: these rules clearly conflict in our data set, because $\sqrt{26} = 5.1$, which is definitely less than 6. So common sense also comes into play, as does trying different numbers of bins and bin widths: if the choice drastically changes the appearance of the data, further investigation is in order.

Bivariate Charts

Charts that display information about the relationship between two variables are called *bivariate charts*: the most common example is the scatterplot. Scatterplots define each point in a data set by two values, commonly referred to as *x* and *y*, and plot each point on a pair of axes. Conventionally the vertical axis is called the *y*-axis and represents the *y*-value for each point, and the horizontal axis is called the *x*-axis and represents the *x*-value. Scatterplots are a very important tool for examining bivariate relationships among variables, a topic further discussed in Chapter 9.

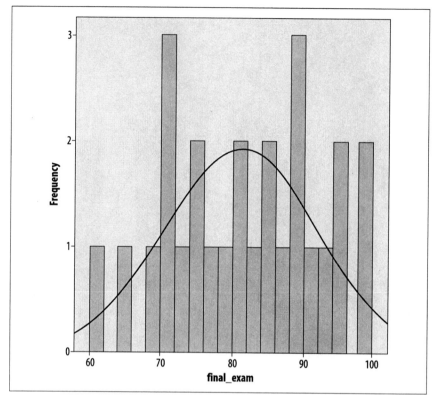

Figure 4-13. Histogram with a bin width of two

Univariate, Bivariate, Multivariate

People sometimes get confused about the meaning of terms like *univariate* and *bivariate*. However, it's easy to keep them straight if you recall that *uni-* means one and *bi-* means two: think of a *uni*cycle, which has one wheel, and a *bi*cycle, which has two. *Multi* means many and in statistics is often used to mean "more than two." *Univariate* statistics such as the mean therefore describe characteristics of one variable, and the bar chart and histogram are examples of univariate graphic displays. *Bivariate* statistics such as Pearson's correlation coefficient describe the relationship between two variables, and bivariate graphs such as the scatterplot display the relationship between two variables. *Multivariate* statistics such as the multiple correlation and multivariate regression describe the relationship between more than two variables.

Scatterplots

Consider the data set shown in Table 4-9, which consists of the verbal and math SAT (Scholastic Aptitude Test) scores for a hypothetical group of 15 students.

Table 4-9. SAT scores for 15 students

Math	Verbal
750	750
700	710
720	700
790	780
700	680
750	700
620	610
640	630
700	710
710	680
540	550
570	600
580	600
790	750
710	720

Other than the fact that most of these scores are fairly high (the SAT is calibrated so that the median score is 500, and most of these scores are well above that), it's difficult to discern much of a pattern between the math and verbal scores from the raw data. Sometimes the math score is higher, sometimes the verbal score. However, creating a scatterplot of the two variables, as in Figure 4-14, with math SAT score on the y-axis (vertical axis) and verbal SAT score on the x-axis (horizontal) makes their relationship clear.

Despite some small inconsistencies, verbal and math scores have a strong linear relationship: people with high verbal scores tend to have high math scores and vice versa, and those with lower scores in one area tend to have lower scores in the other. Not all relationships between two variables are linear, however: Figure 4-15 shows a scatterplot of variables that are highly related but for which the relationship is quadratic rather than linear.

In the data presented in this scatterplot, the x-values in each pair are the integers from −10 to 10, and the y-values are the squares of the x-values. As noted above, scatterplots are a simple way to examine the type of relationship between two variables, and patterns like the quadratic are easy to differentiate from the linear pattern.

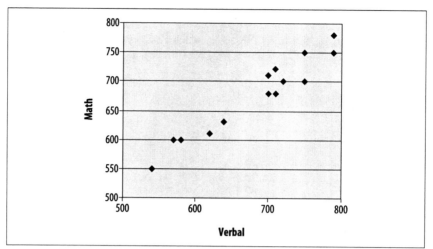

Figure 4-14. Scatterplot of verbal and math SAT scores

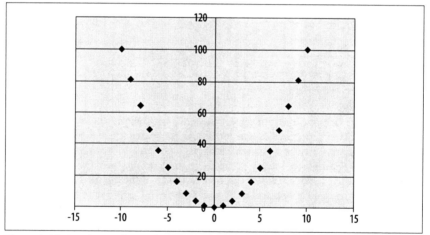

Figure 4-15. Quadratic relationship among variables

Line Graphs

Line graphs are also often used to display the relationship between two variables, often between time on the *x*-axis and some other variable on the *y*-axis. One requirement for a line graph is that there can only be one *y*-value for each *x*-value, so it would not be an appropriate choice for data such as the SAT data presented above. Consider the data in Table 4-10, from the U.S. Centers for Disease Control and Prevention (CDC), showing the percentage of obesity among U.S. adults, measured annually over a 13-year period.

Table 4-10. Percentage of obesity among U.S. adults, 1990–2002 (source: CDC)

1990	11.6
1991	12.6
1992	12.6
1993	13.7
1994	14.4
1995	15.8
1996	16.8
1997	16.6
1998	18.3
1999	19.7
2000	20.1
2001	21
2002	22.1

What we can see from this table is that obesity has been increasing at a steady pace; occasionally there is a decrease from one year to the next, but more often there is a small increase (1–2 percent). This information can also be presented as a line chart, as in Figure 4-16.

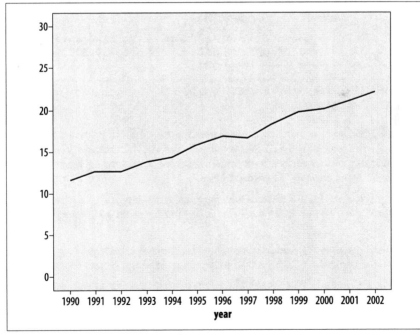

Figure 4-16. Obesity among U.S. adults, 1990–2002 (CDC)

Although the line graph makes the overall pattern of steady increase clear, the visual effect of the graph is highly dependent on the scale and range used for the y-axis (which in this case shows percentage of obesity). Figure 4-16 is a sensible representation of the data, but if we wanted to increase the effect we could choose a larger scale and smaller range for the y-axis (vertical axis), as in Figure 4-17.

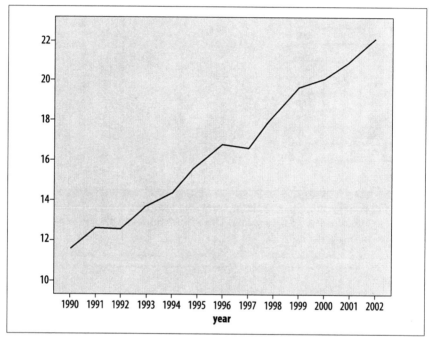

Figure 4-17. Obesity among U.S. adults, 1990–2002 (CDC), using a restricted range to inflate the visual impact of the trend

Figure 4-17 presents exactly the same data as Figure 4-16, but a smaller range was chosen for the y-axis (10%–22.5%, versus 0%–30%). The narrower range makes the differences between years look larger: choosing a misleading range is one of the time-honored ways to "lie with statistics."

The same trick works in reverse: if we graph the same data using a wide range for the vertical axis, the changes over the entire period seem much smaller, as in Figure 4-18.

Figure 4-18 presents the same obesity data as Figures 4-16 and 4-17, with a large range on the vertical axis (0%–100%) to decrease the visual impact of the trend.

So which scale should be chosen? There is no perfect answer to this question: all present the same information, and none strictly speaking are incorrect. In this case, if I were presenting this chart without reference to any other graphics, the scale would be 4–16 because it shows the true floor for the data (0%, which is the lowest possible value) and includes a reasonable range above the highest data point. One principle that should be observed is that if multiple charts are compared to each other (for instance, charts showing the percent obesity in

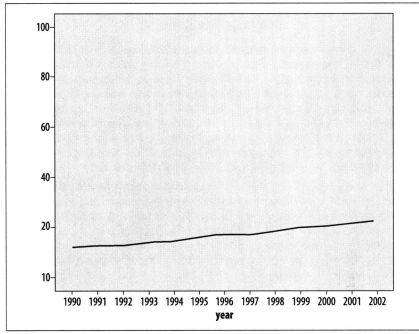

Figure 4-18. Obesity among U.S. adults, 1990–2002 (CDC), using a large range to decrease the visual impact of the trend

different countries over the same time period, or charts of different health risks for the same period), they should all use the same scale to avoid misleading the reader.

Exercises

Like any other aspect of statistics, learning the techniques of descriptive statistics requires practice. The data sets provided are deliberately simple, because if you can apply a technique correctly with 10 cases, you can also apply it with 1,000.

My advice is to try solving the problems several ways, for instance, by hand, using a calculator, and using whatever software is available to you. Even spreadsheet programs like Excel have many simple mathematical and statistical functions available, and now would be a good time to investigate those possibilities. In addition, by solving a problem several ways, you will have more confidence that you are using the software correctly.

Most graphic presentations are created using software, and while each package has good and bad points, most will be able to produce most if not all of the graphics presented in this chapter, and quite a few other types of graphs as well. So the best way to become familiar with graphics is to investigate whatever software you have access to and practice graphing data you work with (or that you make up). Always keep in mind that graphic displays are a form of communication, and therefore should clearly indicate whatever you think is most important about a given data set.

How to Lie with Statistics

Darrell Huff was a freelance writer who also worked as an editor at *Look* magazine, *Better Homes and Gardens*, and *Liberty*, among other publications. His greatest claim to fame, however, is the classic *How to Lie with Statistics*, first published in 1954: some say it is the most widely read statistics book in the world. Huff was not a trained statistician, and his presentation can be charitably described as informal (and some illustrations would be quite offensive if they were included in a contemporary book). Yet this slim volume has retained its popularity over the years, remains in print, and has been translated into many languages, including a Chinese edition published in 2003.

Huff draws his examples of "lies," by which he means the misleading presentation of information, from the contemporary media and political and commercial discourse. Some of his most insightful examples are in the chapters on graphic presentation, from the use of a deliberately misleading scale to the lack of any axis labels at all. One reason for the continuing popularity of *How to Lie with Statistics*, unfortunately, is that many of the misleading techniques he identified in 1954 are still in use today.

Question

When is each of the following an appropriate measure of central tendency? Think of some examples for each from your work or studies.

Mean
Median
Mode

Answer

The mean is appropriate for interval or ratio data that is continuous, symmetrical, and does not contain significant outliers.

The median is appropriate for continuous data that may be skewed (asymmetrical), based on ranks, or contain extreme values.

The mode is most appropriate for categorical variables, or for continuous data sets where one value dominates the others.

Question

What is the median of this data set?

1 2 3 4 5 6 7 8 9

Answer

5: The data set has 9 values, which is an odd number; the median is therefore the middle value when the values are arranged in order. To look at this question more mathematically, since there are n = 9 values, the median is the $(n + 1)/2$th value, and thus the median is the $(9 + 1)/2$th or fifth value.

Question

What is the median of this data set?

12345678

Answer

4.5: The data set has 8 values, which is an even number; the median is therefore the average of the middle two values, in this case 4 and 5. To look at this question more mathematically, the median for an even-numbered set of values is the average of the $(n/2)$th and $(n/2)$th + 1 value; $n = 8$ in this case, so the median is the average of the $(8/2)$th and $(8/2)$th + 1 values, i.e., the fourth and fifth values.

Question

What is the mean of the following data set?

123456789

Answer

The mean is:

$$\bar{x} = \frac{1}{n} \sum_{i=1}^{n} x_i$$

In this case, $n = 9$ and

$$\sum_{i=1}^{n} x_i = (1 + 2 + 3 + 4 + 5 + 6 + 7 + 8 + 9) = 45$$

so $\bar{x} = 45/9 = 5$.

Question

What are the mean and median of the following (admittedly bizarre) data set?

1, 7, 21, 3, −17

Answer

The mean is $((1 + 7 + 21 + 3 + (-17))/5 = 15/5 = 3$.

The median, since there are an odd number of values, is the $(n + 1)/2$th value, i.e., the third value. The data values in order are (−17, 1, 3, 7, 21), so the median is the third value or 3.

Question

What are the variance and standard deviation of the following data set? Calculate this using both the population and sample formulas.

1 3 5

Answer

The population formula to calculate variance is:

$$\sigma^2 = \frac{1}{n} \sum_{i=1}^{n} (x_i - \bar{x})^2$$

And the sample formula is:

$$s^2 = \frac{1}{(n-1)} \sum_{i=1}^{n} (x_i - \bar{x})^2$$

In this case, $n = 3$, $\bar{x} = 3$, and the sum of the squared deviation scores = $(-2)^2 + 0^2 + 2^2 = 8$. The population variance is therefore 8/3 or 2.67, and the population standard deviation is the square root of the variance or 1.63. The sample variance is 8/2 or 4, and the sample standard deviation is the square root of the variance or 2.

5

Research Design

When trying to make sense of the world using statistics, it's important to consider which type of research design will provide you with the most accurate answer for the type of question you are asking. Typically, the selection of a particular design involves an amount of horse trading between the extent to which a research environment is able to be controlled versus only being able to be observed, and whether the research goal involves determining some underlying cause for a phenomenon (science), or whether the goal is to optimize the yield or output for a specific process while minimizing effort (technology). Various other impediments to the use of certain designs—such as human research ethics committees—may require you to carry out your research in ways that are less than ideal in a statistical sense, but ensure that your work is consistent with community rather than just scientific values.

The two main types of research design are *experimental* and *observational* studies. In an experiment, some degree of manipulation is involved, since the intention is that the researcher should maximize control over as many aspects of the environment as possible. All variable measurements and manipulations are under research control, including the allocation of experimental units (or subjects or participants). An observational study implies that no change of the environment is necessary, and that the allocation of experimental units to groups is outside the control of the researcher. Indeed, the goal of an experiment is to control the environment in such a way that manipulation of a *treatment* variable yields a direct, corresponding change in one or more *response* variables. In practice, it may be very difficult to control the confounding effects of all variables in an environment, since some *intermediate* variables may not even be known to exist, yet still exert an influence. Note that a group or category of treatments is known as a *factor*. You may also see treatments referred to as *independent* or *predictor* variables, and responses may also be known as *dependent* variables. While some experimental designs attempt to measure the effect of a single treatment or factor on a response, a single experiment can actually involve multiple factors, each with multiple levels. The goal here is to establish the main effect of each factor on the response, but also possibly their interaction. For example, a study examining the effect of seasonal factors and time of drug administration on mood might have two factors; season (summer, fall, winter, spring) and medication administration (morning, night). Thus, the goal would be to determine a main effect for season (e.g., are you happier in summer), a main effect for medication administration (e.g., is morning more effective than night), and any interactions (e.g., are you happiest in summer when medication is taken in the morning). Any or all of these main effects and interactions may form part of your research hypothesis.

An observational study is less invasive than an experimental study, but has correspondingly less effectiveness than an experiment, in terms of inference; the major difference between an observational and an experimental study lies in the strength of the explanation that can be drawn from the results of each. In order to minimize *systematic error* (or *bias*, a subject discussed at greater length in Chapter 1), an experimenter will randomly allocate experimental units to groups, whereas in observational studies, this is generally not possible, as you can only observe what is present in the environment during an observation. There may be very good reasons why you would choose an observational approach over an experimental design: for example, while there is a large body of work in social psychology in which the influence of many intermediate variables is controlled in the laboratory, the ultimate ecological validity of the work is clearly most accurately demonstrated in the real world. Conversely, the variables measured during observation may have had their construct validity determined in controlled laboratory experiments. So, over the long term, a research plan might set out a combination of experimental and observational work to be performed.

Observational Studies

Not all observational studies (also known as *quasi-experimental studies*) are the same in terms of their effectiveness in answering research questions. The least biased observational studies are "forward-looking" (*prospective*) and focus on a randomly selected group (a *cohort*). Thus, a *prospective longitudinal study*, also known as a *cohort study*, observes people forward in time from their entry into the study. Cohort studies often start from birth or another common point in time, such as the start of school. Systematic error can be reduced by ensuring that all participants have as much in common as possible, for example, by selecting on birthdate, social class, etc. In contrast, a *retrospective longitudinal study* reviews participants backward in time from their entry into the study. The goal of this type of *case control study* might be to determine, for example, what social activities in the clinical group have led to catching a specific disease. However, if there is a long gap in time between the study and the events in question, then *recall bias* may influence the results.

Case control designs are also the only ethically supportable designs where a treatment may be harmful, as is the case in disease; for example, in order to link the AIDS syndrome with HIV infection, it would be clearly unethical to administer the virus to an experimental treatment group and compare their immune responses and clinical outcomes with a control group. For some case control studies, even if an experimental manipulation was possible, there simply may not be enough potential participants—particularly for very rare disease cases—so an observational design in preferable. Experimental designs are usually only appropriate where the treatments are either known to be all not harmful or all beneficial. For example, a researcher investigating the effectiveness of a new form of pain relief might use baseline responses for other pain relief medications that are known to be safe in the quantities administered.

Two other types of observational study are worth mentioning. A *cross-sectional design* involves a single observation (e.g., a questionnaire or an interview), which may be useful if an immediate response to a specific question is required (e.g., what donut flavors are popular in New York City today), but have clear limits on their *generalizability* (since donut flavor preferences may change seasonally). Another technique is so-called *secondary analysis*, where data from many different sources is combined to investigate a particular problem. In this case, the investigator does not exert any influence over the data collection, and the analysis is generally retrospective, since data is usually discovered post-hoc from a range of sources. While some researchers have questioned the validity of relying on other people's data, secondary analysis can be very useful for developing new leads when investigating difficult or complex questions. For example, cohort studies undertaken in different countries may have systematic bias related to geographical or social factors, and secondary analysis can be used to trace and/or eliminate this bias by examining whether the relationship between variables is consistent across these different countries. Note that the only source of control in a secondary analysis is the selection of variables, although it may be possible for a secondary analysis to specify new variables to be measured in a future prospective study.

While observational studies are generally considered weaker in terms of statistical inference, they have one important characteristic: response variables (like human behavior) can often be observed within the natural environment, enhancing their *ecological validity*, or the sense in which what is being observed has not been artificially constrained by engaging in a narrowly defined experimental paradigm. Going one step further, some observational studies use participant observation methods, where a researcher becomes involved in the activity under study. If this participation is hidden from the actual participants, then ethical issues may arise around the use of deception. However, in other cases, there may be ethical objections to withholding an intervention in which investigator participation is required. For example, in clinical settings in speech language pathology, the clinical investigator would typically play a very active role in eliciting responses to various treatments, since these would not normally be forthcoming from the participants. Observational studies may be the only solution to investigate a research problem where ethical considerations prevent the use of randomized experimental trials.

Observational studies potentially suffer from a number of biases, including biases in selection, which are either known or suspected. One way of removing these biases is to make adjustments for those that are known (perhaps using a covariance correction), and at least making clear those that are suspected to exist. Often, even if the magnitude of a bias is not known, the direction of bias can be easily determined. For example, members of a conservative political party may be reasonably assumed to hold conservative social values, even if the strength of these beliefs or their extent is unknown (as an illustration, all conservatives may favor jailing of drug dealers, but a proportion may favor the death penalty over imprisonment).

Case control studies match patients with nonpatients based on covariates to eliminate bias, without having to use an experimental design. A covariate in the context of observational studies is a variable that is unaffected by the administration of a treatment (e.g., collected prior to a study), such as age, sex, or IQ, whereas a variable that is predicted to be affected by a treatment is known as an outcome. This type of matching is also common in experimental designs (i.e., a *matched pair design*).

For some observational studies, such as clinical populations, the treatment group may be relatively small, but in order to match as closely as possible on all covariates, a large potential pool of controls is generally useful. Indeed, it may also be possible to further eliminate biases by matching a single member of the treatment group to several controls who are all matched on key covariates. You can imagine, though, that as the number of covariates increases, and/or as the number of controls per treatment unit increases, the likelihood of finding "perfect matches" decreases. If you are concerned that bias may be creeping in, you can calculate a *propensity score*, which is the probability of being assigned to a treatment or control group based on the covariate values. Think of it as a form of validation for the assignments that you have made during case control.*

So far, I've focused on known biases, but what about biases that are unknown? In experiments, randomization takes care of both identified and unknown sources of bias—if the selection is performed randomly, both types of bias will be controlled for. But in an observational study, there may simply be a confounding, hidden source of bias that is unknown, such as a hidden covariate. In this situation, it may be possible to use a sensitivity analysis to determine whether there is a source of hidden bias, especially one of great magnitude.

Experimental Studies

There are three different elements to an experimental study, and the configuration of the design can range from the very simple to the very complex:

Experimental units
> The objects under examination. In human experiments, units are generally referred to as participants, given their active engagement in the experimental process.

Treatments
> The procedures applied to each unit in the experimental setting, which are either qualitative or quantitative (with contrasting, well-defined levels of interest), depending on whether the variables are real values or categorical.

---

* For more information, see Rosenbaum, P. R., and Rubin, D. B., (1983), "The Central Role of the Propensity Score in Observational Studies for Causal Effects," *Biometrika* 70, 41–55.

Responses
These form the criteria on which the effect of the experimental treatments can be compared.

The term *treatment* implies a fairly active process of performing some transformation on the unit, whereas the status of the unit may already be known, and no transformation is necessary. A random allocation to a specific treatment group (control or experimental) may be determined on the basis of variables such as age or sex, for control purposes, or the variables may form an integral part of the hypothesis. For example, a psychological study of reaction times in driving might attempt to control for sex differences by balancing the number of males and females in the control and experimental groups, or the hypothesis might explicitly predict that sex has an effect on reaction times. In the latter case, sex has an explanatory role in the study.

In some experimental designs, a comparison is made between a *baseline* measurement for each unit before treatment, and the measurement for the unit after treatment (also known as pre-test and post-test responses). This type of design is known as a *within-subjects* design, and provides a high degree of experimental control, since measurements on units are only ever made with themselves, i.e., participants act as their own controls. Conversely, in a *between-subjects* design, comparisons are made between units that are matched on as many variables as possible, to ensure the least confounded comparison of the treatment on units from the control and experimental groups. Note that this terminology is mainly used in social and health sciences, and may differ in other disciplines. I will discuss these commonly used designs further in the "Blocking and Common Designs" section later in this chapter.

Ingredients of a Good Design

The goal of an experiment is to clearly show differences, if they exist, when a treatment is applied to a group of experimental units. Differences can be expressed in terms of magnitude, as well as through limits placed on the *confidence* of the analysis, which may arise from random error. Good procedures for allocating experimental units to treatment and control groups are an important step here, and indeed, separate experimental from observational studies.

A major goal of any experimental design is to minimize or preferably eliminate *systematic errors* or *biases* in the data collected.

For many reasons—including ethical and resource considerations—the amount of data collected should be minimally sufficient to answer a particular research question. The use of effective sampling and *power* calculations ensures that the smallest number of experimental units is subjected to experimentation, and that a result can be achieved with the least cost and least effort.

An effective research design makes analysis much easier later on. For example, if you design your experiment in such a way that you will not encounter missing observations, then you will not need to worry about coding missing data and any limitations of interpretation from your results that may subsequently arise (a topic explored further in Chapter 3). Where missing data arises because of the loss of experimental units due to mortality (in the human case), then a mortality bias may exist in your data.

Statistical theory is very flexible to the extent that many sophisticated types of designs are mathematically possible, but in practice, most statistics (and therefore designs) are structured according to the requirements of the *general linear model*. This makes analysis easy, since many techniques such as correlation and regression are based on this model. But to make valid use of the general linear model, an experiment must be designed with several important factors in mind, including *balance* and *orthogonality*.

Balance means that treatments are administered in equal numbers within each experimental block, meaning that they will occur with the same frequency. A lack of balance indicates that some bias has arisen in the allocation of units to the treatment or control group. Randomization of group allocation, blinding, and identifying biases are all mechanisms for ensuring that balance is maintained; these are discussed later in this chapter.

Orthogonality means that the effects of different treatments can be independently estimated without interfering with each other. For example, if you have two treatments in an experiment, and you build up a statistical model that measures their effects on experimental units, you should be able to remove either treatment from the model, and get the same answer for the remaining treatment.

None of this is as complicated as it first sounds, and if you stick to some well-known recipes and templates for factorial design, you won't need to worry about more exotic exceptional cases.

Gathering Experimental Data

So, you want to run an experiment, but where do you start? I will outline some general guidance below, roughly in the order in which you need to carry out each step, but my advice is to stand on the shoulders of giants. That is, if you are running experiments in a scientific discipline, then look at some articles from research journals in that specific discipline, and ensure that the designs and analyses that you carry out are consistent with what others are using in the field. The process of peer review ensures that the methodology used has been vetted by at least two experts—so, how can you go wrong? In industry or manufacturing, it may be harder to find guidance, but company technical reports and previous analyses should provide some previous examples—even if they have not been peer-reviewed—that you might find instructive.

Having said that, you will be surprised at just how much variation and urban mythology surrounds experimental design, so let's walk through the steps one by one. I'll focus primarily on science in this example:

1. Identify the experimental units that you want to measure something from.
2. Identify the treatments that you want to administer, and the controls that you will use.
3. Specify treatment levels.

4. Identify the response variables that you will measure from the experimental units.

5. Generate a testable hypothesis that predicts what effect the treatment will have on the response variables.

6. Run the experiment.

7. Analyze the results.

That was easy, right? Design (steps 1–5) can be easy, but let's look at each step in detail to see what's really involved.

Identifying Experimental Units

Recall that statistics are estimates of population parameters drawn from a sample. In order to ensure that these estimates are accurate (i.e., representative), the analyses that you later use to determine the *significance* of experimental effects rely on the assumption that you have truly selected these units randomly. *Bias* can very easily creep into a design at this first stage, and yet circumstances may dictate that this cannot be easily avoided. For example, many research studies in psychology solely use undergraduate psychology students as participants. This serves two purposes: firstly, as part of their coursework, students are exposed to a wide variety of experimental designs and get to experience, first-hand, what is involved in running an experiment; and secondly, the participant group is easily accessible for psychological researchers. In some ways, control may be achieved in this group because the participants may be the same age, have an even split in terms of sex, may come from the same home town, listen to the same music, dress with the same fashion styles, have above average intelligence, etc. However, in selecting from a limited sub-sample of the population at large, the inferences that you can make about the broader population are limited. Thus, some research papers tell us more about the behavior of college students at this specific point in time, rather than the population at large. It's a difficult area, and different fields will have different expectations about what is acceptable practice in terms of wider *inference*.

What is meant by random selection in this context? Imagine a lottery in which every citizen of a country receives a ticket. All of the tickets are entered into a large box, which is mixed by rotation through many different angles. An assistant is then asked to pick one ticket by placing his hand in the box and selecting the first ticket that he touches. In this case, every ticket has an equal chance of being selected. If you needed 100 members for a control and 100 for an experimental group, then you could select them using a similar process, where the first 100 selections are allocated to a control group, and the next 100 are allocated to the treatment group. Of course, you could alternate selection by allocating the first ticket to a control group, the next to the treatment group, the third to a control group, and so on. However, if the sampling is truly random, then the two techniques will be equivalent. Importantly, for the selection to be random, the allocation of any particular individual must be truly independent of the selection of any of the others.

Are there any situations where random allocation is not possible? Certainly—and this is a major limitation for applying inference in experimental studies. There are relatively few sources of true random variation; if, for example, you use a computer program to generate a set of random numbers as the basis for allocating participants to treatment groups, the program will almost certainly be using a pseudorandom number generator, which approximates the characteristics of truly random numbers—but the numbers are not truly randomly generated. This is because such programs are generally seeded with a small set of numbers. Some strategies to increase the randomness, such as using a timestamp as a seed, rather than a fixed value, produce better numbers. Fortunately, there are techniques— such as the Pearson chi-square test or Kendall and Smith's randomness hypothesis tests—that can be used to determine the randomness of any set of numbers that might be used for treatment allocation.

Random selection is not always possible in real-world applications. Indeed, in many areas of science, there are quite structured (and potentially biased) mechanisms for identifying groups of interest on which an experiment is conducted, where inference may have to be limited to a population that is not perhaps the largest possible population.

Imagine that you are a microbiologist interested in examining bacteria present in hospitals. If you use a filter with pores of diameter one μm, any bacteria less than this will not be part of the population that you are observing. This sampling limitation will introduce systematic bias into the study; however, as long as you are clear that the population about which you can make inferences is bacteria of diameter > one μm, and nothing else, your results will be valid.

One way to ensure that you make valid inferences is to limit the population you are experimenting on. Concise experimental results that make valid inferences are important. Over time, accumulating evidence from many well-designed studies will be more reliable in identifying population characteristics. For example, carrying out tests of reaction time to English words may be used to make inferences about the perceptual and cognitive processing performance of English-speaking people. However, it may only be used to suggest further hypotheses about non-English speakers. Subsequent experiments aimed at increasing the generalizability of the finding might include the same experiment but with German words displayed to German speakers, and Japanese kanji to Japanese speakers. Indeed, this is the way that more general results are built up in science.

Identifying Treatments and Controls

Treatments are the manipulations that you want to perform in order to demonstrate an experimental effect. For example, a pharmaceutical company has spent millions of dollars on a new "smart" drug, and after many years testing in the lab, they now want to see if it works in practice. So, they set up a clinical trial, where they select 1,000 participants by randomly selecting names from a national phone book, giving a truly representative sample of the population on significant parameters, such as age, gender, etc. Luckily, they have a 100% success rate in recruiting participants for the study (everyone wants to be smarter, right?), so they don't have to worry about noncompliance.* All participants are going to be tested on the same day in identical experimental conditions (exactly the same location, temperature, lighting, chair, desk, etc.). At nine in the morning, participants are administered an intelligence test via computer; at noon, they are given an oral dose of the "smart" drug with water; and at three p.m., they sit for the same intelligence test again. The results showed an average increase in intelligence of 15%! The company is ecstatic, and they release the results of the test to the stock exchange, resulting in a large increase in the company's share price. But what's wrong with the treatments administered in this experiment?

Firstly, since everyone was tested in exactly the same place, and under exactly the same experimental conditions, the result cannot be automatically assumed to apply to other locations and environments. If the test was administered under a different temperature, the results might be different. In addition, there might have been some aspect of the testing facility that biased the result—say, the chair or desk used, or building oxygen levels—and it's difficult to rule these confounding influences out.

Secondly, the fact that the baseline and experimental conditions were always carried out in the same order will almost certainly have been a contributing factor in the 15% increase in intelligence, i.e., there will have been a learning effect from the first time that participants undertook the test to the second time, given that the questions were exactly the same (or even if they were of the same general form).

Thirdly, there is no way that the researchers can be sure that some other confounding variable was not responsible for the result, since there was no experimental control in the overall process; for example, there could be some physiological response to drinking water at noon (in this paradigm) that increases intelligence levels in the afternoon.

Finally, participants could be experiencing the placebo effect, where they expect that having taken the drug, their performance will improve. This is a well-known phenomenon in psychology, and requires the creation of an additional control group to be tested under similar circumstances, but with an inert rather than active substance being administered.

There are numerous such objections that could be made to the design as it stands, but fortunately, there are well-defined ways in which the design can be strengthened by using experimental controls. For example, if half of the randomly selected sample was then randomly allocated to a control group, and the remaining half allocated to an experimental group, then an inert control tablet could be administered to the control group, and the "smart" drug to the experimental group. In this case, the learning effect from taking the test twice can be estimated from the control group, and any performance differences between the two groups can be determined statistically after the treatment has been applied.

___

* Noncompliance is a major issue in experimental designs, since if participants are deleted from a truly random sample, the deletion will cause the sample to be nonrandom, and the limits of inference must be correspondingly reduced. In practice, many analytical procedures make allowances for missing data, and missing cases should always be included where possible in the analysis.

Of course, in real clinical drug trials, the research designs are structured quite differently, and investigations are staged in phased trials that have explicit goals at each step, starting with broad dose-response relationships, investigations of toxicity, and so on, with controls being tightened at each stage until an optimal and safe dosage can be identified that produces the desired clinical outcome.

Specifying Treatment Levels

In practice, you may not be specifically interested in determining whether some factors are influencing the experimental result—you may simply wish to cancel out any systematic errors that may be arising. This can often be achieved by balancing the design, to ensure that equal numbers of participants are tested in different levels of the treatment. For example, if you are interested in whether the "smart" drug increases intelligence in general, your sampling should ensure that there are an equal number of male and female participants, a spread of testing times, etc. However, if you are interested in determining whether sex or time of drug administration influences the performance of the drug treatment, then these variables would need to be explicitly recognized as experimental factors, and their levels specified in the design. For categorical variables like sex, the levels or categories (male and female) are easy to specify. However, for continuous variables (like time of day), it may be easier to collapse the levels to hourly times (in which case there will be 24 levels, assuming equal dosage across the 24-hour day), or simply morning, afternoon, and evening (3 levels). Once again, the research question guides the selection of levels and the experimental effects that you are interested in. Otherwise, counterbalancing and randomization can be used to mitigate error arising from bias. Indeed, replication of the result but extending or being able to generalize across spatial and temporal scales is important for establishing the generalizability of the result.

Once treatment levels have been determined, researchers generally refer to the treatments and their levels as a formal *factorial* design, in the form $A_1(n_1) \times A_2(n_2) \times \ldots A_x(n_x)$, where $A_{1..x}$ are the treatments and $n_{1..x}$ are the levels within each treatment. For example, if you wanted to determine the effect of sex and time of drug administration on intelligence, and you had a control and experimental group, then there would be three treatments, with their levels as follows: SEX: male/female, TIME: morning/afternoon/evening, DRUG: smart/placebo. Thus, the design can be expressed as SEX (2) $\times$ TIME (3) $\times$ DRUG (2), which can be read as a "2 by 3 by 2" design. We will deal with the analysis of *main effects* within and *interactions* between these treatments in the analysis chapters.

Specifying Response Variables

Sometimes, the response variable will be fairly obvious, but in other cases, more than one response variable may need to be measured, depending on how precisely the variable can be operationalized from some abstract concept. Intelligence is a very good example: the abstract concept may appear to be fairly straightforward to the layperson, and yet there is no single test that directly measures intelligence. Instead, many different measures of general ability across different skills (numerical, analytical, etc.) are measured as response variables, and may be combined to form a single number (an intelligence quotient, or IQ), representing some latent

Treatments or Characteristics?

One important difference arises between physical and social sciences in the definition of treatments. The word treatment implies an active process of applying a process that is transformative, e.g., administering a drug to improve intelligence. However, in social sciences, treatments are quite often made up of fixed characteristics, such as sex. Should such characteristics be regarded as treatments since no transformation takes place? Are designs that use such treatments experimental, quasi-experimental, or actually observational? The issue is fundamental to demonstrating a causal relationship between treatments and responses, since using a nontransformative treatment leaves open the question of what characteristic of the experimental units is actually responsible for any differences observed between treatment levels. Ultimately, the type of inferences that can be made from any research design are limited by such considerations. In technological research, an experiment may have a more explicit optimization goal, such as the estimation of an effect size, to determine the optimal combination and proportions of different treatments and levels that will maximize the value of the response variable.

In some fields, a distinction is made between "independent" variables (or set characteristics) and treatments, which seems to be a sensible compromise.

structure amongst the correlated responses. There are advanced techniques (covered in Chapter 16) that describe how to combine and reduce the number of response variables to a smaller, more meaningful (in the sense of interpretation) set.

Indeed, it is likely that for a concept as problematic to define as intelligence, the safest bet might be to use a number of different instruments to obtain response variables, and then determine how much they agree with each other. Indeed, techniques for determining the mutual consistency of response variables play an important role in validating experimental designs.

There are three main types of response variables: *baseline*, *response*, and *intermediate*. In the previous section, we saw how a baseline measure of intelligence was used to estimate a direct experimental effect on a response variable (intelligence). An intermediate variable is used to explain the relationship between the treatment and response variable where it is indirect, but controllable. If you're interested in establishing a causal relationship as part of an explanatory model, then you will clearly want to be aware of all of the variables involved in a process.

In some designs, the distinction between treatments and intermediate variables may not be important. For instance, if you are a chemist and you are interested in the chemical properties of water, you may be happy to work at the level of atomic particles (protons, neutrons, electrons) rather than the subatomic level in your analysis.

In very complicated systems, unanticipated interventions (or unobservable intermediate variables) may influence the result, especially if such variables are highly correlated with a treatment, or where the act of performing the experiment changes the behavior of that which is being observed. Thus, it may be hard to causally draw out whether a treatment is specifically responsible for a change in response. Another general principle is that the longer the delay between a treatment being administered and a response being observed, the greater the likelihood of some intermediate variable affecting the result, and possibly leading to spurious conclusions. Seasonal factors, such as temperature, humidity, and so on, exert a very strong influence on the outcomes of agricultural production, for example, perhaps more than a treatment looking at fertilizer productivity.

Inference and Threats to Validity

The choice of research design—experimental or observational—is usually governed by constraints over how data of interest can be collected, and the type of *statistical inference* required; that is, inferring characteristics of a population using statistics calculated on a sample considered to be representative of that population. Before I review the structure of these research designs, I will firstly discuss inference, and why it is the cornerstone of statistics.

Sometimes, data is collected for a very specific purpose, without any desire to understand, characterize, or make predictions about a broader phenomenon. For example, clinicians in a hospital ward dedicated to the treatment of hypertension may be interested in how different anti-hypertensive medications affect each of the patients individually—these effects are difficult to predict, since so many physiological factors are involved. The selection of safe and effective medication for each patient is the primary motivation. In each of the patient cases, data is collected and stored for the primary clinical purpose. However, after a number of years, clinicians begin to notice some patient factors that appear to predict which drug will have the best clinical outcome for certain patients. Drug A appears to be most effective for men, while Drug B appears to be most effective for women. Rather than administering A and B to every patient—since drug administration is inherently risky—the clinicians decide that they would like to determine whether the results that they have observed for individual patients are true for the wider population. Thus, based on the samples that they have obtained in the past (clinical) context, they wish to make inferences about the parameters of a larger population. This will assist in both quantitatively characterizing and predicting the effects of the drugs on patients.

Who is the population in this instance? The population is the group of persons who suffer from hypertensive illness. Since it is infeasible to test the effects of Drug A and Drug B on all hypertensive patients worldwide, a sample—a representative subset of the population—is usually selected for such a study. A number of research designs could be used to investigate the two hypotheses for the study, i.e., that Drug A is most effective for men, and that Drug B is most effective for women. Case control designs or clinical trials could be used to test the effects of Drug A and Drug B on men and women, perhaps using matched samples for the control and experimental conditions. You will learn more about these techniques and strategies later in this chapter.

It's important to distinguish between the estimates obtained from a sample and the numerical characteristics of a population that would be determined if every member of the population were measured. For example, a parameter (from the population) might be characterized as the percentage of male patients whose systolic blood pressure reached 120 after administration of Drug A, while the statistic (from the sample) might be the percentage of male patients whose systolic blood pressure reached 120 after administration of Drug A at Hospital X. If the responses to the treatment by patients at Hospital X were truly representative of the population, then the statistics computed would be considered true estimates of the population parameters.

In real-life situations, you rarely encounter a variable with zero variance, i.e., where every experimental unit responds in exactly the same way. In this case, statistics from random samples are treated as random variables, and the responses gathered take on the form of a probability (or sampling) distribution. The properties of these distributions and associated theorems (such as the Central Limit Theorem*) mean that you can make valid comparisons between experimental and control groups using the designs described in this chapter, and using the analytical tools described in Chapter 7 and the balance of this book.

Validity means being sure that what you are measuring is what you intend to measure or claim to have measured, as described in Chapter 1. In experiments, the validity of an observed treatment difference between responses (between an experimental and control condition, for example) is the extent to which the result cannot be attributed to error in sampling or measurement. Continuing the hypertensive drug example, if case-control and subsequent experimental studies were undertaken by the same clinical team inside the same hospital, to what extent would the results be valid? The main concerns about greater interpretation vis-à-vis the population would be the fact that some consistent bias in measurement may be giving rise to the result, and/or that the clinical population being tested was too small and/or not sufficiently representative of the broader hypertensive population. Thus, two easy techniques to improve the validity of a research result are to test across different laboratories, research teams, and facilities, and to sample with sufficiently large sample sizes to observe an experimental effect. The actual sizes required to test the statistical significance of differences between treatments depends on the specific test being used (power and sample size calculations are further discussed in Chapter 18).

In some fields, such as psychology, general notions of validity have been refined to develop typologies of validity. The American Psychological Association, for instance, originally classified validity into four categories: content validity, construct validity, concurrent validity, and predictive validity. In more recent times, this classification has itself become more refined into the following major types: construct validity, content validity, internal validity, and statistical validity. I will review each of these below, and discuss threats to validity in each case.

* Where the normal distribution accurately represents the actual distribution of responses, since the Central Limit Theorem predicts that sample means will approximate the normal distribution.

Construct Validity

Construct validity refers to the extent to which an abstract concept that is operationalized as a variable is both sound and measurable. Thus, an invalid construct is one in which there is no general agreement in the field as to whether the construct is useful and/or correct, or simply disagreed upon. Even if a theoretical construct is widely accepted, there may be little agreement on how it should be measured, which means an assessment of convergent validity is required. A classic example is intelligence, where there are divergent views on what the construct actually means, as well as how each of the specific constructs are operationalized. For example, even if you accepted that tests of general cognitive ability were valid operationalizations of intelligence, the many batteries of tests of general ability are not always correlated with each other.

Internal consistency for such test batteries means that there is consistency within the set of items for a given scale, while the agreement between different batteries can be assessed through criterion validity. Measures have been developed to estimate internal consistency (further discussed in Chapters 1 and 19), including *Cronbach's alpha*, which will be higher if a correlation is consistent for large samples when compared to small samples.

Criterion validity for these sorts of tests indicates how well the responses for the particular battery of tests matches others used in the field, or responses for different tests that (based on theoretical assumptions) should be correlated. For example, if you develop a new test of verbal ability, you would expect the distribution of results to be correlated with the results of similar tests, or other measures of linguistic performance. On the other hand, if you feel that the criterion is wrong, you could create an entirely new test and consider some other method of determining validity (such as predictive validity).

Predictive validity can be achieved when a response can be shown to be accurately predictive of the intended phenomenon. A classic example is whether performance on exams at the end of high school—used to determine entrance into specific courses—is actually predictive of performance in those courses, and/or the professions for which they are intended to prepare students (in many studies, they are not).

Another phenomenon, known as *regression towards the mean*, predicts that students at either end of the spectrum (e.g., those scoring 0% or 100%) would be more likely on re-test to score closer to the mean for purely statistical reasons, because of measurement error. In this example, the phenomenon arises because some responses (correct for the 100% group, and incorrect for the 0% group) were no doubt the result of chance (or guessing) as much as knowledge, skill, or intelligence. Thus, you would expect on re-test that—given that the chances of all your guesses being right or wrong a second time around are low—low scorers would score higher, and high scorers would score lower. The phenomenon is observed throughout the natural world and the business world, and may lead to spurious inferences being made about current weak performance because previous performance was high. For example, many companies that report strong, above-average earnings in their first year may see a subsequent reduction in their second year, which is independent of any treatment effect. You can see that in nonexperimental (naturalistic) settings, it's easy for these sorts of inappropriate inferences to be made; the main strategy for dealing with such biases is to be aware of them!

External validity is usually the most significant issue when attempting to draw inferences about population parameters from sample statistics. Where there are many successful ways to replicate experiments in different conditions, by different teams, then external validity is enhanced. Any study that claims to make valid inferences about a population, based on the statistics from a single study, is usually treated with skepticism.

Content Validity

Content validity is concerned with establishing acceptance that a construct measures what it claims to. Typically, an operationalized variable might be criticized as being too broadly or too narrowly defined, in the sense that it measures too much or too little information, or where the variable scope is just not clearly defined. For example, intelligence tests that focus solely on tests of verbal and/or numerical ability may neglect other critical facets of intelligence, such as emotional intelligence.

Internal Validity

Internal validity can only be established when there are no biases that may lead to the misidentification of explanatory variables in a study. Internal validity can be threatened by:

- Systematic biases in selection
- Intentional bias inherent in a researcher's genuine desire to establish causal relationships
- Self-serving bias in responding to solicitation for participation in a study on the part of participants
- Giving answers to questionnaires and other studies involving self-ratings that make the participant "look good"

Where an experimental design is used, issues involving comparisons between treatments can also become a validity issue (in addition to raising ethical issues). For example, is it ethical to withhold a treatment from the control group simply for the purpose of establishing experimental control? If blinded designs are not implemented—and sometimes this is simply not practical—the potential for rivalry between treatment groups must be considered a source of bias.

Statistical Validity and Hypothesis Testing

Achieving statistical validity ensures that correct conclusions are drawn from any statistical analyses performed on data. Validity in this sense can be achieved by carrying out hypothesis testing—using predictions about data that have been made *a priori*—and by determining the reliability of the results through repeated independent experimentation, known as *replication*. Reliable results may then be reproduced for different factor levels, variable combinations, etc., and can be accepted as generalizable, in the sense that they can be used to make inferences about broader populations than those that have been directly tested.

While hypothesis testing will be discussed in detail in Chapter 7, at the stage of selecting a research design, it's important to consider whether you have sufficient information to make a prediction about the effect of a treatment on a variable. If you do, then conducting an experiment is certainly appropriate, and a clear hypothesis statement can be generated. However, if you don't, it may be better to consider further observation, or using an observational study to further clarify your thinking vis-à-vis treatment effects. This is particularly important in research involving animal or human subjects, who may suffer distress during experimentation.

There are inherent risks in hypothesis testing when the hypothesis is ambiguous, not specified in sufficient detail to make predictions concerning operationalized variables, or completely absent. A *Type I error* occurs in hypothesis testing when a researcher finds that, on the basis of experimentation, a treatment has a statistically significant effect on an experimental group, when, in fact, it does not. The probability of committing a Type I error is estimated using p values, with $p < 0.05$ or $p < 0.01$ being commonly accepted values in different fields. This simply means that you have a 1 in 20 or 1 in 100 chance, respectively, of committing a Type I error—not very reassuring, especially if you do 20 experiments per year, as you could expect a significant result to occur simply by chance. Conversely, a *Type II error* means that you have missed a significant effect of the treatment on an experimental group. Again, there could be many explanations: power may be too low to demonstrate the effect, or the sample size (or a nonrandomly selected sample) is simply too small for the effect to be observed.

A *hypothesis* is a clearly formulated statement that makes a prediction about the effect of a treatment on an experimental unit that forms the basis for your experimental design. For example, a general prediction like "fertilizer increases crop growth" is not sufficient to be a testable hypothesis; more likely, a hypothesis would be formulated predicting the effect of a specific fertilizer on a specific crop, and even specific concentrations. For example, does 100 mg/L of phosphate increase the size of sunflowers by 10% after 5 days? In this instance, you could randomly select a treatment and control group of sunflowers, with the treatment group being sprayed with 100 mg/L of liquid phosphate and the control group being sprayed with water. The growth of the flowers could then be measured from both groups, and the hypothesis could be supported or not supported.

One important aspect of hypothesis testing vis-à-vis research design is the limitation of inference inherent in one experiment. If the results from the sunflower experiment were taken that the hypothesis was "proven" 100%, then there would still be a strong possibility that one or more factors were responsible for the result—even if there was a control group. For example, the concentration of the phosphate may have actually been 110 mg/L due to a lab error, there may have been excess fertilizer landing by chance only on the experimental group but not the control group, the control group may have been placed further into the shade than the experimental group, and so on. This is why any one experiment is only used to provide support for a hypothesis, rather than *proof*. Indeed, many scientists believe it is impossible to "prove" anything; rather, hypotheses gain support over many years, as results become more generalizable, and any null results are taken into account. Indeed, if the result had failed to produce any difference in crop size, no one would have suggested that fertilizers don't enhance crop yield! Once again, you must always be alert to potential sources of random or systematic error in the design and conduct of experimental research.

A related issue is *reliability* of experimental results, which is the likelihood that the same experiment will yield identical results if performed many times over, by the same team or by other teams. If the goal of generalizability is considered to increase the breadth of knowledge concerning an experimental effect, then reliability is more concerned with the confidence that the result is repeatable. The extent to which a result is reliable for individual units versus a population is an important issue in experimental research. Determining reliability in this sense depends on the extent to which individual experimental units are the focus of the research, or whether the objective of the research is to test a hypothesis about idealized units, represented by response means, rather than individual unit responses. For example, in well-characterized systems, it is possible to make very precise predictions about the expected behavior of all experimental units: two hydrogen atoms and one oxygen will always combine in exactly the same way, for example, and this can be predicted accurately on a per-molecule basis. In this instance, the responses for all units are assumed to be equal.

However, in many physical systems, there is a naturally occurring distribution of responses in the population. Thus, the result of an experiment must be able to distinguish between variation in population responses that may be measurement error, and error due to the individual (or group) differences that result from a treatment. So, the hypothesis is often couched in terms of an average effect, rather than an effect that is consistently produced for every unit. By focusing on average effects, which may be biased by outliers of significant magnitude, you may find that a significant number of experimental units may actually experience no effect, or even a negative effect.

It may be possible to reduce the variation among a particular population by *a priori* identification of features that allow them to be reduced into further distinct populations, using categories that can be accurately determined from the population. For example, you might classify and group trees naturally according to trunk size, color, leaf density, etc., until the experimental units have similar response properties. At the same time, generalizability will be limited by focusing on very specific subpopulations.

Eliminating Bias

I have already discussed how systematic errors can be reduced through effective design; however, there are many additional design techniques that a researcher can use to minimize bias in research results. You've seen how randomization can be used to ensure that treatment and control groups are equally comparable, but there are many other influences that can potentially confound an experiment.

Hypothesis Testing Versus Data Mining

Given that a statistical significance $p < 0.05$ implies that 1 in 20 experiments will result in a Type I error, the onus is on the researcher to construct experiments that are consistent with, or attempt to explain, phenomena based on a model or theory. However, some researchers may collect a large amount of data on many different response variables, and try to relate these to explicit treatments of known characteristics of the sample. When undertaken on a large scale, this approach is known as *data mining*. Data mining—as a form of secondary analysis—is incredibly useful in exploring large, existing data sets, usually collected through observation or aggregated from different sources. At its simplest, the purpose of data mining is to determine correlations between many different variables, which may later form the basis for an experimental prediction. Alternatively, in industrial settings, it may be used to create decision rules in production systems, based on relationships observed in the data. For example, a financial database might reveal that all bank customers with income > \$100,000 and living at an address > 3 years never default on a home loan. Thus, the bank may decide to offer loans to customers who meet these requirements, and who currently do not have a loan. But generally there is no causal inference made; the decision rules are pragmatic in nature. Contrast this with the scenario where no *a priori* hypotheses are formed when undertaking experiments; if $p < 0.05$, a Type I error may occur 5% of the time, and you can imagine that the incidence will be higher where unexpected results, not predicted, are taken as experimentally "proven" relationships. That's not to say that there's not a place for unexpected findings (or serendipity) in science, but they usually occur at the observational or exploratory stage of research.

Blinding

You may have heard of the so-called *placebo effect*, in which participants in an experiment who have been allocated to a control group appear to exhibit some of the effects of the treatment. This effect arises from many sources, including an expectancy effect (since in drug trials, for example, the experimental substance and its known effects and risks would be disclosed to participants), as well as bias introduced by the behavior of the treatment allocators or response gatherers in an experiment. For example, if a treatment allocator knows that a participant is going to receive the treatment, they may act more cautiously than if they were administering a control. Conversely, the response gatherer (i.e., the person responsible for observing and measuring data in an experiment) may also be influenced by membership knowledge of the treatment and control groups.

Using single-, double-, or triple-blind experimental methods can effectively control these sources of error:

Single-blind

> The participant does not know whether he or she has been allocated to a treatment or control group.

Double-blind

Neither the participant nor the treatment allocator knows whether the participant has been allocated to a treatment or control group.

Triple-blind

Neither the participant, the treatment allocator, nor the response gatherer knows whether the participant has been allocated to a treatment or control group.

In small laboratories, the roles of treatment allocator and response gatherer may be carried out by the same individual, and thus triple-blind status can often be as easily achieved as double-blind status. While blinding is highly desirable, it may not always be possible to achieve at one or more of the levels. For example, a control tablet may be physically different from the experimental tablet, and a researcher may be familiar with it. Good experimental control can be achieved only by trying to hold all variables constant across the two groups, except for the treatment(s).

Retrospective Adjustment

In the previous section, I mentioned the potential bias of the response gatherer, arising from not being blind to the treatment status of participants. Another potential source of bias arises when there are multiple response gatherers, or when different instruments are used to gather response data, making essentially independent judgments of responses in either control or experimental treatments. Bias can be reduced in a number of ways: responses from multiple judges could be averaged, for example, to reach a "consensus" value, but what about situations in which there was only one judgment made for each unit by one of a group of judges? It may be possible to examine the overall set of decisions made by each judge, and attempt a retrospective adjustment for perceived bias; however, the calibration of such post-hoc adjustments can be difficult, since inter-rater reliability must be established for the pool of judges.

Blocking and Common Designs

The purpose of *blocking* is to set up experiments in such a way that comparable (and preferably identical) responses can be elicited from the same treatment. The idea is to use as much *a priori* information as possible about experimental units to allocate them to experimental blocks, so that all units in a specific block give the same response to a treatment. Perhaps the most famous example of blocking is the use of identical twins in psychological research to examine the effect of "nature versus nurture," since the twins have exactly the same genetic makeup. In circumstances where the twins have been separated at birth, for example, or sent to different schools, the impact of differences in environment can be determined while controlling for genetic factors. The advantage of blocking with identical twins is that variation due to one factor (genetics) can be tightly controlled because the responses will be very closely matched. The disadvantage with identical twin research is that the subject pool is limited, and the numbers of separated identical twins are even fewer.

However, the level of control for blocking to be effective does not have to be so very tight to use a *matched pair design*. The differences in responses *between subjects* can be controlled by matching on as many potentially confounding (or unit-treatment correlated) factors as possible. In psychological research, this typically means matching on factors such as age, sex, and IQ, but may also include quite specific controls, such as visual acuity or color blindness in perceptual experiments. In some cases, these factors will be *nuisance factors*, but in other cases, they are just not central to the research question.

It may not be possible to match participants on all possible sources of influence extraneous to the research question, but most scientific fields have a set of well-known criteria on which matching has been shown to be effective. The advantage of matched pair designs is that, on a per-unit basis, you can establish more confidence that an experimental effect genuinely occurs for all units, rather than hoping randomization will iron out any differences. However, by introducing structure into the treatment allocation process, the benefits of randomization may be lost. In a *randomized block design*, though, it may be possible to allocate treatments to matched units in a random way, to preserve the reduction in bias achieved therein.

 A rule of thumb in research design is to block wherever possible, and where you can't block, randomize.

Recall that a matched pair design attempts to control as many extraneous factors as possible by matching experimental and control treatment units as closely as possible. Further control can be achieved by allowing units to act as their own controls in a *within-subjects design*, although it may not always be physically possible or practical to do this. For example, if the treatment irreversibly alters the unit in some way, then a within-subjects design provides a very high level of experimental control. This is because you can observe the true experimental effect for all units, excepting error due to any changes in the environment between the control and experimental session.

Within-subjects designs are used extensively in psychology; however, since many of the experiments involve some modification to behavior or cognition, you may wonder whether there isn't a possible confounding "learning" effect. If all units were given the control treatment first and then administered the experimental treatment (or vice versa), there certainly would be potential for a learning effect (or *maturation bias*) to influence the results.

However, randomization again provides an antidote, in the form of a *Latin Square*, which provides an unbiased way to randomize the allocation of participants to treatments. In any design where y conditions are presented to each participant $(T_1, T_2, ..., T_y)$, the trials for each participant are grouped together and randomized, using a Latin Square, to ensure that no sequence is ever repeated twice for different subjects. For example, if the reaction time to five objects is measured with trials T_1, T_2, T_3, T_4, and T_5, and there are five participants, then a randomized Latin Square would produce the design shown in the following table, governing the order of stimulus presentation.

T_1	T_5	T_2	T_3	T_4
T_3	T_2	T_4	T_5	T_1
T_4	T_3	T_5	T_1	T_2
T_5	T_4	T_1	T_2	T_3
T_2	T_1	T_3	T_4	T_5

Using a Latin Square in this way ensures that any between-subjects variation affects all treatments in an equal way. Note that there are 161,279 other possible randomizations of the 5×5 Latin Square that would retain their characteristic property of no orthogonal (row or column) having the same number more than once. If your design required that at least one instance of the ordinal presentation of treatments (i.e., T_1, T_2, T_3, T_4, and T_5) then the reduced form could be used—because the first row and column would preserve ordinality—but yielding only 55 possible randomizations.

Example Experimental Design

In this section, I review a real example of an experiment and discuss the design decisions made, comparing how it could have been conducted using two common experimental designs, and provide examples that highlight the relative strengths and/or weaknesses of each strategy.

Martin and Siddle (2003)[*] set out to investigate the main effects of alcohol and tranquilizers on reaction time, P300 amplitude, and P300 latency, as well as their interaction. P300 amplitude and latency are measures derived from event-related potentials in the brain at 300ms. All three responses are related to different information processing mechanisms in the brain.

The research question was based on previous studies that had independently demonstrated the impact of alcohol or tranquilizers on these response variables, but not their interaction. Also, studies investigating the effect of alcohol on the response variables tended to use large doses, and studies looking at tranquilizers focused on strong and not weak ones (hence, temazepam, a mild tranquilizer, was selected). Thus, three questions were posed: (1) would alcohol have a significant main effect on any of the response variables; (2) would temazepam have a significant main effect on any of the response variables; and (3) would alcohol and tempazepam interact?

The experiment used a within-subjects design, so that participants acted as their own controls. The factorial design was 2 (alcohol, control)×2 (tranquilizer, control); thus, every participant performed the same experiment four times, having either no alcohol or temazepam, alcohol only, temazepam only, or both alcohol and temazepam.

The results indicated a significant main effect for temazepam on P300 amplitude (i.e., with or without alcohol), and significant main effect for alcohol on P300

[*] F. Martin and D. Siddle (2003). "The interactive effects of alcohol and temazepam on P300 and reaction time." *Brain and Cognition*, 53(1), 58–65.

latency and reaction time. However, there was no significant interaction between the two factors at all. Given that alcohol and temazepam have different main effects, and since they don't interact, the study supports the idea that alcohol and temazepam independently affect different information processing mechanisms in the brain.

If you were designing this experiment, what would you have done? Would you have selected a matched pair design instead of a within-subjects design? Possibly, since this would have reduced the number of trials that each participant had to complete, but in this instance, using a within-subjects design also allowed for smaller participant numbers to be used (N = 24), whereas a larger sample may have been needed to demonstrate an effect between subjects. No doubt, you would have randomized the selection of participants, perhaps by selecting names from a phone book using page numbers and columns generated by a random number generator. Content validity would not be a concern, since the response variables used are widely accepted in the field as reflecting information processing characteristics of the brain. You would also have ensured blinding of the researcher administering the alcohol or temazepam, ensuring that the control for each was physically the same in appearance. Would you have chosen to increase the number of factors rather than having a 2×2? For example, perhaps there would only be an interaction between alcohol and temazepam at high respective dosages, so perhaps a 3×3 design would have been more appropriate? The question here is not necessarily experimental but ethical; you want to limit the amount of tranquilizer being administered to each participant, and in the absence of a compelling theoretical reason (or clinical evidence or observation) to suspect otherwise, I think that a 2×2 study was correct.

6

Critiquing Statistics Presented by Others

This chapter explains how to read and critique statistics presented by someone else, including those contained in published research articles and workplace presentations, and will teach you to evaluate the research design. You'll also learn to critique the statistics chosen and their presentation, and common ways authors and presenters try to cover up weaknesses in their data.

The Misuse of Statistics

Broadly, the misuse of statistics falls into two very distinct categories: ignorance and intention. The ignorant use of statistics arises when a person attempts to use descriptive or inferential statistics to support an argument, where the technique, test, or methodology is inappropriate. The intentional misuse of statistics arises when a person attempts to conceal, obfuscate, or over-interpret results that have been obtained. Intuitively, you may think that ignorance arises mostly with complex statistical procedures such as multivariate analysis—and it certainly does—but even basic descriptive statistical procedures are routinely misused.

Intentional misuse is rife in descriptive statistics as well; witness the infamous "stock charting" techniques, in which graph axes are typically labeled with an ordinal scale that manipulates intervals to make a stock appear to rise faster in price than it actually does. The assumptions of inferential testing that make the calculation of statistical tests valid are routinely ignored, as they represent an "inconvenient truth" regarding the analysis that is being performed.

In this chapter, some examples are drawn from the contemporary debate surrounding climate change and global warming, since the public mood in most countries has clearly changed over the past few years. Has research progressed sufficiently that most governments and citizens are now convinced of its truth? Has experimental evidence been obtained that more strongly confirms specific hypotheses about climate change that were previously predicted? If "statistically significant" changes in climatic systems are detected, are they meaningful changes? Or is the population weary of sensational headlines, simply accepting that if a statement is repeated often enough by credible sources (such as newspapers, TV news and former vice presidents) then it must be true?

There is no simple answer to any of these questions, but as with all scientific debates, the pattern of observation, prediction, hypothesis testing, and analysis should remain unchanged.

Common Problems

If you are presented with a set of dazzling statistics that are meant to "prove" or support some argument, theory, or proposition, start with the following checklist to start asking the tough questions:

Truly representative sampling
> If the investigator is attempting to make inferences about a population by using a sample, how was the sample selected? Was it truly randomly selected? Were there any biases in the selection process? The results of any inferential tests will only be valid if the sample is truly representative of the population that the investigator wants to make inferences about. In some cases, samples may be maliciously constructed to prove a particular fallacious argument. Alternatively, there may be a self-selection bias that arises when some members of a population respond to a sampling request while others do not. For inferences about a population to be valid, the sample must be truly representative with all sources of bias removed.

Response bias
> Respondents may be tempted to tell you exactly what you want to hear, simply because they are asked a particular leading question, so as not to offend.

Conscious bias
> Are arguments presented in a disinterested, objective fashion? Or is there a clear intention to report a result at any cost?

Missing data and refusals
> How is missing data treated in the analysis? If participants were selected randomly, but some refused to participate, how were they counted in the analysis?

Small sample sizes
> Were the sample sizes selected large enough for a null hypothesis to be rejected? Was the sample size selected on the basis of a power calculation?

Large sample sizes
> A sample that is too large is overly sensitive to small differences that may not be important from a subject-matter perspective.

Effect sizes
> If a result is statistically significant, was an effect size reported? If not, how was the importance of the result established? Was it meaningful in the context of the phenomenon under investigation?

Parametric tests

Was the data analyzed using a parametric test when a nonparametric test may have been more appropriate?

Test selection

Was the correct inferential test used for the scale of variable used? Different techniques are used for different DV (dependent variable) and IV (independent variable) combinations of categorical, ordinal, and/or interval/ratio data.

Association and causality

Is the only evidence for a causal relationship between two variables a measure of association, such as correlation? In this situation, it is incorrect to assert a causal relationship, even if one variable is labeled "dependent" on an "independent" variable.

Training and test data

Has a model been developed using one data set and then tested using the same data set? This problem occurs frequently in pattern recognition applications.

Operationalization

Is the variable selected to measure some particular phenomenon actually measuring it? This is a common problem in psychology, where latent variables (such as intelligence) are measured indirectly by performance on different cognitive tasks.

Assumptions

Have the assumptions that underlie the validity of the test been met? How has the investigator ensured that they have been met? For example, if a test assumes that a population is normally distributed, and it is in fact bimodal, then the results of the test will be meaningless.

Testing the null hypothesis

To determine whether two groups are drawn from the same or different populations, it is common practice to test the null hypothesis that they are drawn from the same population. This derives from basic scientific methodology, where theories are supported by numerous and reliable sets of tests of null hypotheses that are rejected, rather than the (apparently) more straightforward approach of testing the hypothesis directly. Beware of any piece of research that attempts to "prove" a theory by a single experiment.

Blinding

Was the study single-, double-, or triple-blinded? For example, could the participants or investigators have introduced some bias by having knowledge of the treatment or control conditions in an experiment?

Controls

If the effect of a treatment is demonstrated in a pre/post-treatment model, are matched controls receiving a placebo within the same experimental paradigm to control for the placebo effect? *A designed experiment is the only common way to be able to make causal inferences.*

Quick Checklist

Investigations supported by statistics follow a surprisingly standard lifecycle. If you are reviewing a piece of work, try and determine what the sequence of events was during the investigation. Did the investigators start off with one hypothesis and change their minds once the results were in? Did they try numerous different tests with various post-hoc adjustments to make sure that they could report a "significance test" result? Asking searching questions about the research process is like a detective asking questions about movements at a certain date and time—inconsistencies and story-changing can be very revealing!

In short, investigations based on statistics should proceed along the following lines:

- Assuming that a period of observation and exploration has preceded the start of an investigation, research questions should be stated up front. Investigators must have formulated hypotheses (and the corresponding null hypotheses) well before they begin to collect data. Otherwise, the use of hypothesis testing is invalid, and the investigation may take on the flavor of a "fishing expedition." Given that a $p = 0.05$ result represents a 1 in 20 chance of making a Type I error, and since many thousands of studies are published each year in the scientific literature alone, many "facts" must surely be open to question. This is where independent repeatability and reliability are critical to the integrity of the scientific method.

- The relationship between the population of interest and the sample obtained must be clearly understood. It's not sufficient to make inferences about the entire human population based on a sample of highly educated, healthy, middle-class college students from one college. Honestly.

- Hypotheses must relate to the effect of specific IVs on DVs. Thus, it's critical to know as much about the DVs as possible, especially every source of variation in the DVs. This is particularly important where DVs are thought or known to be highly correlated (i.e., multicollinearity). The DVs must be measurable and must operationalize underlying concepts completely.

- In complex designs, where there are both main effects and interactions to consider, all of the possible combinations of main effects and interactions and their possible interpretations must be noted.

- Procedures for random sampling and handling missing data or refusals must be formalized early on, in order to prevent bias from arising. Remember that a truly representative sample must be randomly selected. Where purely random sampling is not feasible, it may be possible to identify particular strata within the population and sample those in proportion to their occurrence within the population. For example, since the proportion of males and females is generally known for most populations, sampling can be performed appropriately and "in proportion."

- Always select the simplest test that will allow you to explore the inferences that you need to examine. Multivariate techniques are incredibly important, but if you only need to make simple comparisons, they may be inappropriate.

- Selection of tests must always be balanced against known or expected characteristics of the data. For example, if testing mean differences, and only small samples may be available, then use a two-sample *t*-test in preference to an ANOVA. Although the two only differ in terms of the number of group means being tested, a designed experiment is the only common way to be able to make causal inferences.

- Don't be afraid to report deviations, nonsignificant test results, and failure to reject null hypotheses—not every experiment can or should result in a major scientific result!

Research Design

Generally, the design of an investigation of a question of interest needs to follow the guidelines presented in Chapter 5 if meaningful inferences are eventually to be made. However, many investigations do not follow these types of guidelines at all—especially if you have a newspaper to sell that relies on sensational headlines to grab the attention of an inattentive reader. Other investigations are based on a single sample or event whose significance is then extrapolated to indicate some more fundamental shift.

One of the major problems in the climate change debate is that there is no experimental apparatus available that completely represents the complexity of the earth's climatic systems. Indeed, to thoroughly test the various hypotheses that have been developed, you might need to have a number of different planets, identical in composition to earth, which can be assigned to control and treatment conditions to test different hypotheses. For example, does global warming arise from greenhouse gases? Or does it arise from increases in solar activity? Do both factors contribute to global warming? Do the factors have main effects on global warming, or do they interact to produce global warming, or both?

Clearly, it is not possible to obtain such samples for physical objects like planetary systems, so investigators may resort to using models that have been demonstrated to have predictive value. However, even though computer models of the climate have improved dramatically over the years, thanks to advances in computer processing power and model refinement, they are still unable to predict the weather accurately beyond a few days. If the models have predictive validity, how well do they rate on explanatory value? A classic example is the development of supervised learning algorithms such as artificial neural networks; these generalized function approximators were used to predict many different phenomena in psychology and neuroscience, based on the *backpropagation* algorithm, which attempts to minimize errors between an output array of units, and an expected value supplied for training, where hidden unit values are adjusted to compensate for error. While the networks do have predictive validity, real neural networks rarely provide for the backpropagation of potentials in the way that the model implies. Thus, you would not want to base the development of new neurosurgical procedures on this type of artificial neural network, even though it has a level of predictive validity.

Variation

Understanding *variation* is critical to all systems. Variation can arise from legitimate sources of a population in question, but also from measurement error. Variation may be cyclical, so cross-sectional designs may not always correctly identify that local minima may be perfectly acceptable in the lifecycle of a system. In climatic systems, for example, variation in temperature occurred prior to the industrial revolution and the consequent increase in the release of greenhouse gases; how do you partition the variation expected due to normal cyclical effects from that which can be directly attributed to human activity? This is one of the critical issues facing environmental science, since the atmosphere definitely warmed since the last ice age, without any human interference, until the industrial revolution.

Population

Scope in defining a *population* is critical in accurately specifying the limits of inference that can be made from a particular study. If all members of a population are measured in some way, and there is no missing data or refusals, then you don't need statistics at all, as you can directly calculate parameters of interest. Part of the problem in defining a population is when there is some fundamental misunderstanding of the population in question. Imagine that a survey of attitudes undertaken at a census in Utah is taken as the same population as California; both states are located in the same country and are relatively close in geographical terms, and so it may be tempting use Utah as a sample for California. And it may well be the case that many attitudes in Utah would be predictive of those in California—but there may well be very significant differences. A better design would be to sample in both states and determine if there are significant differences between attitudes, with confidence intervals and other techniques used to ensure the validity of the result.

Sampling

There are two keys aspects of *sampling*: size and randomness. A truly representative sample must be both "large enough" and randomly selected to give an accurate estimate (statistic) of any population parameters. Being sufficiently large to represent the population is a difficult problem—calculations of statistical power certainly provide a basis for this, in terms of inferential testing—but more sophisticated sampling schemes will attempt to identify all sources of variation in the population that might introduce bias, and sample within those appropriately. Sampling bias can occur when participants are able to self-select into a study. Alternatively, sampling can be biased where there is a confounding factor at work.

Random sampling is very difficult to achieve for cross-sectional or short-term studies. Long-term cohort studies with high retention rates that were originally sampled randomly provide the best basis for studies in social or health sciences. Being able to track all sources of variability in an individual's life, and examining how similar they are to other members of the sample, provides a reliable basis for estimating population parameters.

Controls

A recent study indicated that the administration of antidepressant medication to a large number of participants in a clinical study was no more effective than a placebo. Thus, the expectation of receiving a cure resulted in the same improvement in depressive symptoms as receiving a tablet with the active ingredient. The placebo effect is very powerful in humans, and most studies should provide some type of explicit control where the effect of a treatment is intended to be demonstrated. In clinical and pharmaceutical sciences, the methods and processes for controls are well established. However, in areas like climatic modeling, finding suitable controls is difficult if not impossible.

The Power of Coincidence

When statistical significance is measured at the $p = 0.01$ or 0.05 level, this means that there is a 1 in 100 or a 1 in 20 chance respectively of a Type I error being committed. Thus, in the case of $p = 0.05$, a repetition of the experiment would lead to 19 out of 20 cases being significant, and 1 out of 20 being insignificant. This is why independent replication and repeatability are so important. In addition, the world is full of coincidences, and experiments are subject to measurement error, and the interaction of the two can lead to some downright wacky and unexpectedly "significant" findings, to which no actual significance should be attached. Imagine that there are 20 earths surrounding the sun, and you choose one to examine the effects of global warming. You find a correlation between increases in industrial activity and temperatures for the last 200 years. Since you know that there is a 1 in 20 chance of committing a Type I error, you would check out at least some of the other planets, or perform an experiment on them all, with half acting as matched controls for the others.

You can see the difficulty here in understanding the causal sources of global warming: there are no other 19 planets that you can experiment with, or verify your model against—but at the same time, you know there is a strong possibility of committing a Type I error. There is also the insidious effect of the *Poisson distribution*, which can be used to model rare events that cluster together, and shows that 25% of intervals will have multiple events, while others will be empty. The effect can be seen from some cancer cluster cases that do not appear to have any particular relationship with each other.

Descriptive Statistics

The issues surrounding the appropriate interpretation of inferential tests are complex and prone to error. However, the use of descriptive statistics also has enormous potential to introduce errors in reasoning and understanding. Some of these errors are deliberate attempts to misguide and mislead. Others are simply poor choices. In this section, you will learn about some common problems associated with descriptive statistics, especially measures of central tendency and graphing.

Measures of Central Tendency

The choices for deception here are endless: if researchers wanted to overemphasize average temperature increases, they would select the mean as the measure of central tendency, or conversely, the median if they wanted to underemphasize the effect. Over a 10-year period, if there are 8 years of 70 degree averages, and 2 years of 80 degree averages, then selecting the mode would also underemphasize the effect. Unscrupulous investigators will often choose the measure of central tendency that best fits their desired outcome.

Measures of central tendency can also be very misleading when the sample and/or the population changes from measurement to measurement. Average house prices are a classic example: these are based solely on sales in a particular period, such as one year. From year to year, the sample from which the average is calculated will almost certainly change, unless all houses sold in one year are resold the next, and no other houses are sold. This would surely be a very unlikely event. And yet eager homeowners often take a "10% average rise in house prices" to mean that theirs has increased by the same proportion. Where the population itself changes—such as where many new homes are built and sold in one year—the median will almost certainly rise. And yet existing houses may sell for exactly the same price (or less) than the year before. A more valid method of determining the average house price would be to sample amongst the population so that each house has an equal chance of being valued and added to the sample. Furthermore, since the proportion of existing houses to new builds is known, the sample could be further stratified, so that average prices for both types of houses could be reported and/or aggregated afterward.

The fair solution is usually to eliminate cases from analysis that lie two standard deviations above or below the mean. This also helps to minimize measurement error effects; in reaction time experiments, for example, it's not uncommon for participants to become incredibly bored and miss a stimulus. If the computer program waits only for two seconds to accept a response, but the stimulus is missed, then a reaction time that is usually on the order of 20–80ms is now recorded as 2,000ms, which is up to two orders of magnitude greater. If this case is not culled, then the mean would be greatly overestimated.

Note that by removing outliers, you are reducing the generalizability of your results, so you should never remove outliers casually.

Standard Error and Confidence Intervals

Given that measurement error exists when trying to gather data in most disciplines, the standard error and/or probable error should be reported, especially when trying to compare the means of two different groups. The standard error is an estimate of the variation of error in making a particular observation, i.e., it represents the gap between the measured value and the "true" value. Since it's not always possible to understand what factors may influence measurement error, standard error is always estimated.

Normally, the standard error is estimated by using standard deviation divided by the square root of n; thus, as the sample size increases, the standard error generally decreases, as observations become more reliable.

Where the standard errors overlap between the means for two groups (i.e., standard errors of the mean), it doesn't make sense to attempt to distinguish between them, since there is no way to achieve certainty in measurement within that range. But where the null hypothesis of equality of means is tested, it's useful to examine the standard errors to see if they do overlap.

Standard errors of the mean are used to calculate confidence intervals. For example, where the upper and lower 95% confidence intervals are to be presented, these can be calculated by using the following formula, where x is equal to the sample mean, y is equal to the standard error of the sample, and z is the 0.975 quantile of the normal distribution:

```
Upper 95% Limit=x+(y*z)
Lower 95% Limit=x-(y*z)
```

For confidence intervals at these levels, if a single population is repeatedly sampled, then 95% of the samples should capture the true population mean.

Be wary of any study that doesn't provide confidence intervals, especially if the sample size is small.

Graphing

Graphs are often provided in the absence of statistics to provide an accessible way of understanding how variables are related. However, graphs can be misused in a number of ways; for example, axes may be unlabeled, meaning that they cannot be correctly interpreted, or axes can be manipulated to obscure or enhance the real relationship between variables.

The old adage "a picture tells a thousand words" is certainly true, but the "thousand words" can change dramatically depending on the choice of scale. Figure 6-1 shows a fictional set of temperature increases ranging between 70–77 degrees over a 50-year timespan. The rise in temperature is strongly correlated with the year, $r = 0.96$. Figure 6-1 certainly shows this almost perfectly linear rise.

However, by stretching out the Year axis, suddenly the visual effect is of an overall slower rise in temperature, as shown in Figure 6-2.

Note that if the Temperature scale is now adjusted to start at 1 degree rather than 68, then the relationship is even further flattened, and the two variables visually appear to be uncorrelated, as shown in Figure 6-3.

If this was not enough, then converting the Temperature scale to the log has the effect of flattening the line even further, as shown in Figure 6-4.

Of course, if you took the opposite view, you could always stretch the Temperature axis vertically, and make it appear as if the temperature rise was, in fact very steep, as shown in Figure 6-5.

Another common tactic is to use "2-D" graphs, where a doubling in size is shown by doubling both the x- and y-dimensions, thereby exaggerating the increase.

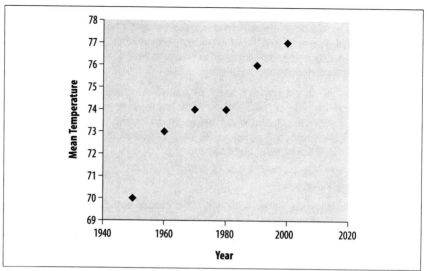

Figure 6-1. Graph manipulation

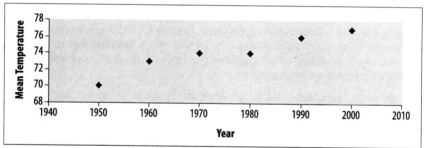

Figure 6-2. Graph manipulation

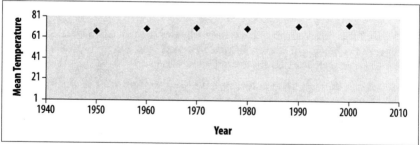

Figure 6-3. Graph manipulation

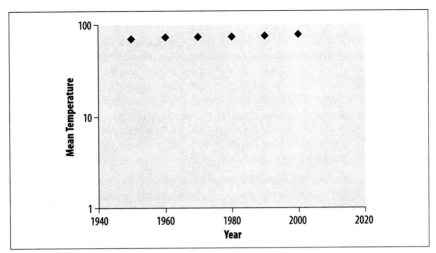

Figure 6-4. Graph manipulation

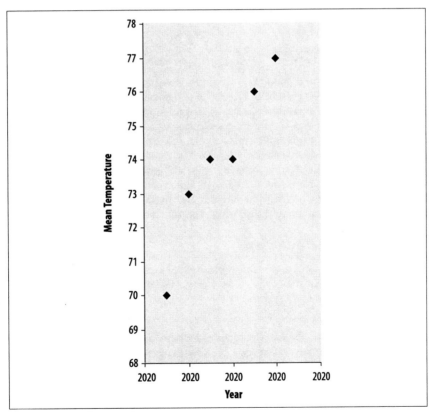

Figure 6-5. Graph manipulation

Extrapolation and Trends

A common tool used in marketing is *extrapolation* of a known relationship between two variables, outside a measured range, to form a trend. For example, if the S&P 500 index has increased by 10 points for the last 10 weeks, a gambler might feel some confidence in betting that the index might increase by 10 points during the following week. In this case, using simple linear interpolation provides the best estimate possible—but because the stock market is subject to a lot of random variation, the index will not always rise in accordance with previous experience. If the system is not a linear system, then linear interpolation is not appropriate.

For nonlinear systems, polynomial extrapolation and conic extrapolation may be appropriate, if the system in question can be modeled by using the functional form in each respective case.

Trends can be useful when smooth data is extrapolated by smooth functions, or nonsmooth data is extrapolated using a nonsmooth function. But when the system under study is not deterministic, subject to random error, or chaotic, the usefulness of trending is limited and may give wildly inaccurate and potentially misleading results, especially if there is branching or bifurcation in the data.

Inferential Statistics

So far, you have learned about key problems in research design and descriptive statistics that are often present in reports of statistical work performed. In some cases, deception may be behind the incorrect presentation of an analysis, and the omission of key statistics should raise your suspicions. With inferential testing, however, not only does deception play a role, but also the incorrect or inappropriate use of some tests is a major problem. The most significant problem is that the assumptions of multivariate tests are routinely ignored, and yet the results of these tests are extremely sensitive to any violation of the assumptions. Investigators can and should be proactive in determining whether their data actually meets all of the underlying assumptions for using a specific multivariate test.

Assumptions of Statistical Tests

Typical violations of some statistical tests are given below, and mechanisms to test whether the assumptions are violated are also provided.

t-tests

Two-sample *t*-tests assume that the samples are unrelated; if they are related, then a paired *t*-test should be used (*t*-tests are discussed further in Chapter 8). Unrelated here means independent—you can test for linear independence by using the correlation coefficient. Serial correlation may become an issue if data is collected over a period of time.

t-tests are also influenced by outliers; these should be removed when they are two or more standard deviations above or below the mean. Alternatively, they may be visually detected by using a boxplot or a normal Q-Q plot. Use caution with outlier removal, as removal of any data will reduce the generalizability of your results.

Note that discarding outliers on the basis of sound statistical measures—such as the standard deviation—is an entirely separate activity from discarding data that happens to be unfavorable. For example, when there is a 5% chance of committing a Type I error, then discarding the 19/20 experiments that do not meet your favored conclusion would not be statistically valid (or ethical).

t-tests assume that the underlying population variances of the two groups are equal (since the variances are pooled as part of the test); if they are not, then the *Welch-Satterthwaite t-test* should be used, since this provides a direct means to adjust for the inequality. An *F*-test could be performed to directly test the equivalence of variances, or a side-by-side boxplot comparison could be used.

Normality of the distributions of both variables is assumed, although for the small samples that a *t*-test is often used to test, this may be difficult to establish—a histogram of the distribution should reveal any significant lack of symmetry (or skew). In this case, a nonparametric or "distribution-free" test (discussed in Chapter 11), such as the *Wilcoxon rank-sum test*, may be more appropriate. The lack of balance in sample sizes may result in biased estimation of the population parameters in one of the groups; certainly, the standard error of the mean will be greatest in the smaller group.

Note that the *t*-test is often used with small sample sizes. Using small samples in any design may result in a lack of power, meaning a true difference may not be determined. Unless variances are small, testing within small samples may produce a nonsignificant result, even if there truly is a significant difference. Relaxation of the alpha level will increase power, as will increasing the sample size and/or reducing variance.

ANOVA

ANOVA has a large number of assumptions that need to be met, which usually requires directly determining whether the assumption is met (rather than hoping that it is met, or ignoring it). ANOVA (discussed further in Chapter 12) assumes independence and normality—again, the impact of outliers needs to be considered if these are the main cause of the nonnormality, and attempting to screen them may radically change the result of the *F*-test, but at least the result would then be valid. The *most important* assumption, from a practitioner's perspective, is the equality of variances.

ANOVA is most reliable when sample sizes are balanced and when the population variances are equal. Skewed distributions and unequal variances may make the interpretation of the *F*-test unreliable. A side-by-side boxplot comparison may be very helpful; if data is sampled from a truly normal distribution, then there should be symmetry in the boxplots. If there is no attempt to establish normality, ask why. While it's true that—if the population data is normally distributed—increasing the sample size will bring about a greater approximation to normality, if the population is not normal, then increasing the sample size won't help. And yet many studies rely on large numbers to claim reliability, putting great faith in the Central Limit Theorem. Levene's test and Bartlett's test are very useful for determining whether the assumption of equal population variances has been met from a sample.

If samples are both nonnormal and population variance is thought to be unequal, and/or there is a lack of balance in sample sizes, it might be best to use a nonparametric test, such as the Kruskal-Wallis. Alternatively, if sample sizes are unequal, but the other assumptions are met, then a Tukey-Kramer adjustment may be made.

MANOVA

In addition to the assumptions underlying univariate ANOVA, MANOVA assumes the equality of variance-covariance matrices (more on MANOVA can be found in Chapter 13). This assumption can be tested using the Box test, and significance levels are often provided. Data is also assumed to be multivariate normal; unfortunately, there is no direct test available for multivariate normality, but univariate normality tests should at least be undertaken.

MANOVA is also sensitive to outliers, and these should be removed before analysis, again noting that removing any cases from your analysis may reduce the generalizability of your results. Tests for linear relationships (to exclude nonlinear relations) should be performed; however, where multicollinearity arises, reducing redundancy for dependent measures (through principal component analysis or similar) should be considered—usually where $r > 0.80$, as a rule of thumb.

Linear regression

Like the other techniques described here, *linear regression* assumes the independence of errors in the IV and DV: if a seasonal effect is present, then examining the residuals should indicate that a more complex model is required (look for any pattern other than a random distribution). Linear regression is covered in depth in Chapters 12 and 14. Time series analysis, for example, provides methods to remove seasonal or cyclical trends from data before performing linear regression.

Examining residuals is more an art than a science. However, by becoming familiar with residual analysis, you will be better able to assess the regression analyses presented by others and pinpoint any problems.

Table 6-1 shows average wholesale coffee prices per pound for the past 10 years. As you can see, the rise in prices is strongly correlated with the year, $r = 0.991$. There is some random variation present in the data—perhaps some prices were transcribed incorrectly, or perhaps some growers were slightly more or less greedy each year. But generally, the relationship is linear.

Figure 6-6 shows the residuals from the model fit, with an overlaid normal distribution. Although there are some deviations for such a small sample, it's actually a good fit.

Table 6-1. Average wholesale coffee prices

Year	Price
1998	2.40
1999	2.89
2000	3.75
2001	4.00
2002	4.20
2003	4.82
2004	5.19
2005	5.98
2006	6.36
2007	7.31

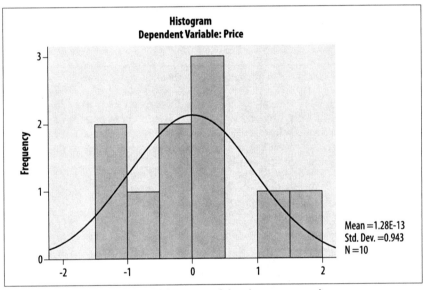

Figure 6-6. Residuals and overlaid fit to normal distribution—no outliers

However, what if coffee prices had spiked in 2002 to \$10.86? The correlation would then be $r = 0.572$, resulting in only 32% of the variation being accounted for in the DV by the IV, rather than 98%. The residual plot in Figure 6-7 shows 9 cases clustered around standardized residuals of 1, while there is only one case with a residual of approximately 3. If that single case had been removed as an outlier, using the ±2 SD criterion, then the almost-perfect fit observed in Figure 6-6 would have been maintained.

Imagine a seasonal effect (shown in Table 6-2) that reflects government policy to run a subsidization program every second year to ensure that growers can remain

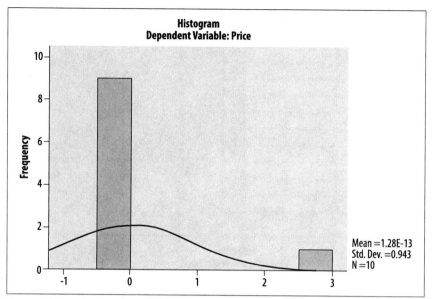

Figure 6-7. Residuals and overlaid fit to normal distribution—single outlier

competitive in a global market. In this case, there is an increasing linear trend overall ($r = 0.74$), but you can see a repeating pattern where there are serial clusters that are above and below zero. You wouldn't see this from the histogram, which is why, especially with regression through time, it's useful to examine the serial order of residuals. Figure 6-8 illustrates this.

Table 6-2. Average wholesale coffee prices with cyclical effect

Year	Price
1998	2.51
1999	1.97
2000	2.63
2001	1.91
2002	2.66
2003	2.12
2004	2.86
2005	2.94
2006	3.48
2007	3.25

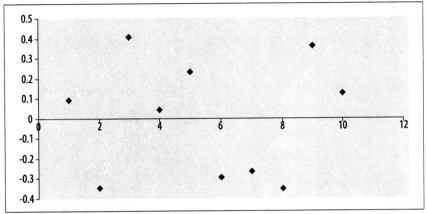

Figure 6-8. Residuals plotted serially: cyclical effect

In other situations, there may be an observed expansion in the divergence at positive and negative parts of the cycle—perhaps the government increases spending on the subsidy in the first year, and then has to decrease the subsidy because it has less money. In this situation, you may see a bifurcation, as shown in Figure 6-9. Again, the correlation is still high, at $r = 0.79$, but the residuals, shown in Figure 6-10, clearly show the oscillation between successive residuals, as well as their increase in magnitude.

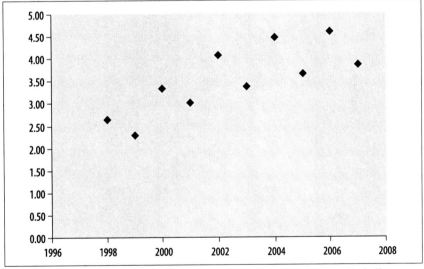

Figure 6-9. Residuals and overlaid fit to normal distribution: increasing cyclical effect

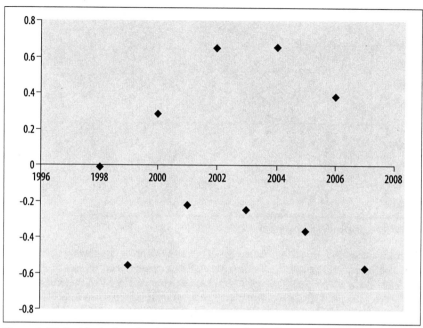

Figure 6-10. Residuals plotted serially: increasing cyclical effect

7

Inferential Statistics

Statistical inference is the science of characterizing or making decisions about a population using information from a sample drawn from that population. Most of the practice of statistics is concerned with inferential statistics, and many sophisticated techniques have been developed to facilitate this type of inference.

The name "inferential statistics" derives from the term "inference," given two definitions by the Merriam-Webster online dictionary (*http://www.m-w.com/ dictionary/inference*):

a) the act of passing from one proposition, statement, or judgment considered as true to another whose truth is believed to follow from that of the former

b) the act of passing from statistical sample data to generalizations (as of the value of population parameters) usually with calculated degrees of certainty

The second meaning, which is specific to statistics, is clearly related to the first. Inference in general is a method of making suppositions about an unknown, drawing on what is known to be true. Statistical inference is a refinement of ordinary inference, and is a process of making generalizations about unmeasured populations using data calculated on measured samples. Statistical inference has the additional advantage of quantifying the degree of certainty for a particular inference.

People sometimes get confused about the difference between descriptive statistics (covered in Chapter 4) and inferential statistics, partly because in many cases the statistical procedures used are identical while the interpretation differs. For instance, the same formula is used for calculating a mean whether the data represents a population or a sample: add up all the data values and divide by the number of values. There are differences in the notations of the formula, however, such as the use of the Greek letter μ to represent the population mean (which is properly called a *parameter* since it is a number that describes a population) and the Latin letter x with a bar over it ($\bar{x}$), pronounced "x-bar," to represent a sample mean (properly called a *statistic* since it is a number that represents a sample), and the use of the uppercase *N* for population size versus the lowercase *n* for sample size. In other cases, the formula is different: for instance to calculate a population standard deviation we divide by *N*, while for the sample standard deviation we divide by $n - 1$.

So it can make a difference, even before you get to the interpretation stage, whether you are working with descriptive or inferential statistics. To answer this question, think about the purpose of your study: is it merely to describe the specific people or entities that provided the data upon which you will perform the calculations? Or is it to generalize to a larger group of which the study objects are considered representative? The basic rule is this:

> Any time you want to generalize your results beyond the specific cases that provided your data, you should be doing inferential statistics.

To look at the same question from another side:

> Any time the cases that provided your data do not represent the entire population of interest, you should be doing inferential statistics.

Probability Distributions

Statistical inference frequently relies on making assumptions about the way data is distributed, or requires performing some transformation of the data to make it better fit some known distribution. Therefore we will begin the topic of statistical inference with a presentation of the concept of a theoretical probability distribution, and a review of two commonly used distributions.

A theoretical probability distribution is defined by a formula that specifies what values can be taken by data points within the distribution, and how common each value will be (or, in the case of continuous distributions, how common a given range of values will be). Graphic presentations of theoretical probability distributions are often used to present statistical concepts: the well-known "bell curve" of the normal distribution is a good example.

Theoretical probability distributions are useful in inferential statistics because their properties and characteristics are known. If the actual distribution of a given data set is reasonably close to that of a theoretical probability distribution, many calculations may be performed on the actual data using assumptions drawn from the theoretical distribution. In addition, thanks to the Central Limit Theorem, we can assume that the distribution of means of samples of a sufficient size is normal, even if the population from which the samples were drawn is not normally distributed.

Probability distributions are commonly classified as *continuous*, meaning the data can take on any value within a specified range, or *discrete*, meaning the data can only take on certain values. We examine the normal distribution as an example of a continuous distribution, and the binomial distribution as an example of a discrete distribution.

The Normal Distribution

We will use the normal distribution for our example of a continuous distribution because it is arguably the most commonly used distribution in statistics. This is due in part to the fact that the normal distribution is a reasonable description of how many continuous variables are distributed in reality, from industrial process variation to intelligence test scores. A second reason for the widespread use of the normal distribution is that under specified conditions we may assume that sampling distributions of statistics such as the sample mean are normally distributed even if the samples are drawn from populations that are not normally distributed. This is discussed further in the section on the Central Limit Theorem later in this chapter. The normal distribution is also referred to as the "bell curve" due to its characteristic shape, and as the "Gaussian distribution" in honor of the 18th-century physicist and mathematician Karl Gauss, who used this distribution to analyze astronomical data.

There are an infinite number of normal distributions, all of which have the same basic shape but differ according to their mean μ (the Greek letter *mu*) and standard deviation σ (the Greek letter *sigma*). Examples of three normal distributions with different means and standard deviations are displayed in Figure 7-1.

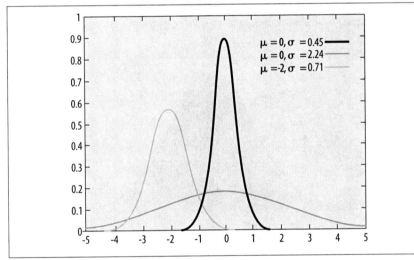

Figure 7-1. Three normal distributions

The normal distribution with a mean of 0 and standard deviation of 1 is known as the *standard normal distribution* or *Z distribution*. Any normal distribution can be transformed to the standard normal distribution by converting the original values to standardized scores (a process discussed below and further in Chapter 19), which facilitates comparison among populations with different means and variances.

All normal distributions, regardless of the mean and variance, share certain characteristics. These include:

- Symmetry
- Unimodality (a single most common value)
- A continuous range from $-\infty$ to $+\infty$ (negative infinity to positive infinity)
- A total area under the curve of 1
- A common value for the mean, median, and mode

As we noted above, there are an infinite number of normal distributions, but they all share certain properties. For the sake of convenience, we often describe normal distributions in terms of units of standard deviation rather than raw numbers, because that allows us to apply the same description to any normal distribution.

Because all normal distributions have the same basic shape, we can make some assumptions about how data is distributed within any normal distribution. The empirical rule states that for any normal distribution:

- About 68% of the data will fall within one standard deviation of the mean
- About 95% of the data will fall within two standard deviations of the mean
- Over 99% of the data will fall within three standard deviations of the mean

This is illustrated in Figure 7-2, which expresses values in units of standard deviation.

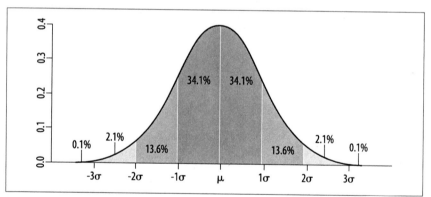

Figure 7-2. Percent of data falling into specified ranges of the normal distribution

Knowledge of these properties of the normal distribution gives us a way to judge whether a particular value is typical or atypical compared to other values in the population. This comparison is facilitated by converting raw scores (scores in their natural metric, for instance weight measured in pounds or kilograms) into Z-scores, which express the value of the score in terms of units of the standard deviation for their population. Converting all data points to Z-scores is equivalent to transforming a normally distributed population to the standard normal distribution. For this reason, Z-scores are sometimes referred to as *normalized scores*, the process of computing them as *normalizing* the scores, and the standard normal distribution is sometimes called the *Z distribution*.

A Z-score is the distance of a data point from the mean, expressed in standard deviations. The formula to calculate a Z-score for a population with known mean and standard deviation is:

$$Z = \frac{x - \mu}{\sigma}$$

If the variable x is distributed normally with mean of 100 and standard deviation of 5, i.e., $x \sim N(100, 5)$, a value of 105 has a Z-score of 1, because:

$$Z = \frac{105 - 100}{5} = 1$$

Similarly, a value of 110 from this population has a Z-score of 2, and a value of 85 has a Z-score of –3. Using the empirical rule cited above, we classify the value 105 as above average but not remarkable among the population (about 15.9% of the population would be expected to have higher Z-scores). A Z-score of 2 is more unusual (about 2.5% of the population would be expected to have higher Z-scores) and –3 is quite unusual (less than half of one percent of the population would be expected to have scores this low or lower).

The advantage of Z-scores is that they facilitate comparison among populations with different means and standard deviations. For instance, comparing our population $x \sim N(100, 5)$ with another population $y \sim N(50, 10)$, we can't immediately say whether a score of 95 among the first population is more or less unusual than a score of 35 among the second population. However, this comparison is easily made using Z-scores:

$$\frac{95 - 100}{5} = -1$$

$$\frac{35 - 50}{10} = -1.5$$

Conversion to Z-scores places both populations on the same metric, and we can see that the second score is more extreme because –1.5 is further from 0, the mean of the standard normal distribution, than –1.

The Binomial Distribution

We will use the binomial distribution as an example of a discrete distribution, i.e., a distribution for a variable in which only certain values are possible. Consider the case of flipping a coin five times: the number of times the coin comes up heads can take the values 0, 1, 2, 3, 4, or 5, but not the values 3.2 or 4.6. The variable "number of heads in five coin flips" is therefore a discrete variable.

The binomial distribution applies to many types of real-life data with dichotomous outcomes (outcomes that can take only two values), from machine parts that are either defective or acceptable to students who either pass or fail a class.

Events in a binomial distribution are generated by a *Bernoulli process*. A single trial or experiment within a Bernoulli process is called a Bernoulli trial or Bernoulli experiment. The binomial distribution describes the number of successes in n trials of a Bernoulli process. "Success" doesn't necessarily mean something good, just that the outcome we are looking for occurred. For instance, if we were describing how many machine parts out of a sample of 10 were defective, each part would be considered a separate trial and would be classified as a success if it were defective. The binomial distribution describes how likely it is that a particular number of parts from the sample of 10 will be defective, given some estimate of the overall rate of defective parts.

Data represented by the binomial distribution must meet four requirements:

1. The outcome of each trial is one of two mutually exclusive outcomes.
2. Each trial is independent, so the result of one trial has no influence on the result from any other trial.
3. The probability of success is constant for every trial.
4. There are a fixed number of trials, denoted as n.

Common examples of data described by the binomial distribution include the number of heads in 10 flips of a coin, where the probability of heads on any toss is known to be 50%; the number of males in a sample of 5 drawn from a large population known to be 55% male (the population must be large enough for its makeup to not change appreciably by the removal of five members); and the number of defective items in a sample of 20, drawn from a large population whose defect rate is known to be 1%.

The formula to calculate the probability of a particular number of successes on a particular number of trials is:

$$\binom{n}{k} p^k (1-p)^{n-k}$$

where:

$$\binom{n}{k} = \frac{n!}{k!(n-k)!}$$

$$\binom{n}{k}$$

is a combination, discussed in Chapter 2, which expresses the number of ways k items can be chosen from a set of n objects.

The symbol ! in this equation means factorial: $n! = (n)(n-1)(n-2)...(1)$. For instance, $5! = 5 * 4 * 3 * 2 * 1 = 120$.

n is the number of trials: if we are flipping a coin 10 times, $n = 10$.

k is the number of successes: for instance, if we want to know the probability of 5 successes in 10 trials, $k = 5$.

p, a number between 0 and 1, is the probability of success: for instance, if we are flipping a fair coin and the event is heads, $p = 0.5$.

The binomial formula can be used to calculate the probability of getting a particular number of successes given a fixed probability of success per trial and a fixed number of trials.

Figure 7-3 shows the graph for three binomial distributions (each combination of p and n will produce a different distribution).

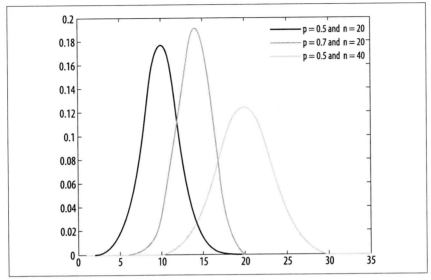

Figure 7-3. Three binomial distributions

Note that as n increases, the binomial distribution comes to resemble the normal distribution. A common rule of thumb is that if both np and $n(1 - p)$ are greater than 5, the binomial distribution may be approximated by the normal distribution. In Figure 7-3, the distribution ($p = 0.5$, $n = 40$) qualifies for the normal approximation because:

$np = 40(0.5) = 20$
$n(1 - p) = 40(1 - 0.5) = 20$

However, a distribution with $p = 0.1$ and $n = 40$ could not use the normal approximation because:

$np = 40(0.1) = 4$

Complex calculations based on the binomial distribution are usually done using computer software, but a simple example will demonstrate how the formula works. Suppose we are flipping a fair coin five times: what is the probability that we will get exactly one head? We will define "heads" as a success and use the binomial formula to solve this problem. In this example:

$p = 0.5$ (the definition of a "fair coin" is that heads and tails are equally likely)
$n = 5$ (because we are conducting five trials)
$k = 1$ (because we are calculating the probability of exactly one success)

The probability of exactly one success in five trials, given a probability of success on each trial as 0.5, is:

$$P(n = 1) = \binom{5}{1}0.5^1(1 - 0.5)^{5-1} = 0.156$$

Breaking down the steps:

$$\binom{5}{1} = \frac{5!}{1!(5-1)!} = \frac{5 \times 4 \times 3 \times 2 \times 1}{1(4 \times 3 \times 2 \times 1)} = 5$$

And therefore:

$$P(n = 1) = 5 \times (0.5) \times (0.5)^4 = 0.156$$

Independent and Dependent Variables

There are many ways to characterize variables: one of the most common is by the roles they play in a research design or data analysis. Using those criteria, a simple way to describe variables is as either *dependent*, if they represent some outcome of the study, or *independent*, if they are presumed to influence the value of the dependent variable(s). Many study designs include a third category, *control variables*, which may influence the dependent variable but are not the main focus of interest.

Note that the labels "independent," "dependent," and "control" relate to the roles played by the variables in a given design or experiment. This is because a variable, for instance weight, could easily be an independent variable in one study, a dependent variable in another, and a control variable in a third. In addition, other labels are also used to describe dependent and independent variables, with some authors preferring to reserve specific labels for particular types of studies. Control variables are particularly problematic because many types of control variables have been defined, depending on their relationship to the independent and dependent variables of interest, and the type of study design employed. This discussion will concentrate on independent and dependent variables.

We will use the example of a regression equation to illustrate the concept of independent and dependent variables. This is just a brief introduction: the topic of regression is covered in detail in Chapters 12 and 14.

In a standard linear model such as an OLS regression equation (OLS means Ordinary Least Squares; if not otherwise specified, this is what is meant by a regression equation), the outcome or dependent variable is customarily indicated by the letter Y, while the independent variables are indicated by X. Subscripts are used to identify each individual X variable: X_1, X_2, and so on.

This should be clear from the conventional way of notating a regression equation:

$$Y = \beta_0 + \beta_1 X_1 + \beta_2 X_2 + \beta_3 X_3 + \dots + e$$

The *e* in this equation means "error" and refers to the fact that we don't expect any regression equation to perfectly predict Y. Note that each X in the equation is preceded by a β, which is called its *regression coefficient*: β_1 is the regression coefficient for X_1, β_2 is the regression coefficient for X_2, and so on. These coefficients are determined through a mathematical process in order to make the best possible equation for predicting the value of Y from the value of the Xs.

Because of this notational convention, the dependent variable is also referred to as the "Y variable" and the independent variables as the "X variables." Other terms used for the dependent variable include the *outcome variable*, the *response variable*, and the *explained variable*. Other names for independent variables include *regressors*, *predictor variables*, and *explanatory variables*.

Some researchers believe that the terms independent and dependent should be reserved for experimental studies, in which case at least the primary independent variables have been manipulated in some way by the researcher, while the values for the dependent variable are merely observed and recorded. In this interpretation, the terms "independent" and "dependent" imply causality, i.e., that the value of the dependent variable *depends* at least in part on the values of the independent variables, a statement that is impossible to establish in many nonexperimental designs. This may be illustrated by comparing a randomized controlled trial (an experimental design) with a cross-sectional survey (an observational design).

In a randomized clinical trial of the effects of a new drug on hypertension, if the correct procedures are followed and significant results are achieved, the researcher can be comfortable (or as comfortable as one can ever be when dealing with inferential statistics, whose conclusions are inherently probabilistic rather than absolute) in asserting that changes in blood pressure observed were caused or influenced by the new drug.

In a cross-sectional survey of juvenile delinquency and drug use, however, it is impossible to establish a causal effect because either variable could cause the other, and any relationships found could be due to other variables. For instance, children who use drugs may be more likely to become delinquent, or delinquents may be more likely to use drugs. Even if this issue is resolved by including temporality in these questions (it might be possible to determine which came first, drug use or delinquency) the explanation cannot be discarded that those who use drugs (a self-selected group) differ in other ways from those who do not. For instance, the drug users may be less intelligent, or more intelligent, than the nondrug users, or may have different family circumstances, and either of those variables could influence delinquency independently of drug use.

Populations and Samples

The concept of populations and samples, discussed briefly in Chapter 4, is crucial to understanding inferential statistics. The process of defining the population and selecting an appropriate sampling method can be quite complex (in fact, many doctoral-level statisticians specialize in just that type of work) and requires more study than can be covered here. Instead, the basic issues and concepts will be discussed, and the reader interested in further information on the subject should consult a specialized textbook (several are listed in Appendix C).

The population of interest (often called merely "the population") consists of all the people or other units (for instance, airplane parts or Atlantic salmon) that the researchers would like to study, if they had infinite resources. To put it another way, the population of interest is all the units to which the researchers would like to be able to generalize their results. Defining the population of interest is the first step in drawing a sample: it may be very broad, such as everyone living in the United States in 2007, or narrow, such as Canadian men aged 65–75 with a diagnosis of congestive heart failure.

Almost all research is based on a study sample drawn from a population, rather than the population itself, for practical reasons. The rare exceptions are studies such as those based on the U.S. census, which intends to collect data from every individual living in the United States in a particular year.

Nonprobability Sampling

There are many ways to draw a sample. Unfortunately, some of the most convenient (described in this section) are based on nonprobability sampling, which leaves them highly subject to sampling bias. This means there is a high probability that the sample drawn using a nonprobability method will not be representative of the population the sample is meant to represent, and therefore may lead to incorrect conclusions about that population. The convenient methods are popular because they allow the researcher to bypass the steps of defining the population of interest, assigning a probability of selection to each member, drawing a sample, and applying sampling weights. The drawback is that with this type of sampling no information is available about how the sample relates to the population of interest, so little faith may be laid in conclusions about that population based on results from the sample.

Volunteer samples are a commonly used type of nonprobability sample. An example would be if a researcher advertises in the newspaper for study subjects and accepts those who answer the ad and who volunteer to be part of the study. Unfortunately, people who volunteer for studies can't be assumed to be representative of any general population. Volunteer samples are particularly common in circumstances when it would be difficult to randomly select from a population, for instance in a study about people who use illegal drugs. Much useful information can be gained from volunteer samples, particularly in the early stages of a project: for instance, you might use volunteer subjects to gather information about drug use within a community, which you could use to construct a questionnaire that would be administered to a random sample from the community. But results from volunteer samples have limited usefulness if the goal is to generalize beyond the sample.

Convenience samples are another type of nonprobability sampling that may be used to collect initial information, but like volunteer sampling have limited usefulness if the goal is to generalize beyond the sample. For instance, you might collect information about the shopping habits of people in a geographical area by interviewing the first 30 people you saw at a particular shopping mall, and use that information to construct a study that would use a randomized design. But it would not be valid to conclude, for instance, that because 75% of your convenience sample favored shopping at The Gap over Old Navy, that 75% of the people living in the area would do likewise.

Quota sampling is a type of nonprobability sampling in which a data collector is ordered to find a certain number of subjects within broad classifications: in the shopping example, the data collector might be instructed to collect data from a sample of 15 men and 15 women. Quota sampling is a slight improvement in accuracy over convenience sampling because it can specify the makeup of the sample: without the quota requirements the shopping mall sample might be 25 women and 5 men. But it does not get around the main problem of all nonprobability sampling, which is that you have no way of knowing if the people sampled were representative of the population of interest. You may have an even representation of men and women in a quota sample, for instance, but are they representative of all the men and women who shop at the mall?

Probability Sampling

In probability sampling, every member of the population has a known probability of selection to be in the sample. Although more difficult to execute than nonprobability sampling, it has the benefit of allowing the researcher to generalize the results obtained to the population of interest.

Drawing a sample from a population requires devising some type of sampling frame, which allows the researcher to identify and sample members of the population. Sometimes an obvious sampling frame exists: if the population is students enrolled at a particular school, a list of all enrolled students could serve as the sampling frame. Other times a less optimal sampling frame must be used: for instance, a telephone directory or block of phone numbers in use may be employed for a survey carried out by telephone. A problem with either frame is that people without phone service are not included in the population from which the sample is drawn, although they may be included in the population of interest. Weighting and other procedures can be used during analysis to make results from the study sample more applicable to the population of interest.

The most basic type of probability sampling is *simple random sampling* (SRS). In SRS all samples of a given size have an equal probability of being selected. Suppose you wanted to draw a random sample of 50 students attending a particular school. You obtain a list of the students and select 50 at random from the list, using a random number table or random number generator. Because the list represents an enumeration of the entire population and the choice of who to include in the sample is completely random, every student has an equal probability of being selected for the sample, as does every combination of students up to the size of the sample.

In most cases, SRS has the most desirable statistical properties of any kind of sampling, including the smallest confidence intervals around parameter estimates, and requires the least complex procedures to analyze. However, SRS is impossible or prohibitively expensive to execute in some contexts, so other methods of probability sampling have been developed to deal with situations where SRS is not possible or practical.

Systematic sampling is very similar to SRS. To draw a systematic sample, you need a list or other enumeration of your population. You then choose a start number between 1 and n at random, and include in that sample the nth object and every nth object following, n being chosen to produce the sample size desired. Suppose you want to draw a random sample of 100 objects from a population of 1,000. The steps to draw a systematic sample are:

1. Set $n = 10$, because $1000/100 = 10$.
2. Choose a number at random between 1 and 10.
3. Select the object with that number, and every 10th object thereafter.

If the number chosen at random was 7, your sample would include the 7th, 17th, 27th, and so on, up to the 997th object.

Systematic sampling technique is particularly useful when the population accrues over time and there is no predetermined list of objects. For instance, if you want to survey people who will be making court appearances in the upcoming year, at the start of the study you will not know who those people will be. So you could make an estimate of n based on the court caseload in the previous year, keep an ordered list of people making court appearances, and then survey every nth person who appears in court. If you determine that n is 14, you would then survey the 14th person, 28th person, 42nd person, and so on.

One caution when using systematic sampling is that you must ensure that the data is not cyclic in a way that corresponds with n. For instance, if particular hours or days in court were reserved for particular types of cases, and your choice of n meant that people whose court dates were scheduled for those times had no possibility of being selected, then your sample would not be random.

There are many types of *complex random samples*, an umbrella term for probability sampling methods that impose one or more layers of complexity beyond that of SRS. In a *stratified sample*, the population of interest is divided into nonoverlapping groups or *strata* based on common characteristics. For people, these characteristics might be gender or age, for cities they might be population size or type of government, for hospitals they might be type of organization or number of beds. If comparing different strata is a primary goal of the study, stratified sampling is a good choice because it can be designed to ensure adequate sampling from each strata of interest. For instance, using SRS might not produce sufficient elderly people to accurately compare their results with middle-aged people, while a stratified sample can be designed to oversample the elderly to ensure sufficient sample size, then correct statistically for the oversampling.

In a *cluster sample*, the population is sampled making use of pre-existing groups. This technique is often used in national surveys that require in-person interviews or the collection of physical specimens (e.g., blood samples), because it would be prohibitively expensive to send survey personnel to interview one person in Ruckersville, Virginia, one in Chadron, Nebraska, one in Barrow, Alaska, and so on. A more economical procedure is to create a sampling plan that incorporates several levels of random selection. On a national level, this could be executed by selecting geographic regions, then states within regions, cities within states, and so on down to individual households and individuals within households. Precision is decreased with cluster sampling because objects that are clustered within units (for instance, households within cites and cities within states) tend to be more similar than objects selected through SRS. Offsetting this loss of precision is the fact that the cost savings of cluster sampling are usually substantial, so a larger sample can be collected.

Cluster sampling can be combined with the technique of *sampling proportional to size*. For instance, you may wish to draw a sample of elementary school students. There is no national list of all elementary school students, but you could compile a list of all elementary schools, and each school would have a list of their students. So you could select schools at random, possibly in a multistage process, then draw a random sample from the selected schools. Because schools enroll different numbers of students, you might want to include this information in your sampling plan so you don't have a disproportionate number of students from small schools (which are more numerous but contain fewer students). Then you could select a different number of students from each school according to the number enrolled in the school: twice as many from a school with 400 students as from one that enrolled 200, for instance. In this way, your final sample will have a representative proportion of students from both large and small schools.

The Central Limit Theorem

The Central Limit Theorem states that the sampling distribution of the sample mean approximates the normal distribution, regardless of the distribution of the population from which the samples are drawn, if the sample size is sufficiently large. This fact enables us to make statistical inferences using tests based on the approximate normality of the mean, even if the sample is drawn from a population that is not normally distributed.

The Central Limit Theorem may be stated as follows with regard to the sample mean:

> Let $X_1, \ldots X_n$ be a random sample from some population with mean μ and variance σ^2. Then for large n,
>
> $$\bar{X} \dot\sim N(\mu, \frac{\sigma^2}{n})$$
>
> even if the underlying distribution of individual observations in the population is not normal.
>
> The symbol $\dot\sim$ is used to represent "approximately distributed" and the formula can be read as: "the mean of X is approximately normally distributed with mean μ and variance σ^2/n".[*]

The application of the Central Limit Theorem in practice can be seen through computer simulations that repeatedly draw samples of specified size from a nonnormal population. Figure 7-4 displays a histogram for a population of randomly generated data (100 cases) with a uniform distribution of values ranging from 0 to 100.

The distribution in Figure 7-4 is decidedly nonnormal. However, the Central Limit Theorem says that when samples of sufficient size are drawn from a nonnormal population, the means of those samples tend to assume a normal distribution. Note that the theorem does not state what constitutes a "sufficient size." Analysts have developed rules of thumb regarding this issue, such as the

---

[*] Bernard Rosner, *Fundamentals of Biostatistics*, Fifth Edition; Brooks/Cole, 2000, p. 174.

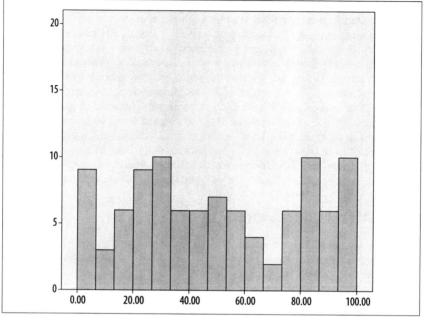

Figure 7-4. Histogram of a uniformly distributed population (N = 100) with range 0–100

often-repeated rule that the sample size should be 30 or larger, but no absolute rule applies in all cases. For data that is approximately normal, normality of the sampling distribution of the sample mean may be normal with a sample size as small as 10 or 15, while with highly skewed distribution the sample size required may be 40 or greater. I've seen one example in which a sample size of 200 was required!

The influence of sample size on the distribution of sample means can be seen by comparing Figures 7-5 and 7-6. Figure 7-5 displays the distribution of the means of 100 samples of size $n = 2$ drawn from the population shown in Figure 7-4, while Figure 7-6 displays the distribution of the means of 100 samples of size $n = 25$ drawn from the same population. Figure 7-5 looks more like a uniform than a normal distribution, indicating that a sample size of 2 is not sufficient to invoke the Central Limit Theorem for this population.

Figure 7-6 displays the distribution of 100 means calculated from samples of size 25 drawn from the uniform distribution displayed in Figure 7-4. It looks very much like a normal distribution, so a sample size of 25 appears to be sufficient to invoke the Central Limit Theorem for this population.

Figures 7-7 to 7-9 demonstrate the same principles with samples drawn from a skewed population. Figure 7-7 shows the distribution of values for a data set of size 100 with a skewed distribution.

Figures 7-8 and 7-9 demonstrate how the distribution of sample means drawn from the population displayed in Figure 7-7 changes with the size of the sample. Figure 7-8 shows the distribution of means calculated from 100 samples of size 2,

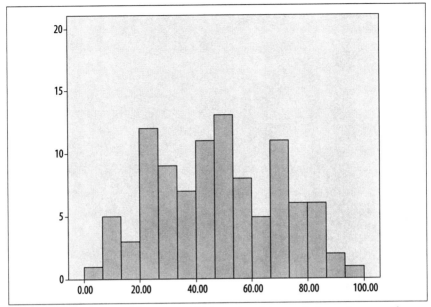

Figure 7-5. Distribution of the means of 100 samples of size n = 2, drawn from a uniform distribution

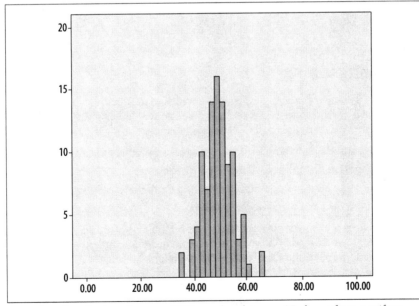

Figure 7-6. Distribution of means of 100 samples of size n = 25, drawn from a uniform distribution

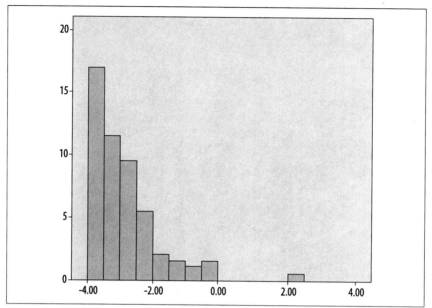

Figure 7-7. Histogram of skewed population (N = 100)

while Figure 7-9 shows the distribution of means from 100 samples of size 25. As with the uniform data example, a sample of size $n = 2$ is not sufficient to invoke the Central Limit Theorem for this data, while a sample of 25 seems to be sufficient.

Hypothesis Testing

Hypothesis testing is fundamental to inferential statistics because it allows us to use statistical methods to make decisions about real-life problems. There are several conceptual steps involved in hypothesis testing:

1. Develop a research hypothesis that can be tested mathematically.
2. Formally state the null and alternative hypotheses.
3. Decide on an appropriate statistical test and do the calculations.
4. Make your decision based on the results.

Take the example of a new medication to treat high blood pressure. The manufacturer wants to establish that it works better than current standard treatments for the same condition, so the research hypothesis might be something like "Hypertensive patients treated with drug X will show greater lowering in their blood pressure than hypertensive patients receiving the standard treatment." If we use μ_1 to signify the mean blood pressure in the group treated with drug X, and μ_2 as the mean blood pressure in the group receiving standard treatment, our null and alternative hypotheses could be formally stated as:

$H_0: \mu_1 \geq \mu_2$
$H_A: \mu_1 < \mu_2$

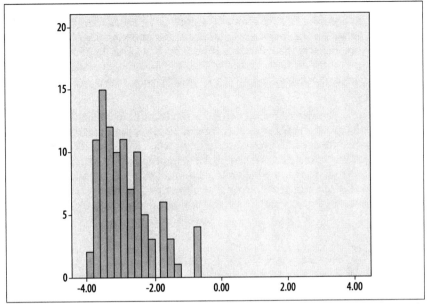

Figure 7-8. Distribution of means from samples of size n = 2, drawn from a population with skewed distribution

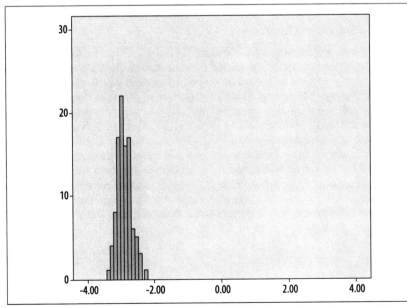

Figure 7-9. Distribution of means from samples of size n = 25, drawn from a population with skewed distribution

H_0 is called the null hypothesis: in this example, the null hypothesis states that drug X is no improvement over standard treatment. H_A, sometimes written as H_1, is called the alternative hypothesis: in this case, the alternative hypothesis is that drug X is more effective than standard treatment. Note that the null and alternative hypotheses must be both mutually exclusive (no results could satisfy both conditions) and exhaustive (all possible results will satisfy one of the two conditions).

In this example, the alternative hypothesis is *single-tailed*: we state that the blood pressure of the group treated with drug X must be lower than that of the standard treatment group for the null hypothesis to be rejected. We could also state a *two-tailed* alternative hypothesis if that were more appropriate to our research question. If we were interested in whether the blood pressure of patients treated with drug A was different, either higher or lower, than that of patients receiving standard treatment, we would state this using a two-tailed alternative hypothesis:

$H_0: \mu_1 = \mu_2$
$H_A: \mu_1 \neq \mu_2$

Normally the first two steps would be performed before the experiment is designed or the data collected; the statistic to be used for hypothesis testing is also sometimes specified at this time, or is implicit in the hypothesis and type of data involved. We then collect the data and perform the statistical calculations, in this case probably a *t*-test or ANOVA, and based on our results make one of two decisions:

- Reject the null hypothesis and accept the alternative hypothesis, or
- Fail to reject the null hypothesis

The first case is sometimes called "finding significance" or "finding significant results." The process of statistical testing involves establishing a probability level or *p*-value (a topic treated in greater detail below) beyond which we will consider results from our sample strong enough to support rejection of the null hypothesis. In practice, the *p*-value is commonly set at 0.05. Why this particular value? It's an arbitrary cutoff point and dates back to the early twentieth century, when statistics were computed by hand and the results compared to published tables used to determine whether a result was significant or not. The use of $p < 0.05$ as the standard for significant results has been challenged (see the upcoming sidebar, "Controversies About Hypothesis Testing") but still remains common practice for published research. Alternative lower values are sometimes used, such as $p < 0.01$ or $p < 0.001$, but no one has been successful in legitimizing the use of a higher cutoff, such as $p < 0.10$.

Note that failure to reject the null hypothesis does not mean that we have proven it to be true, only that the experiment or study did not find sufficient evidence to reject it.

Inferential statistics allows us to make probabilistic statements about the data, but the possibility of error is inherent in the process. Statisticians have classified two types of errors when making decisions in inferential statistics, and set levels for error rates that are commonly considered acceptable. The two types of error are displayed in Table 7-1.

Controversies About Hypothesis Testing

Despite the ubiquitousness of hypothesis testing in modern statistical practice, and the canonical place that the $\alpha = 0.05$ significance level seems to have achieved, many researchers have criticized both practices. One of the chief critics is Jacob Cohen, whose arguments are presented in, among other places, his 1994 article "The World Is Round (p < 0.05)." (*American Psychologist*, 49:2, December 1994, 997–1003). Valid as many of these criticisms are, both hypothesis testing and the 0.05 significance level don't seem to be going away anytime soon. And some level must be set at which results are considered significant, to avoid attributing significance to differences due to chance or to sampling error. A sensible compromise is to realize that there's nothing magical about 0.05, even if it is sometimes treated as such, and that the significance level of a sample calculation is affected by many factors, including the size of the sample involved. It's a common saying among statisticians that if you have a large enough sample, any little difference will be statistically significant. The take-home message is that statistical methods don't relieve the researcher of the need to apply common sense.

Table 7-1. Type I and Type II errors

		True state	
		H_0 true	H_A true
Decision based on sample statistic	**Accept H_0**	Correct decision: H_0 true and H_0 not rejected	Type II error or β
	Reject H_0	Type I error or α	Correct decision: H_0 false and H_0 rejected

The diagonal boxes represent correct decisions: H_0 is true and was not rejected in the study, or H_0 is false and was rejected in the study. The other two boxes (often referred to as the off-diagonal boxes) represent Type I and Type II errors. A Type I error, also known as alpha or α, represents the error made when the null hypothesis is true but is rejected in a study. A Type II error, also called beta or β, represents the error made when H_0 is false but is not rejected in a study.

The level of acceptability for Type I error is conventionally set at 0.05, as noted above. Setting alpha at 0.05 means that we accept the fact of a 5% probability of Type I error. To put it another way, we understand when setting the alpha level at 0.05 that we accept the fact that in our study we have a 5% chance of rejecting the null hypothesis when we should have accepted it.

Type II error has received less attention in statistical theory because historically it has been considered a less serious error to fail to make an inference that is true (Type II error) than to make an inference that is false (Type I error). Conventional levels of acceptability for Type II error are $\beta = 0.1$ or $\beta = 0.2$. If $\beta = 0.1$, that means the study has a 10% probability of a Type II error, i.e., there is a 10% chance that the null hypothesis will be false but will be accepted in the study.

The reciprocal of Type II error is *power*, defined as $1 - \beta$. The importance of setting an appropriate power level has become more appreciated in recent years, particularly in the medical field. Researchers and funding agencies have become concerned with power, and thus with Type II error, in part because they don't want to spend the time and effort to conduct a study unless it has a reasonable probability of finding significant results if they do exist. Power calculation is discussed in more detail in Chapter 18.

Confidence Intervals

When we calculate a single statistic, such as the mean, to describe a sample, that is referred to as calculating a *point estimate* because the number represents a single point on the number line. The sample mean is a point estimate, and is a useful statistic as the best estimate of the population mean. However, we know that the sample mean is only an estimate and that if we drew a different sample, the mean of the sample would probably be different. We don't expect that every possible sample we could draw will have the same sample mean. It is reasonable to ask how much the point estimate is likely to vary by chance if we had chosen a different sample, and in many professional fields it has become common practice to report both point estimates and *interval estimates*. A point estimate is a single number, while an interval estimate is a range or interval of numbers.

The most common interval estimate is the *confidence interval*, which is the interval between two values that represent the upper and lower *confidence limits* or *confidence bounds* for a statistic. The formula used to calculate the confidence interval depends on the statistic being used and will be included in the relevant chapters: this section is meant to convey the concept of the confidence interval. The confidence interval is calculated using a predetermined significance level, often called α (the Greek letter *alpha*), which is most often set at 0.05, as discussed above. The *confidence coefficient* is calculated as $(1 - \alpha)$ or, as a percentage, $100(1 - \alpha)\%$. Thus if $\alpha = 0.05$, the confidence coefficient is 0.95 or 95%. The latter usage is more common; for instance, professional journals often require that you report the 95% confidence interval for your statistics.

Confidence intervals are based on the notion that if a study was repeated an infinite number of times, each time drawing a different sample from the same population and constructing a confidence interval based on that sample, x% of the time the confidence interval would contain the unknown parameter value that the study seeks to estimate. For instance, if our test statistic is the mean and we are using 95% confidence intervals, over an infinite number of repetitions of the study, 95% of the time the confidence interval constructed from the study would contain the mean of the population. For this reason, the confidence interval is sometimes described as presenting a plausible range of values for the mean.

The confidence interval conveys important information about the precision of the point estimate. For instance, suppose we have two samples of students and in both cases the mean IQ score is 100. In one case, however, the 95% confidence interval is (95,105), while in the other case the 95% confidence interval is (80,120). Because the former confidence interval is much narrower than the latter, the estimate of the mean is more precise for the first sample.

p-values

It is a fact of life when working with inferential statistics that we are always trying to estimate something that we can't measure directly. For instance, we don't have the ability to collect data from every hypertensive adult in the world, but we can select a sample of hypertensive adults, design an experiment involving them, and analyze the data we thus collect. Because we understand that sampling error is always a possibility in studies based on samples, we want to know the probability that the results obtained from our sample were not due to chance. If we had the means to draw repeated samples from the population and repeat the experiment, how likely is it that we would obtain similar results most of the time?

A *p-value* usually expresses the probability that results at least as extreme as those obtained in a sample were due to chance. The phrase "at least as extreme" is necessary because most statistical tests involve comparing the test statistic to some hypothetical distribution (such as the normal distribution, as illustrated below) where scores closer to the center of the distribution are most common and scores become less likely as they are further from the center of the distribution.

This may be clearer by considering a simple illustration. Suppose we are engaged in an experiment flipping a coin that we believe to be fair, i.e., a coin for which heads or tails are equally likely outcomes for any single flip. We can express this formally as $P(H) = P(T) = 0.5$. We will call each flip a trial. Our expectation is that we will get 5 heads on 10 trials, although we know that on any particular set of 10 trials we may get a different number of heads. So we flip the coin 10 times and 8 times it comes up heads. We want to know the *p-value* of this result, i.e., how likely is it that a coin with a probability of 0.5 for heads on any single trial would produce 8 heads in 10 trials?

Using a binomial table, computer software, or the binomial formula, we find that the probability of this exact result (8 heads in 10 trials) is 0.0439, meaning that less than 5% of the time would we expect to get exactly 8 heads in 10 flips with a fair coin. The probability for 9 heads in 10 trials is 0.0098, and for 10 heads in 10 trials is 0.0010. This demonstrates that as results move further away from the expected result of 5 heads in 10 trials, they become less likely.

If we are evaluating the probability that the coin truly is fair, results that are far from our expectation give us strong evidence that it in fact is not fair. For this reason, we usually calculate the probability not just of the result we obtained in our experiment, but of results at least as extreme as those we obtained. In this case, the probability of getting 8, 9 or 10 heads in 10 flips of a fair coin is 0.0439 + 0.0098 + 0.0010, or 0.0547. This is the *p-value* for the result of at least 8 heads in 10 trials using a coin where $P(\text{heads}) = 0.5$.

p-values are commonly reported for most research results involving statistical calculations, in part because intuition is a poor guide to how unusual a particular result is. For instance, many people might think it is unusual to get 8 or more heads on 10 trials using a fair coin. In this case, the binomial probability of such a result has a *p-value* of 0.0547. This result does not allow us to reject the null hypothesis that the coin is fair, i.e., $P(\text{heads}) = 0.5$, using the standard rule of thumb that a *p-value* must be less than 0.05 for results to be considered significant.

Data Transformations

Many of the most common statistical procedures are what are known as *parametric* statistics, meaning that they make certain assumptions about the distribution of the population from which the sample analyzed was drawn. If the raw data does not meet these assumptions, the researcher has several options for analyzing the data. One is to use alternate, *nonparametric* statistical procedures, which make fewer or no assumptions about the data distribution, but are also frequently less powerful than their parametric counterparts. Nonparametric statistics are discussed in Chapter 11. Another possibility is to *transform* the data in some way so that the assumptions of the desired parametric statistical procedure are met. There are many ways to transform data, depending on the distribution involved and the assumptions that are violated. Because the most common transformation problem is to make a data set closer to a normal distribution, the most common transformation for improving the normality of data will be discussed here. For more detail about data transformation, the reader should consult a more advanced textbook such as that by Mosteller and Tukey listed in Appendix C.

The first step in using a data transformation is to evaluate the data set in question and decide which, if any, transformations are appropriate. Two approaches are recommended to evaluate a data set. One is to graph the data, for instance by creating a histogram with a superimposed normal curve. This allows a visual evaluation of the general shape of the data and an opportunity to identify outliers. The shape of the data also aids in suggesting what transformation to apply. The second method is to compute one of the statistics provided in most statistical computing packages to test whether the data fits a particular distribution. Two commonly used tests for this purpose are the Anderson-Darling and the Kolmogorov-Smirnov. One or both statistics are included in many statistical packages, and a statistical calculator to compute the Kolmogorov-Smirnov test is available at *http://jumk.de/statistic-calculator/*.

Data that is right-skewed may be made more normal by application of the square-root or log transformations. The square-root transformation computes the square root of each value. If the data value is 4, the transformed value is 2 because $\sqrt{4} = 2$. The log transformation computes the natural log of each value, so if the data value is 4, the transformed value is 1.386 because $ln(4) = 1.386$. Either transformation may be accomplished easily using statistical software such as SPSS or SAS.

Figure 7-10 displays a right-skewed data set. Figure 7-11 shows the same data after a square-root transformation (the values graphed are the square roots of the data in Figure 7-10) and Figure 7-12 shows the same data after a log transformation (the values graphed are the natural logs of the data displayed in Figure 7-10).

Comparing the three graphs visually, Figure 7-10 is definitely right-skewed, meaning that most of the data values are relatively low, with a few higher values creating a long "tail" to the "right." It does not seem to fit the superimposed normal distribution curve. Figure 7-11 seems to be a much better fit to the normal distribution, and Figure 7-12 seems to have replaced the right-skew with a left-skew, so it is also nonnormal. For a second source of information, we calculate the Kolmogorov-Smirnov statistic (using SPSS, although it is available in other programs as well) to statistically evaluate how well each data set fits the normal distribution. Results for the three data sets and the results shown in Table 7-2.

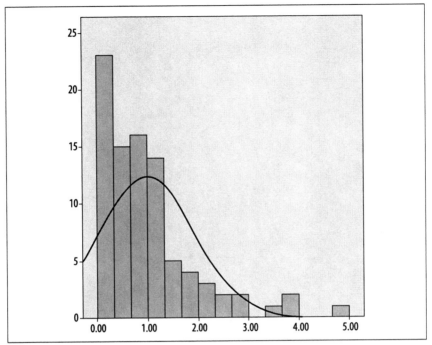

Figure 7-10. Right-skewed data set (raw values)

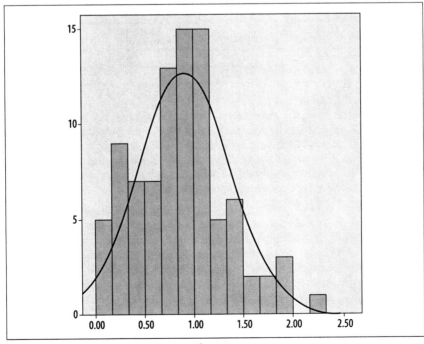

Figure 7-11. Data after square-root transformation

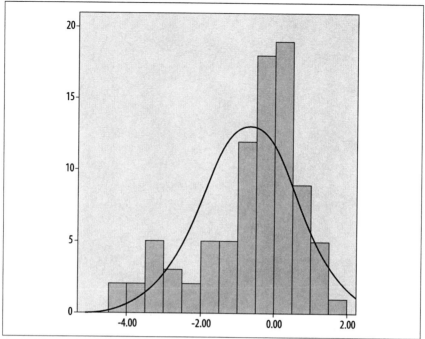

Figure 7-12. Data after natural log transformation

Table 7-2. Kolmogorov-Smirnov test for normality for a data set and two transformations

	Raw data	Square-root transformation	Natural log transformation
Kolmogorov-Smirnov Z	1.46	0.66	1.41
p	0.029	0.78	0.04

The null hypothesis for the Kolmogorov-Smirnov test is that the data is normally distributed, so if the test indicates that the null hypothesis should be rejected, this means that the data is not normally distributed. Using the rule that we reject the null hypothesis if $p < 0.05$, we conclude that the raw data and the natural log transformation data are not normally distributed, while the square-root transformation data can be accepted as normally distributed.

If a variable has a left or negative skew, you can reflect the data and then apply the square-root or log transformation. To reflect a variable, take the largest value in the data set, add 1, then subtract each value of the variable from the new number. If the largest value in the data set is 35, we would subtract each value from 36 (i.e., 35 + 1) to get the reflected values. Reflection changes a left-skewed distribution to one with a right skew, and the square-root and log transformations can then be applied to see if they improve normality.

Note that data transformation is not a guaranteed solution to a distribution problem: sometimes it makes the problem worse or introduces a new problem!

For this reason, the transformed data should also be evaluated for normality, as we did above, to see if the transformation resulted in data that has the desired distribution. Note also that a transformation changes the unit of the data: for instance, if you apply the log transformation to a population of blood pressure scores, your unit is then the log of blood pressure rather than the original scores. If you reflect a variable, this reverses the values (what was the highest score is now the lowest) and so the interpretation of any statistic based on those values is also reversed. The effects of other transformation must be kept in mind when interpreting the statistical results.

Exercises

Here's a quick review of the concepts explained in this chapter.

Conceptual Questions

1. In each of the following sets of variables, which are likely candidates to be treated as independent and which as dependent within a research study?
   - Gender, alcohol consumption, and driving record
   - High school GPA (grade point average), university freshman year GPA, choice of major, race/ethnicity, and gender
   - Age, ethnicity, smoking habits, use of hormone replacement therapy, and occurrence of breast cancer
   - Accuracy on a coding task, type of instructions given, practice time, and anxiety level

2. Why is the Central Limit Theorem of primary importance to the practice of inferential statistics?

3. What type of sampling is described by the following scenarios?
   - The goal is to collect information on HIV status, obtained through blood tests, on the U.S. population. Because it is expensive to send researchers to many locations, a sampling plan is devised using successively smaller regions of the country, beginning with Census Region (Northeast, South, Midwest, and West) and ending with census block groups.
   - The goal is to find out how elementary school students are reacting to a recently appointed principal. The researcher wants to include equal numbers of male and female students in the sample, so the interviewer is sent to the school with instructions to interview 10 male and 10 female students from among those on the playground after school one day.
   - The goal is to learn more about the domestic life of policemen working in a major city, including how home life is affected when the policeman's spouse is employed outside the home. A complete list of all men and women working as policemen in this city is available, and a computer is used to draw a random sample of 200 from this list. The sample is then interviewed by telephone.

- A factory supervisor is concerned that the quality of parts produced may not be equal on all shifts or at all times within a shift (the factory operates 24 hours per day). A sampling plan is devised to collect samples of 30 parts at 9 times during the work day: within the first 2 hours, within the 6 hours, and within the last 2 hours of each of the 3 daily shifts.

4. How are hypothesis testing, confidence intervals, and the p-value related?

Problems

1. Calculate the Z-scores of the following data values, assuming they came from a normal population with $\mu = 100$ and $\sigma^2 = 4$.
   - 108
   - 95
   - 98

2. Which of the following raw scores has a more extreme Z-score, i.e., a Z-score further (in either a positive or negative direction) from 0?
   - A score of 190, from a population with $\mu = 180$ and $\sigma^2 = 16$
   - A score of 175, from a population with $\mu = 200$ and $\sigma^2 = 25$

Solutions

1. Calculating Z-scores using a specified normal distribution.

$$Z(108) = \frac{108 - 100}{\sqrt{4}} = \frac{8}{2} = 4$$

$$Z(95) = \frac{95 - 100}{\sqrt{4}} = \frac{-5}{2} = -2.5$$

$$Z(98) = \frac{98 - 100}{\sqrt{4}} = \frac{-2}{2} = -1.0$$

2. In the second example, the score is more extreme, because -5 is further from 0 than 2.5.

$$\frac{190 - 180}{\sqrt{16}} = \frac{10}{4} = 2.5$$

$$\frac{175 - 200}{\sqrt{25}} = \frac{-25}{5} = -5.0$$

8

The t-Test

The purpose of the *t*-test is to make inferences about single means, or inferences about two means or variances, where sample sizes are small and/or the population distribution is unknown. While not always used in practice—since the one-way Analysis of Variance (ANOVA) is mathematically equivalent to the *t*-test, and since most researchers attempt to gather a reasonable number of samples to avoid Type II errors—understanding the logic and outcomes of the *t*-test (and its distribution) will make it much easier for you to follow ANOVA and more sophisticated analytical techniques, especially where your sampling is necessarily limited.

The t Distribution

In Chapter 7, you learned how to use the normal (or *Gaussian*) distribution, which is a continuous probability distribution, to assist in making inferences about a population. Recall that the known mathematical properties of the distribution can be used to determine probabilities of characteristics occurring within the population, even when the population mean is unknown. Thus, hypothesis testing can be carried out using limited sampling, and correct inferences drawn, if the population is normally distributed. In many natural systems, populations are normally distributed, but sometimes they are not, and thus, the normal distribution cannot be used as a model.

However, if you have gathered enough samples, it may still be possible to use the properties of the normal distribution, since the sampling distribution of averages is likely to be normal, according to the Central Limit Theorem (or at least have some of the key characteristics of a normal distribution, such as being unimodal and symmetrical). Thus, irrespective of the underlying population distribution, the normal distribution can be used to estimate probabilities when samples are sufficiently large: the sample variance can be used to estimate the population variance, and inferences drawn with the assistance of the normal distribution.

This strategy may not always be appropriate for answering your specific research question, especially if you can only obtain limited samples because of financial, physical, or time constraints. Indeed, such a situation faced industrial statistician William Gosset in the early twentieth century, when he worked at the Guinness Brewery in Dublin as an industrial researcher with an enviable role—quality assurance for beer brewing. After studying statistics with Karl Pearson at University College, London, Gosset published a paper under the pseudonym "Student," since Guinness did not want their competitors to know that they were employing statistics to improve quality control.

Gosset's key observation was the dependence on sample size for determining the probability that the mean of the population lies within a given distance of the mean of the sample, if a normal distribution is assumed. Through a combination of mathematical argument and numerical simulation, Gosset noted that when samples are collected from a normal distribution, and if the number of samples is small, and these are used to estimate the variance, then the distribution (for the variable x):

$$t = \frac{\bar{x} - \mu}{\frac{s}{\sqrt{n}}}$$

is both flatter, and has more observations appearing in the tails, than a normal distribution, when sample sizes are less than 30, and where $s/\sqrt{n}$ refers to the standard error. Since both s and x are random variables, this may not be such a surprise. However, as the number of samples increases, the distribution becomes normal, given the dependence on n, and the corresponding effect on degrees of freedom, since $df = n - 1$. This distribution is known as the t *distribution*, and approximates a normal distribution if n (and by implication df) are large (>30 in practical terms).

Books of statistical tables normally provide critical values of t that can be used at different degrees of freedom to make inferences about the population, with an associated probability of committing a Type I error (α). For example, where $n = 21$ and $df = 20$, then $t = 1.725$ at the $p = 0.05$ significance level, and $t = 2.528$ at the $p = 0.01$ significance level. These relations would usually be expressed as $t_{0.05,20} = 1.725$ and $t_{0.01,20} = 2.528$, respectively. Figure 8-1 shows an example t distribution.

t-Tests

Now that you have seen what the t distribution is, you may be wondering about its purpose. In simple terms, t-tests are the simplest form of *parametric hypothesis testing* for real-valued (rather than categorical) data. Using a t-test allows you to test whether the mean of a sample differs significantly from an expected value, or whether the means of two groups different significantly from each other. Significance here means statistical significance, and is related to the probability (p) of committing a Type I error. Typically acceptable probability values are $p < 0.05$, representing a 1 in 20 chance of committing a Type I error, or $p < 0.01$, representing a 1 in 100 chance of committing a Type I error.

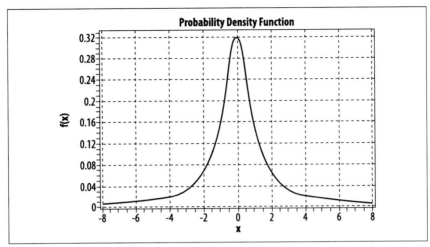

Figure 8-1. Example t distribution

The probability of committing a Type I error relates to two different ways that hypothesis testing is used: in science and in technology. Scientists typically frame their experiments so that they do not directly test the hypothesis, but evaluate a *null hypothesis*. Thus, Type I error here applies to the probability of rejecting the null hypothesis, when in fact it should have been accepted. For example, a scientist has formed two groups, treatment and control, testing the effects of a new weight-loss drug, and her hypothesis is that the weight-loss drug will significantly reduce weight in her treatment group. Weight is measured pre-test and post-test (after six weeks of taking the drug). Age, sex, height, and weight-matched participants are randomly allocated to the treatment and control groups. The null hypothesis in this case—since the two groups were drawn with equal probability from the population—is that there will be no difference between the means of the groups.

Calculating the t statistic, and determining the p value, gives the probability of committing a Type I error. If $p < 0.05$ (or $p < 0.01$), then the experimenter rejects the null hypothesis, which provides support for the hypothesis. If $p > 0.05$, then you would fail to reject the null hypothesis, which provides evidence against the hypothesis. Further replication of the result and/or an enhancement in coverage of different populations increases the generalizability of the finding, and may eventually lead to general acceptance of the result.

In everyday applications of statistics, Type I errors have a more immediate impact; if an engineer, for example, needs to test whether the height of a door is sufficient for 99 in 100 of the population not to bump their heads, then a finding that 1 in 20 will bump his head is quite useful. Or, if the sample size is too small, then the engineer may believe that the height is suitable for 99 in 100 when in fact it is only suitable for 19 in 20. So, in the field, technologists and engineers need to be just as concerned about committing Type I errors, even if they believe that hypothesis testing is something that only scientists do. In safety-critical systems, it's essential that hypothesis testing be used in conjunction with other approaches that are algorithmic in nature, and work perfectly well in simulation.

The value of t can be estimated in practice by using the following formula:

$$t = \frac{(\text{parameter estimate from sample}) - (\text{hypothetical parameter estimate})}{\text{estimator standard error (estimated)}}$$

t-tests are often used industrially since they only require small samples (typically less than 30) for probability values to be calculated. Thus, they are useful in situations where funding is tight, and/or where samples are destroyed during the experimental process.

t-Test Assumptions

Sometimes, in inferential statistics, it seems that assumptions are made to be ignored; the validity of the results from a two-sample t-test, for example, relies on properties of the population distributions being equal, but these assumptions are routinely ignored or violated in practice. Why should you care about these assumptions? Aren't they just trivial mathematical issues that have no relevance to practice? Well, if you're a scientist who makes an amazing discovery of differences between two groups that is accepted for publication in a prestigious journal, only to later have to write a retraction because your analysis was shown to be invalid by a competitor, you'll be very embarrassed.

More seriously, if you're an engineer or technologist, and you designed a new system based on flawed testing, lives could be lost. Always ensure that the assumptions underlying a test are met, and if they are not, use a different technique, or one of the many corrections that exist for such tests. For example, one assumption of t-tests is homogeneity of variance between the two samples; if this assumption is violated, then the unequal variance t-test should be used. Or if a parametric statistic is inappropriate, consider the use of nonparametrics.

Generally, the distributions from which the samples are drawn should be normal. This can be tested directly using the *Shapiro-Wilk test* or *Kolmogorov-Smirnov test*. Equality of variances can be determined by using *Levene's test*, *Brown and Forsythe test*, or *Bartlett's test* for two independent samples. Where normality cannot be established, then nonparametric tests can be used; these include the *Mann-Whitney U test* for independent samples (between subjects), or the *Wilcoxon signed rank test* for dependent samples (within subjects). Chapter 11 discusses the use of these nonparametric techniques.

One-Sample t-Test

As discussed above, inferences about the population mean μ can be made when the population is normally distributed, where the population variance is unknown and a random sample has been selected. The procedure is similar to testing using the normal distribution. Alternatively, the *t* distribution can be used to determine *confidence intervals* for μ, or both approaches can be used together, as you will see in the following example.

An Army recruiter faces the difficult task of selecting new troops based on fairly loose criteria that may bear no resemblance to the tasks that the troops actually need to perform in the field. Suppose the best predictor of performance happens to be response time to Space Invaders, where the accuracy rate for all Army enlisted personnel has been found to be 78%. Thus, the best estimate for the population mean is μ = 78. The Little Rock branch of the Space Invaders Society of America (SISA) has approached the recruiter with a proposal that its members should be considered for rapid recruitment because of their Space Invaders playing skills. The Army recruiter believes that all young recruits being drawn from the same population as the SISA members are not more likely to have a higher performance than the current population, but decides to test the null hypothesis that there is no difference.

Given that (a) there are 100 members of the SISA branch, (b) there is only one Space Invaders machine, (c) the recruiter can only attend for 2 hours, and (d) a game of Space Invaders takes 20 minutes on average, the recruiter decides to randomly select 6 members' names out of a hat. All members of SISA are keen to join the Army, so all respond and attend the testing session.

The recruiter finds that average accuracy $\bar{a}$ = 79% and s = 0.75 for the sample. The number of degrees of freedom *df* = 5, since *n* = 6, and at the 0.05 level of significance, the null hypothesis can be rejected if $t \geq 2.015$ and $t \geq 3.365$ at the 0.01 level.

The value of *t* can be estimated as follows:

$$t = \frac{\bar{a} - \mu}{\frac{s}{\sqrt{n}}}$$

$$= \frac{79 - 78}{\frac{0.75}{\sqrt{6}}}$$

$$= 3.26$$

Thus, the null hypothesis can be rejected at 0.05 but not at the 0.01 level. The recruiter realizes that the chance of committing a Type I error is between 1 in 20 and 1 in 100, and thus believes that a significant increase in performance over the current population could be achieved by recruiting from SISA.

Determining Confidence Intervals

In the case where you genuinely do not know the mean of a population, it's possible to use the t distribution to determine a confidence interval for the population mean μ. Imagine you are a safety engineer whose role is to assign safety ratings to new vehicles produced by automotive manufacturers; as part of your work, it is necessary to determine the impact that crashing a vehicle at constant velocity will have on the front end of the vehicle. The response variable is known as "crunch," and is obtained by measuring the depth of impact: the safety rating will depend, in part, on the crunch measure, with lower values giving a higher rating. While safety testing is critical for marketing purposes—especially if a high rating is achieved—manufacturers are reluctant to sacrifice too many vehicles to testing, since they can no longer be sold. For every new vehicle, the population mean is unknown, and must be tested.

Imagine that 10 vehicles of the new vehicle model called "Superiox" are crash tested at 60 mph, with the results shown in Table 8-1. Given the small number of samples, you decide to use the t distribution to determine the 95% confidence interval for the crunch variable (y). The mean can be computed as follows:

$$\sum y = 24.0$$

$$\frac{\sum y}{n} = \frac{24.0}{10}$$

$$= 2.4$$

The variance can then be computed as:

$$\sum y^2 = 59.12$$

$$s^2 = \frac{\sum y^2 - \frac{(\sum y)^2}{n}}{n-1}$$

$$= \frac{59.12 - \frac{24^2}{10}}{9}$$

$$= 0.169$$

The standard deviation is then $s = \sqrt{0.169} = 0.411$, and the mean's standard error is $s/\sqrt{v} = 0.411 / \sqrt{10} = 0.129$. Given $n = 10$, $df = 9$, thus:

$$CI_{0.95} = \bar{y} \pm t_{0.025,\,9}\frac{s}{\sqrt{n}}$$
$$= 2.4 \pm 2.262 \times 0.129$$
$$= 2.4 \pm 0.292$$
$$= 2.108 \leq \mu \leq 2.692$$

Thus, the estimated range of the population mean is $1.708 \leq \mu \leq 2.292$. To be more conservative, you may decide to use the upper boundary of the estimated mean in the calculation of the safety rating score.

The t distribution could be used for either one- or two-tailed testing, depending on the scenario. A *one-tailed* test is a test of the null hypothesis for one value, or one "tail" of the distribution. A *two-tailed* test, on the other hand, explicitly tests both "tails" of the distribution. The null hypothesis would thus be rejected if a value fell into either the lower or upper tail of the distribution. Whether you use a one- or two-tailed test depends on whether you are testing for a significant positive or negative difference in a parameter, or both. For example, you would use a one-tailed test for the hypothesis that Californian surfers catch more waves than the average surfer. But you may also want to test the hypothesis that Californian surfers catch more *or* fewer waves than the average surfer. If you just test for an effect in one direction, no test of the opposite direction can be implied.

With such a small number of samples, it may be difficult to make sense of a histogram to verify the unimodality of the distribution, as shown in Figure 8-2. On this occasion, there does seem to be a bimodal distribution, but given the relative symmetry, it's difficult to make a judgment call on such a small sample. However, other graphical aids, such as a boxplot, may be used to explore the symmetry of the distribution.

Table 8-1. Crash testing results for Superiox vehicle at 60 mph

2.3	2.6	3.2	1.8	2.2
2.2	2.6	2.5	1.9	2.7

Two-Sample t-Test

The purpose of the two-sample t-test is to determine whether two population means are significantly different. The test is also known as the *independent samples t-test*, since the two samples are not related to each other, and can therefore be used to implement a between-subjects design. In addition to the assumption of *independence*, both distributions must be normal, and the population variances must be equal (i.e., homogeneous). Ensuring that these requirements are met in any experiment can be difficult, and requires solid design work to be undertaken up front, especially in the random selection of samples from both populations.

An age-old physical performance question is whether male football players are fitter than male ballet dancers, so a sports physiologist organizes a study in partnership with a local hospital research team to answer the question. The two

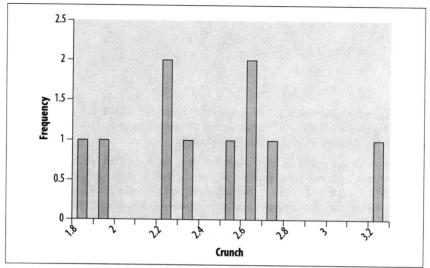

Figure 8-2. Frequency distribution of the crunch variable

groups are independent populations, since no football player is also a ballerina. There are also two lists of ballet dancers and football players located all over the country that are maintained by their respective professional associations, and study members are randomly selected from each group. Since ballet dancers and football players are very busy, only 10 study members from each group can be recruited. All participants will be tested on a range of human performance tasks, including walking, running, and stepping, and corresponding physiological measures associated with fitness, including heart-rate variability, pulse-wave velocity, etc., are then combined to form a single fitness score out of 100.

The participants are all tested in the same facility at the same time of day, and their responses are assessed and combined using the same clinicians. The results for the two groups are shown in Table 8-2.

Table 8-2. Fitness results for football players and ballet dancers

Ballet dancers	Football players
89.2	79.3
78.2	78.3
89.3	85.3
88.3	79.3
87.3	88.9
90.1	91.2
95.2	87.2
94.3	89.2
78.3	93.3
89.3	79.9

Thus, $\bar{y}_{ballet} = 87.95$, $\bar{y}_{football} = 85.19$, $s^2_{ballet} = 32.38$, and $s^2_{football} = 31.18$. The variances of the two samples are within a 5% range of each other, thus, the assumption of equal variance seems reasonable, although a more formal test could be undertaken. Note the skew of the football variable, and the peak frequency for both is quite distinct, as shown in Figure 8-3.

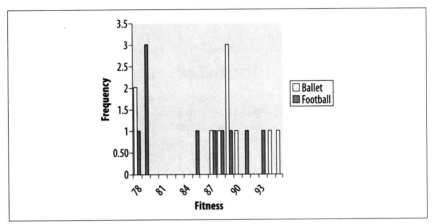

Figure 8-3. Frequency distribution of the fitness variable

The pooled sample variance is given by:

$$s^2_p = \frac{\sum(y_1 - \bar{y}_1)^2 + \sum(y_2 - \bar{y}_2)^2}{(n_1 - 1) + (n_2 - 1)}$$

$$= \frac{(n_1 - 1)s^2_1 + (n_2 - 1)s^2_2}{n_1 + n_2 - 2}$$

$$= \frac{9(32.38) + 9(31.18)}{10 + 10 - 2}$$

$$= 31.78$$

The degrees of freedom in the design are $df = n_1 + n_2 - 2 = 18$. To test the null hypothesis, i.e., that $\mu_{ballet} = \mu_{football}$, t is computed in the following way:

$$t = \frac{(\bar{y}_{ballet} - \bar{y}_{football}) - (\mu_{ballet} - \mu_{football})}{\sqrt{\frac{s^2_p}{n_1} + \frac{s^2_p}{n_2}}}$$

$$= \frac{(87.95 - 85.19) - 0}{\sqrt{\left(\frac{31.78}{10}\right) + \left(\frac{31.78}{10}\right)}}$$

$$= 1.095$$

At $p = 0.05$, $t_{0.95,18} = 1.734$, thus, you would fail to reject the null hypothesis on this occasion, and accept that—for the fitness measure used, and the samples analyzed—there does not appear to be a difference in fitness between the two groups. Looking at the different distributions, you might also argue that the test was not fair and proper, since the shapes were different, and skew was present in one case. Therefore, further experimentation, perhaps with larger and more representative samples, would be required to further test the null hypothesis.

Standard Error

You may be wondering what the $s/\sqrt{v}$ denominator of the formula to compute the various forms of t actually means. $s/\sqrt{v}$ represents the standard error of the mean. Given that the standard error of a random variable x is given by:

$$\sigma_x = \frac{\sigma}{\sqrt{n}}$$

for any sample x, the standard error of the mean of x is given by:

$$s_x = \frac{s}{\sqrt{n}}$$

You can use the standard error to determine how close the mean of a sample is to the population mean for that variable. Using standard error enhances the confidence that your particular sample can be used to say something meaningful about the desired population. It's important to distinguish between the standard deviation of the data, which is a measure of variability or dispersion computed from the probability distribution function, and the standard error, which provides you with certainty about the population mean as estimated from a specific sample.

For two samples, as seen above, the standard error of the difference between two means is given by:

$$\sqrt{\frac{s_p^2}{n_1} + \frac{s_p^2}{n_2}}$$

Again, this indicates the level of confidence you should have that the difference between the two means, as sampled, is indicative of the difference between the population means. In this example, the pooled variance is used, but individual sample variances could also be used.

Repeated Measures t-Test

The purpose of the *repeated measures t-test* is to test the same experimental units under different treatment conditions—usually experimental and control—to determine the treatment effect, allowing units to act as their own controls. This is also known as the *dependent samples t-test*, since the two samples are related to each other, thus implementing a within-subjects design. The other requirement is that sample sizes be equal, which is not the case for a two-sample *t*-test.

A researcher is interested in the effect of a motivational speaker on mathematics exam performance. The experimental condition is this: 30 minutes before an exam being undertaken, high school students will be given a 15-minute talk emphasizing their potential for self-actualization and other concepts believed to be key to enhanced self-esteem, which in turn should lead to higher exam performance. In the control condition, randomly selected newspaper passages will be read to the students. The class size is 10, so a t-test is appropriate to use with this small sample.

The key point is that rather than half of the students being administered the experimental treatment and half being administered the control, all students will participate in both conditions, for two separate exams. To counterbalance any order effects in the design, half of the students (randomly selected) will participate in the control condition first and then the experimental condition, and the other half will undertake the experiment in the reverse order. The results are represented by y.

The results from the experiments are shown in Table 8-3.

Table 8-3. Exam performance after motivational speaker treatment

Experimental	Control	Difference y_d	(Difference)$^2 y^2_d$
65	66	−1	1
75	67	8	64
80	65	15	225
77	68	9	81
74	69	5	25
69	70	−1	1
72	69	3	9
72	68	4	16
71	69	2	4
69	65	6	36

$$\sum y_d = 48 \qquad \sum y^2_d = 462$$

The null hypothesis in this experiment is that $\mu_d = 0$. The mean of y_d is $\sum y_d/n = 48/10 = 4.8$. The variance is then calculated as:

$$s^2_d = \frac{\sum y^2_d - \frac{\left(\sum y_d\right)^2}{n}}{n-1}$$

$$= \frac{462 - \frac{(48)^2}{10}}{9}$$

$$= 25.73$$

The value for t at the $p = 0.05$ probability level is then given by:

$$t = \frac{y_d - \mu_d}{\frac{s_d}{\sqrt{n}}}$$

$$= \frac{4.8 - 0}{\frac{25.73}{\sqrt{10}}}$$

$$= 0.589$$

The number of degrees of freedom $df = 9$, since $n = 10$, and at the 0.05 level of significance, the null hypothesis can be rejected if $t \geq 1.833$, and $t \geq 2.821$ at the 0.01 level. Thus, you would fail to reject the null hypothesis in this experiment.

In this experiment, the null hypothesis is framed as a one-tailed problem, i.e., you are predicting that a treatment (such as motivational speaking) will have a positive effect on performance (in this case, in exams). If you were genuinely unsure whether motivational speaking would have a positive or negative impact on performance, it would be wiser to use a two-tailed version of the null hypothesis, and use the appropriate p value for a two-tailed test. Figure 8-4 illustrates the results.

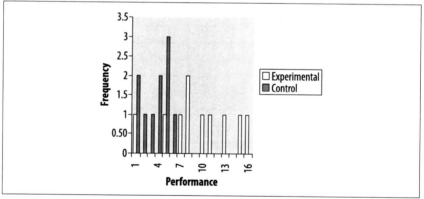

Figure 8-4. Frequency distribution of the performance variable

Unequal Variance t-Test

Recall that the goal of the two-sample t-test is to test whether the difference between the means of two groups is statistically significant. The t-test, in this form, assumes that the variances are the same in the underlying populations, and the test is not reliable (i.e., both Type I and Type II errors may be committed) when these variances are heterogeneous. This is because the variances are pooled to gain the most reliable estimate, and the result of the test would be seriously distorted if they were not equivalent (whilst accepting some difference due to measurement or sampling error). The problem of hypothesis testing between two independent samples where variances are known to be unequal is known as the *Behrens-Fisher problem,* and there have been several proposed solutions.

However, in reality, many researchers use the *t*-test without first verifying that the variances are indeed homogeneous, or use a Mann-Whitney (nonparametric) test as a substitute. In addition, since the *t* distribution approximates a normal distribution for large sample numbers, it is routinely applied to known nonnormal distributions with sufficiently large samples in the belief that large samples can mitigate any concerns about nonnormality (or indeed, *hetereogeneity of variance*). Sufficiently large means greater than 30 in most cases, in the context of the Central Limit Theorem.

The unequal variance *t*-test (or *Welch t-test*) should be used whenever the variances are unknown, and/or the sample size is small, and/or you wish to be conservative in the inferences that you draw. The main difference between the two is that the calculation of the degrees of freedom is more complicated for the unequal variance *t*-test than for the two-sample *t*-test.

If you wish to use the two-sample *t*-test, the best approach is to calculate the homogeneity of variance prior to any *t*-testing, and then decide whether to use the two-sample *t*-test or the unequal variance *t*-test.

Another possibility is to use adjusted degrees of freedom to evaluate the significance of *t*, although this value will almost certainly be more conservative than the *df* computed directly from *n*. The *df* would be computed in the following way for a two independent sample *t*-tests:

$$df = \frac{\left(\dfrac{s_1^2}{n_1} + \dfrac{s_2^2}{n_2}\right)^2}{\dfrac{\left(\dfrac{s_1}{n_1}\right)^2}{n_1-1} + \dfrac{\left(\dfrac{s_2}{n_2}\right)^2}{n_2-1}}$$

Normally, when variances are equal, standard error is minimized when $n_1 = n_2$, but where this is not possible to achieve with the samples available, the robustness of the *t*-test is guaranteed as long as the following equivalence is met:

$$\frac{n_1}{n_2} = \frac{\sigma_1^2}{\sigma_2^2}$$

This equation implies that there should be a proportional relationship between sample sizes and variances. There are many situations in independent-groups *t*-tests where the experimental group are necessarily smaller than the control group. In this case, as the control group increases in size, as long as there is a corresponding increase in variance, the test will be valid. Many of these remedies rely ultimately on the normality assumption being true "in the large," if not for a specific, small sample.

Effect Size and Power

An important question in studies where sample sizes are small and/or limited is determining how many experimental units are required to observe an experimental effect. Recall the crash test example above. The experimenter wants to minimize the number of vehicles destroyed, each vehicle costs a lot of money. On the other hand, the experimenter must be sure that enough cars have been tested to ensure public safety. Alternatively, in animal or human experimentation, it is unethical to apply a treatment to more participants than necessary to see a particular effect.

A statistically significant difference between the mean of one sample and an expected mean, or between two means, does not in itself indicate whether the difference is important. The importance of an observed effect must be determined by the knowledge domain and/or industry standards that are relevant to the problem. For example, in the crash testing scenario, prior experience and/or observation from real crashes may indicate three different thresholds for a crash impact effect at different velocities: one beyond which death will occur, one beyond which death will not occur but injury will occur, and one beyond which neither death nor significant injury will occur. These thresholds might be used to determine a "star rating," for example, so that consumers can make informed choices about car purchasing based on safety. These are examples of important differences, and the differences between them should also be statistically significant. But not all statistically significant differences will be important.

The distance between each of the thresholds in this example corresponds to an *effect size*, or the magnitude of difference between them. The effect size for any test of mean comparisons is given by:

$$\frac{\mu_1 - \mu_2}{\sigma}$$

Population means can be replaced by sample means in any specific experiment, and the standard deviation should be the same for both samples (assuming that the homogeneity of variance assumption for *t*-tests is met), or a pooled estimate can be used. As a concrete example, imagine that the threshold for crash impact at 80 mph, in which death will occur, is 2.5 yards, and 1.5 yards in which serious injury but no death will occur. Assuming equal variance, and if the standard deviation is 0.2, then the effect size will be given by:

$$\frac{2.5 - 1.5}{0.2} = 5.0$$

This is a very large effect size, and so the difference can be considered statistically important. However, if the threshold beyond which no injury will occur is only 1.4 yards, the effect size is much smaller:

$$\frac{1.5 - 1.4}{0.2} = 0.5$$

Here, the difference between the populations is going to be very small indeed, since they differ by 0.5 standard deviations on average. In statistical terms, it's always going to be easier to measure large differences than small differences, when the standard deviation and *n* are equal.

If you know in advance what the expected difference between two means is before you experiment, based on past experience or observation, and you have a reasonable estimate of the standard deviation, you can compute an effect size prior to experimentation. After selecting an appropriate α (e.g., $\alpha = 0.01$), you can compute the number of experimental units required to observe a specific level of power.

Calculation of statistical power *before* you run an experiment is an important step in determining its scope, especially in terms of the likelihood of committing a Type II error. So far, you have learned a lot about Type I errors, but the impact of Type II errors can be quite insidious; imagine signing off on crash tests for a vehicle showing that the sample did not differ from the mean, when in fact the crash performance for the population mean was significantly worse than the acceptable level. Thus, statistical power is best understood as the ability of a test—in this case a *t*-test—to discriminate between two means when in fact they are actually different. Power is formally defined as $1 - \beta$, where β is the probability of committing a Type II error.

Following the crash testing example, if you have an effect size of 4.0, and $\alpha = 0.05$, to achieve power of 0.90 (i.e., where $\beta = 0.1$), then *n* should be at least 4. However, if you have an effect size of 0.5, and $\alpha = 0.05$, to achieve power of 0.90 (i.e., where $\beta = 0.1$), then *n* should be at least 106. That's a very large difference in *n* required to see an experimental effect, but serves to illustrate why effect sizes are critical to understanding the importance of statistically significant results. In practice, because of the conservative nature of scientific hypothesis testing, priority is usually given to conservative α levels (e.g., $\alpha < 0.01$), while β is typically accepted at 0.80 in many fields, especially where a lot of repetition occurs in experimentation. Effect size and power are further discussed in Chapter 18.

Exercises

While you can use a statistical package like Minitab, SPSS, STATA, SAS, or even Excel to compute *t*-tests and their significance, working through some examples yourself will make the underlying concepts easier to understand (especially the difference, say, between standard error and standard deviation). Also, if you consider scenarios from work or school that involve small samples, you may begin to develop a sense of how to approach them inferentially using *t*-tests. If you have understood all of the permutations of *t*-testing as computed by hand, then using a statistical package will be much easier for you. Also, the output generated by many statistical packages is confusing if you don't understand what you should be looking for; e.g., most statistical tests are accompanied by various adjustments and corrections that are usually calculated but may not be relevant to your research question, unless one or more of the assumptions underlying the test have been violated (e.g., homogeneity of variance).

Question

A boutique brewery company is trying to determine the optimal period of fermentation for a new organic ale called Old Sarum, which is free of additives and preservatives that may have been hindering the fermentation process, according to the marketing director. However, given that the organic ingredients have never been used before, the brewer needs to know whether the new recipe will require a different fermentation period from the existing recipe. The average fermentation time for existing brews is 48 hours, so the best estimate for the population mean is $\mu = 48$. The master brewer—skeptical that organic ingredients will make any difference at all to the fermentation process—decides to test the null hypothesis that there is no difference between the population mean and the sample mean of Old Sarum.

However, the pressure from the marketing department means that there is only a limited time available for quality control before the new product is launched, so the brewer is only allowed 20 kegs of beer to be brewed and tested. Since there are 120 kegs, a computer program is used to randomly select 20 from the population.

Answer

The brewer finds that average brewing time is 43 hours and $s = 3.5$ for the sample. The number of degrees of freedom $df = 19$, since $n = 20$, and at the 0.05 level of significance, the null hypothesis can be rejected if $t \geq 1.729$, and $t \geq 2.539$ at the 0.01 level.

The value of t can be estimated as follows:

$$t = \frac{\bar{a} - \mu}{\frac{s}{\sqrt{n}}}$$

$$= \frac{48 - 43}{\frac{3.5}{\sqrt{20}}}$$

$$= 6.39$$

Thus, the null hypothesis can be rejected at both the 0.05 and the 0.01 levels. The brewer realizes that the chance of committing a Type I error is less than 1 in 100 of the population, and thus believes that a significant reduction in brewing times exists between Old Sarum and the existing brews.

Question

The finance department is very unhappy with the brewer, since 20 kegs is a lot of beer to waste on a test. The finance manager decides to conduct a power analysis to determine how many kegs should have been used, taking into account that a difference of only two hours more would have resulted in a cost savings in terms of fermentation.

Answer

The manager begins by computing the effect size:

$$\text{Effect size} = \frac{\mu_1 - \mu_2}{\sigma}$$

$$= \frac{48 - 46}{1.87}$$

$$= 1.06$$

If you have an effect size of 1.06, and $\alpha = 0.05$, to achieve power of 0.90 (i.e., where $\beta = 0.1$), then n should be at least 15. Thus, the finance manager decides to deduct the cost of the five wasted kegs from the brewing department's accounts.

Question

After the success of the Old Sarum in reducing the costly fermentation process, the brewers are under pressure to make sure that it tastes better than other ales. To this end, the marketing department engages a consultant to undertake an expert panel evaluation of the flavor of Old Sarum versus the original ale. The consultant will employ a panel of expert judges, who are expensive to hire, so only 10 will be empanelled to make taste judgments.

Answer

The results from the experiments are shown in Table 8-4.

Table 8-4. Taste test results for Old Sarum

Existing brew /10	Old Sarum /10	Difference	(Difference)^2
6	8	−2	4
7	8	−1	1
8	9	−1	1
7	8	−1	1
7	10	−3	9
8	9	−1	1
6	8	−2	4
6	9	−3	9
7	8	−1	1
7	7	0	0

$$\sum y_d = 15 \qquad \sum y_d^2 = 31$$

The null hypothesis in this experiment is that $\mu_d = 0$. The mean of y_d is:

$$\sum \frac{y_d}{n} = \frac{15}{10} = 1.5$$

The variance is then calculated as:

$$s_d^2 = \frac{\sum y_d^2 - \frac{\left(\sum y_d\right)^2}{n}}{n-1}$$

$$= 0.94$$

The value for t at the $p = 0.05$ probability level is then given by:

$$t = \frac{y_d - \mu_d}{\frac{s_d}{\sqrt{n}}}$$

$$= \frac{1.5 - 0}{\frac{0.88}{\sqrt{10}}}$$

$$= 5.39$$

The number of degrees of freedom $df = 9$, since $n = 10$, and at the 0.05 level of significance the null hypothesis can be rejected if $t \geq 1.833$, and $t \geq 2.821$ at the 0.01 level. Thus, you would reject the null hypothesis in this experiment at both the 0.01 and 0.05 levels of significance and conclude that the panel's judgments clearly favored Old Sarum.

Of course, since the judges were not randomly selected from the population of judges, they may not reflect broader opinion within the expert community, nor would any inference be able to be made about the wider beer-consuming population, who were not judges. The marketing department would be wise to follow up these studies with a much broader set of tests using randomization.

Note that the flavor ratings here should be interpreted as interval data: if the ratings were to be interpreted as ordinal data, then a nonparametric test for differences between groups, such as the Mann-Whitney U test, would be more appropriate.

9

The Correlation Coefficient

This chapter is concerned with measures of *relatedness* between two variables. A simple measure, the *correlation coefficient*, is commonly used to quantify the degree of *association* between two variables. Often, correlations are used during an exploratory or observational stage of research to determine which variables *at least* have a statistical relationship with each other. In experimental designs, correlations are also used to determine the degree of association between independent and dependent (or response) variables. However, the finding of a correlation between two variables does not imply that a change in one variable causes a corresponding change in another—that's why you still need experiments. Indeed, the history of computing correlation coefficients at large, and often without any theoretical or model-based justification, has led to numerous errors in inference being made. In this chapter, you will learn about measures of association, such as Pearson's correlation coefficient, the Spearman rank-order coefficient, the point-biserial correlation coefficient, and phi, and review examples of the appropriate use of each. The key message is that correlations are useful tools, but many variables in nature are correlated; such relationships are not always useful for inference.

Measuring Association

The world is awash with correlations, or statistical associations between two (or more) variables. Often, such relationships are useful to characterize or predict some phenomena. For example, a freshman economics student interested in understanding monthly central bank interest rate movements notices that every quarter, if inflation is greater than an annualized 2%, interest rates always rise by 0.25%. The student reviews 10 years of data and concludes that the association between the two variables (inflation and interest rate increases) is very strong; the student then hypothesizes that inflation could be causing interest rate rises. The student excitedly emails her professor to report the discovery, only to find out that while the two variables are indeed correlated, the relationship is only causal to the extent that central banks use a 2% inflation benchmark as a major factor in their deliberations.

The important lesson here is that a correlation during an observational phase of research has highlighted a very strong association between two variables, one of which always preceded the other. Temporal dependencies like this are often good candidates for causal relationships. In this case, the student was able to determine that while inflation does not have one specific cause, interest rate movements are decided by a committee, based on observation of this variable. Thus, it's entirely appropriate to consider inflation as an independent variable, and the interest rate as a dependent variable. Note that the interest rate settings of the committee constitute an intermediate variable, since the student was able to measure the association between inflation and interest rate movements without any knowledge or observation of the committee's behavior.

It's important to note that, while measures of *association* are often useful in characterizing the broad relationship between two variables, they do not always reflect the underlying and significant variation in the association between specific cases. Thus, while lung cancer is strongly associated with lifelong smoking, nonetheless, some people get lung cancer without ever lighting a cigarette, while some heavy smokers live to 100 and die from other causes.

If you have a very large population, and unless the association between two variables is 100.00%, such cases will always occur. In the case of cancer, there are undoubtedly some other causes of cancer above and beyond direct smoking (such as passive smoking) and also some protective measures for certain individuals that are almost certainly genetic. Therefore, while you might be able to construct a model of the relationships between two variables in the large, it's always possible to refine this model until all variation is explained.

There are a number of techniques that have been developed to quantify the association between variables of different scales (nominal, ordinal, interval, and ratio), including the following:

- Pearson product-moment (with both variables measured on an interval or ratio scale)
- Spearman rank-order (with both variables measured on ordinal scales)
- Phi (with both variables measured on a nominal/dichotomous scale)
- Point biserial (with one variable measured on a nominal/dichotomous scale, and one measured on an interval or ratio scale)

We review examples of each type of correlation in this chapter.

Graphing Associations Through Scatterplots

One of the easiest ways to explore associations between variables is to use *scatterplots*. A scatterplot is a graph that usually takes the form of an explanatory variable (on the horizontal, or x-axis) plotted against a response variable (on the vertical, or y-axis). Each member of a sample corresponds to one data point on the graph described by a set of coordinates (x, y). When many variables are plotted, a sense of the overall pattern of the relationship between the explanatory and response variables often emerges. The relationship may be described as strong or weak. Note that the overall pattern may be influenced by a number of *outliers*, or cases that do not fit the overall pattern.

The form of association can be either *linear* or *nonlinear*: often, statisticians seek to determine the mathematical relationship between two variables (specified by a linear or nonlinear model), whereas engineers may be primarily interested in linearized versions. You will learn more about model building in Chapter 12, but for now, if you can draw a straight line through your (x, y) set of coordinates, you are likely dealing with a linear model.

Quantitative Variables

The association between two quantitative variables can be readily displayed using a scatterplot. Relationships between two variables can be described as signed, additive, multiplicative, or a combination of all three types of relation. For example, a positive multiplicative relationship between interest rates and inflation means that when interest rates increase, inflation will also increase. Conversely, a negative relationship would imply that a decreasing interest rate would be associated with increasing inflation.

Relationships of this type are commonly used in linear algebra; you might think of them as "formulae of straight lines," since they describe exactly the characteristics of straight lines in a two-dimensional plane that can vary. In the general form of the model $y = \pm ax \pm b$:

- y is the dependent variable
- $\pm a$ is the *slope* (i.e., where the value of y when $x = 0$); $-a$ indicates a negative association; $+a$ indicates a positive association
- x is the independent variable
- $\pm b$ is the *intercept* of the straight line

Note that m is sometimes used in place of a in this equation: this is just a different notational convention and does not change the meaning. Given various sources of error, most phenomena that can be described by a linear model actually demonstrate some deviation from expected values. However, by using a scatterplot, you can get some visual clues concerning whether the relationship between two variables is linear or not.

Figure 9-1 shows the association between two variables (x and y) that are strongly positively associated, since for each (x, y) set of coordinates, the values are equal. The model describing this relationship is $x = y$. Using the linear model, $a = 1$, which is positive, so the relationship is positive; and $b = 0$, since the intercept is the origin $(0, 0)$.

However, associations between two variables can also be negative, as shown in Figure 9-2, where each value of y is a negative multiple of x. In this example, the association between the two variables is perfectly negative. Note that the values of the (x, y) coordinates do not need to be the same in order for the association to be strong—the first values plotted are $(1, -2)$, $(2, -4)$, and $(3, -6)$. The model for this

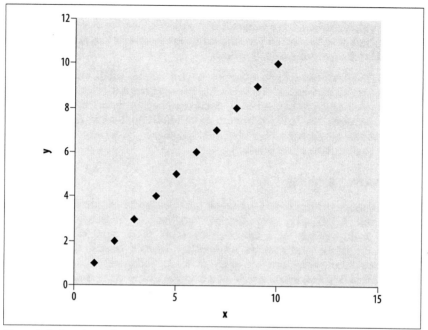

Figure 9-1. Association between two variables described by the model y = x

relationship is $y = -2x$, which is multiplicative in nature. Using the linear model, $a = -2$, which is negative, so the relationship is negative. Again, $b = 0$, since the intercept is the origin $(0, 0)$.

Relationships between two variables can also be additive, or both multiplicative and additive. Figure 9-3 shows the strongly linear relationship between x and y, where the model is multiplicative (by a factor of 2) and additive (with 0.5). The model for this relationship is $y = 2x + 0.5$. Using the linear model, $a = 2$, which is positive, so the relationship is positive.

Sometimes, variables have no relationship, so don't be fooled into thinking that a straight line on a number plane indicates an association. Figure 9-4 shows the situation where the same value of y is related to every possible value of x. In this case, there is no association between the variables. Using the linear model, $a = 0$, and since there is no slope, there is no relationship.

Often, when taking real-world measurements, some amount of random or systematic error is present, as described in Chapter 6. This can obscure the strength of relationship between two variables when you are examining a scatterplot. For example, Figure 9-5 shows exactly the same data as Figure 9-1, except that random error has been added into the model, to better reflect real-world measurements. By looking at the scatterplot, would you have guessed that the model was similar to Figure 9-1? In this example, the model is $x + \varepsilon_x = y + \varepsilon_y$, where ε_x and ε_y represent random error.

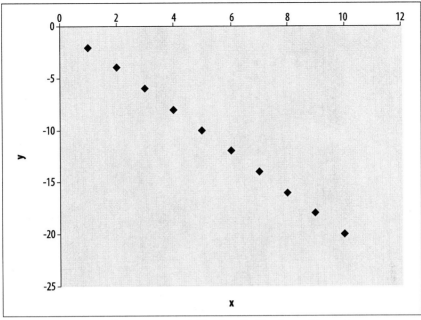

Figure 9-2. Association between two variables described by the model y = −2x

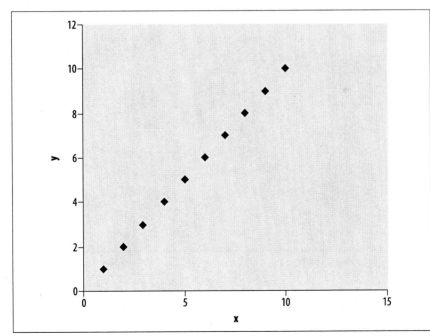

Figure 9-3. Association between two variables described by the model y = 2x + 0.5

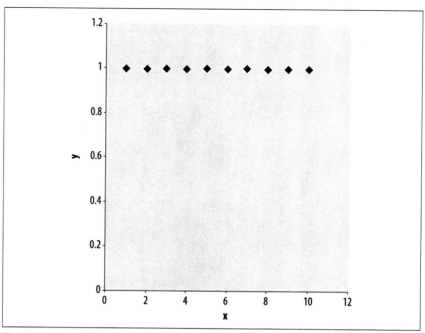

Figure 9-4. Lack of association between two variables described by the model y = 1

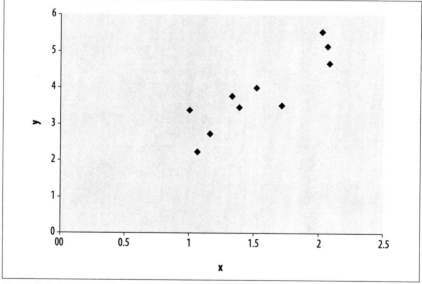

Figure 9-5. Association between two variables described by the model $y+\varepsilon_y = x+\varepsilon_x$

Finally, Figure 9-6 shows a different type of relationship between two variables: an exponential relationship that is described by the model $y = e^x$, where e equals 2.712…, which is the base of the natural logarithm. Exponential and other nonlinear functions mean that variables may not be linearly associated, but are associated in other ways. Indeed, knowing (or being able to predict) the type and class of model required to describe the type of association between two variables is part of the art of being a statistician.

In this example, if you changed the y-axis to be displayed using a logarithmic scale, the relationship between x and y would appear to be strong. Being aware of the underlying linearity either assumed or used explicitly is very important when understanding relationships. Linear associations are the easiest to deal with mathematically, but exponential relationships are very powerful (imagine if Figure 9-6 described the growth in your savings after 10 years of stock market investment!).

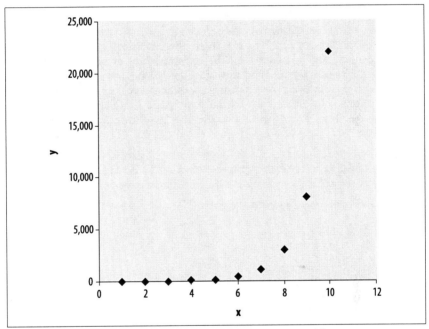

Figure 9-6. Association between two variables described by the model $y = e^x$

Looking forward to Chapter 12, imagine now drawing a straight line through Figure 9-5; you can see there are several possible straight lines that can be drawn through all of the data points. The basis of linear regression goes one step further in quantifying the relationship between two (or more) variables by drawing the line through the plane that best fits the data, in the sense of minimizing the distance between the observed coordinates and their estimate according to a model.

Least-squares linear regression, described in Chapter 12, minimizes the squared deviations from the expected values of each observation, and thus provides the "best fit" to a linear model for the data. The point to note here is that you can use graphical tools as well as mathematical models to determine underlying structure in data.

Pearson's Product-Moment Correlation Coefficient

Using scatterplots is a good visual guide for examining relationships between two variables. However, as you have seen in Figure 9-5, sometimes effects like random error can make it difficult to easily estimate the strength of relationship. Fortunately, there is a single quantitative measure that can be computed to determine the strength and direction of relationship between two variables, *Pearson's product-moment correlation coefficient* for samples, otherwise known as *r*.

In simple terms, the more the relationship between two variables is similar to a straight line (excepting the type shown in Figure 9-4, with a zero slope), the more correlated the variables are said to be. If the values are only weakly scattered around a straight line in the scatterplot, the variables are said to be weakly correlated; weak scattering means that the values are generally clustered around the values predicted by a straight line, but where there is deviation or error from the predicted values. The correlation can be either positive or negative; Figure 9-2 shows a negative relationship, and thus the correlation is also negative. Just like the mean and standard deviation statistics are estimates of their respective population parameters, *r* is an estimate of the population parameter rho (ρ). In the population, if $\rho = 0$ for two variables, the two variables are said to be independent.

The correlation coefficient measures the tendency of two variables to change in value together (i.e., to either increase or decrease). To do this, the sum of products of the two standardized variables are divided by the *degrees of freedom*.[*]

The formula for the correlation coefficient is:

$$r = \frac{\sum\limits_{i=1}^{n}\left(\dfrac{x_i - \bar{x}}{s_x}\right)\left(\dfrac{y_i - \bar{y}}{s_y}\right)}{n-1}$$

Here, you can see that the correlation is derived from the sum of products of standardized deviations from the mean for the (x, y) set of coordinates. Why do you need to standardize the variables? It's the same old question of comparing apples and oranges—if you divide the deviations of each case from the mean by the standard deviation, you can relate variables that have been measured in different units. For example, if you want to measure the relationship between grip strength and weight, measure in pounds per square inch (psi) and pounds, respectively.

To compute r, the following algorithm, corresponding to the formula shown above, is used:

- For each (x, y) set of coordinates, subtract the mean from each observation for x and y.

---

[*] Conceptually, the number of degrees of freedom for error in a sample is the number of chances for change, and is usually defined as the sample size minus one. Mathematically, the difference between the number of observations and the degrees of freedom occurs because the residuals arising from fitting a particular model are always of a smaller dimension: if you know the values of all residuals except one, then you can calculate it, because their sum must be zero.

- Divide by the corresponding standard deviation.
- Multiply the two results together.
- The result is then added to a sum.
- The sum is divided by the degrees of freedom, $n - 1$.

r always ranges in value from -1 to 1, with values close to zero representing weak relationships, and high values representing strong relationships (either strongly negative or strongly positive). A correlation of 1.00 means that the two values are completely or perfectly positively correlated; -1.00 means perfectly negatively correlated; and a correlation of 0.00 means that there is no relationship between two variables.

Let's look at an example. A psychologist is interested in the relationship between net wealth and attractiveness, rated by an expert panel of five judges with a combined score out of 10. The two values of the variables are shown in Table 9-1.

Table 9-1. Measures of wealth and attractiveness

Wealth ($m)	Attractiveness
1.21	2.44
2.24	5.73
1.20	2.93
2.39	5.69
1.10	2.74
1.45	4.26
2.29	5.11
2.33	5.58
1.13	2.42
2.39	5.52

The mean and standard deviation of wealth (x) and attractiveness (y) are 1.773 and 0.593765, and 4.242 and 1.454203 respectively. Table 9-2 shows the result of subtracting the mean from each observation, and division by the standard deviation. Table 9-3 shows the summation of the multiples, and the calculation of the correlation coefficient.

Table 9-2. Computation of deviations and normalization

$x - \bar{x}$	$\dfrac{x - \bar{x}}{s_x}$	$y - \bar{y}$	$\dfrac{y - \bar{y}}{s_y}$
−0.563	−0.94819	−1.802	−1.23917
0.467	0.786506	1.488	1.023241
−0.573	−0.96503	−1.312	−0.90221
0.617	1.039132	1.448	0.995734
−0.673	−1.13345	−1.502	−1.03287
−0.323	−0.54399	0.018	0.012378

Table 9-2. Computation of deviations and normalization (continued)

$x - \bar{x}$	$\dfrac{x - \bar{x}}{s_x}$	$y - \bar{y}$	$\dfrac{y - \bar{y}}{s_y}$
0.517	0.870715	0.868	0.596891
0.557	0.938082	1.338	0.920092
−0.643	−1.08292	−1.822	−1.25292
0.617	1.039132	1.278	0.878832

Table 9-3. Computation of products

Product
1.174961
0.804786
0.87066
1.034699
1.170699
−0.00673
0.519721
0.863121
1.356812
0.913222
$\Sigma = 8.701$

Thus:

$$r = 1/(n - 1) \times 8.701$$
$$= 1/9 \times 8.701$$
$$= 0.967$$

Thus, with a correlation coefficient of $r = 0.967$, the two variables wealth and attractiveness are highly positively correlated, meaning that people with low wealth tend to be less attractive than individuals with high wealth (Figure 9-7). Conversely, wealthy individuals tend to be more attractive than individuals with lower wealth. Note that since we have not performed any experimental manipulation of the variables wealth and attractiveness, it's impossible to say whether wealth causes attractiveness, or indeed, whether attractiveness is a cause of wealth.

Alternatively, even with a strong observed statistical relationship, there may well be an intermediate variable that actually causes the change in a variable's value. You need to be very cautious of over-interpreting correlations, since a surprisingly large number of variables in the physical world are correlated, but the relationship is not meaningful or causative.

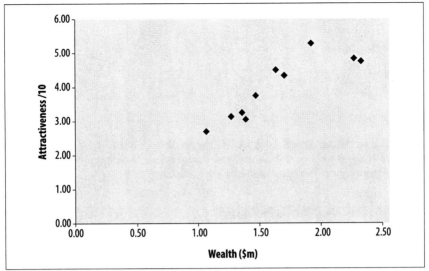

Figure 9-7. Association between two variables (wealth and attractiveness)

However, if you find a highly correlated measure that is fast, easy, and cheap to obtain, and your only interest is prediction, select whichever correlated variable gives you the highest *r*. For example, a psychologist interested in predicting emotional intelligence might administer a test battery of 200 questions that takes three hours to complete. But after obtaining the results from 100,000 participants worldwide, the psychologist discovers that responses to the first question—"Do you love animals?"—have a correlation coefficient of 0.90 with the overall emotional intelligence score. Given the strength of this relationship, in a predictive sense, the psychologist may decide to reduce the number of questions being asked in order to save time and expense, and minimize any discomfort to participants. The relevant correlation for this type of data is called the point-biserial correlation, and is further discussed later in this chapter.

Testing Statistical Significance

The significance of the correlation coefficient can be evaluated by using the *t* statistic. For example, if you believe there is a relationship between two variables in the population, you can test against a null hypothesis that $\rho = 0$, using the *t* statistic and the estimate of *r*:

$$t = \frac{r}{\sqrt{\dfrac{1 - r^2}{n - 2}}}$$

For the association between wealth and attractiveness, we can test the null hypothesis that they are independent, as follows:

$$t = \frac{0.967}{\sqrt{\dfrac{1 - 0.967^2}{10 - 2}}}$$

$$= \frac{0.967}{\sqrt{\dfrac{0.065}{8}}}$$

$$= 10.73$$

According to the statistical tables for t, the result of $t = 10.73$, $df = 8$, is highly statistically significant, $p < 0.01$, and you can reject the null hypothesis. For more details regarding the interpretation of t-tests, see Chapter 8.

Coefficient of Determination

The correlation coefficient alone provides an indication of the strength and direction of relationship. But it doesn't directly tell you the proportion of variation in one variable that can be accounted for by the other. Nor can relative r values be directly compared in proportion, e.g., you can't say that $r = 0.2$ represents double the correlation of $r = 0.1$.

Fortunately, the square of the correlation coefficient provides exactly this measure, and is known as the *coefficient of determination*. Thus, in the emotional intelligence test example above, $r^2 = 0.9 \times 0.9 = 0.81$. Hence, 81% of the variation in wealth can be directly attributed to attractiveness and vice versa. As you will see in Chapters 12 and 14 on linear regression, the unique variance accounted for by multiple explanatory variables can often be combined to account for the majority of variation in a response variable. Thus, if attractiveness uniquely explains 81% of the variation in wealth, and level of education uniquely explains 10% of the variation, then combining the two explanatory variables together to form a model for variation in wealth allows you to explain 91% of the variation in the variable.[*]

Let's look at an example. An environmental engineer working for a government advisory panel on climate change wants to examine the relationship between engine capacity and CO_2 emissions. Since engines are very complex systems, subject to the interaction of thousands of variables in each case, and with numerous variation across cases, the problem of stating the relationship between the two variables is difficult. Also, the number of different environmental and driving conditions under which engines operate varies enormously.

However, the engineer decides that based on studies of driving usage, a representative "driving" pattern can be established that features driving on a 10-mile circuit under different acceleration conditions and velocities. Manufacturers are

---

[*] Note the key term here: "uniquely." Ideally, all explanatory variables should be orthogonal and account for a unique portion of variance. However, this is rarely the case in measured data, unless the variables have been produced by an orthogonal decomposition, such as principal components analysis, described in Chapter 16.

then invited to submit vehicles to be tested, with the relationship between engine size and CO_2 output (grams/mile)—averaged over the 10 miles—being determined by the correlation coefficient, and the coefficient of determination used to determine the proportion of variance in CO_2 accounted for by engine size.

The results of the initial 10 vehicles are shown in Table 9-4, and displayed using a scatterplot in Figure 9-8.

Table 9-4. Results of the initial 10 vehicles

Engine size	CO_2 emissions (grams/mile)
1.9	200
2.2	210
1.8	230
2.5	240
3.2	235
5.4	400
4.3	310
3.3	250
3.2	260
3.3	260

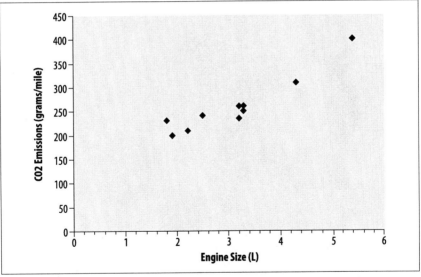

Figure 9-8. Association between two variables (engine size and CO_2 emissions)

The mean and standard deviation of engine size (x) and CO_2 emissions (y) are 3.11 and 1.11201, and 259.5 and 57.9487, respectively. Table 9-2 shows the result of subtracting the mean from each observation, and division by the standard deviation. Tables 9-5 and 9-6 show the summation of the multiples, and the calculation of the correlation coefficient, respectively.

Table 9-5. Computation of deviations and normalization

$x - \bar{x}$	$\dfrac{x - \bar{x}}{s_x}$	$y - \bar{y}$	$\dfrac{y - \bar{y}}{s_y}$
−1.21	−1.08812	−59.5	−1.02677
−0.91	−0.81834	−49.5	−0.8542
−1.31	−1.17805	−29.5	−0.50907
−0.61	−0.54856	−19.5	−0.3365
0.09	0.080935	−24.5	−0.42279
2.29	2.059334	140.5	2.424558
1.19	1.070134	50.5	0.87146
0.19	0.170862	−9.5	−0.16394
0.09	0.080935	0.5	0.008628
0.19	0.170862	0.5	0.008628

Table 9-6. Computation of products

Product
1.117249
0.699027
0.599709
0.184592
−0.03422
4.992975
0.93258
−0.02801
0.000698
0.001474
$\Sigma = 8.466$

Thus:

$$r = 1/(n - 1) \times 8.466$$
$$= 1/9 \times 8.466$$
$$= 0.94068$$

Thus, with a correlation coefficient of $r = 0.94068$, and coefficient of determination can be calculated as follows:

$$r^2 = 0.94068^2$$
$$= 0.88487$$

Hence, 88.487% of variation in CO_2 emissions can be explained by engine capacity.

Spearman Rank-Order Coefficient

Sometimes, you may be interested in determining the relationship between two variables in terms of the ranking of each case within each variable. This is usually the case where you are interested in ordinal relations. For example, in educational testing, it is often difficult *a priori* to predict whether a test or examination will produce a distribution of results that allows for identification and discrimination of certain types of skill levels. If a test is set "too easy," many students will achieve "full marks," while if a test is "too hard," many students will fail, even though—in both cases—there is underlying variation in general ability. Also, when comparing results across tests, you may be more interested in a particular student's *rank* in the class, rather than the student's raw scores. Using ranks rather than raw scores allows these sorts of comparisons to be made fairly and validly.

The *Spearman rank-order coefficient* (or *rank correlation coefficient*) is very similar to the product-moment correlation coefficient, discussed above, except that the ranks are correlated rather than the raw scores. So, a lot of the computational effort involves working out the differences between ranks for individual items on different variables, and then squaring the observed deviations. The formula for calculating the Spearman rank correlation coefficient is:

$$r_s = 1 - \frac{(6\sum d^2)}{n(n^2 - 1)}$$

Let's revisit the relationship between CO_2 emissions and engine size with a view to ranking individual engines by each variable. The first step in calculating r_s is to rank all of the scores based on the two different variables (CO_2 emissions and engine size), as shown in Table 9-7, sorted by the engine rank. Here, you can see that while the ninth and tenth ranked engine by size are the same as the ninth and tenth ranked engine by emissions, the other cases all occupy different relative ranks.

Table 9-7. Ranking of items

Vehicle	Engine size	CO_2 emissions	Engine rank	Emissions rank
A	1.8	230	1	3
B	1.9	200	2	1
C	2.2	210	3	2
D	2.5	240	4	5
E	3.2	235	5.5	4
F	3.2	260	5.5	7.5
G	3.3	250	7.5	6
H	3.3	260	7.5	7.5
I	4.3	310	9	9
J	5.4	400	10	10

Correlation Coefficient

If two variables have the same value, then they occupy a "tied rank," and can be scored as halfway between the two items of interest. If more than two items are tied, the average can be taken of all the ranks concerned.

Now let's compute the differences between the ranks and the squared differences, as shown in Table 9-8.

Table 9-8. Calculation of rank differences

Vehicle	Engine size	CO_2 emissions	Engine rank	Emissions rank	d	d^2
A	1.8	230	1	3	2	4
B	1.9	200	2	1	−1	1
C	2.2	210	3	2	−1	1
D	2.5	240	4	5	−1	1
E	3.2	235	5.5	4	−1.5	2.25
F	3.2	260	5.5	7.5	2	4
G	3.3	250	7.5	6	−1.5	2.25
H	3.3	260	7.5	7.5	0	0
I	4.3	310	9	9	0	0
J	5.4	400	10	10	0	0

Thus, $\Sigma \delta^2 = 15.50$. The correlation can then be computed as:

$$r_s = 1 - (6 \times 15.50) / 10(100 - 1)$$
$$= 0.90606$$

Testing Statistical Significance

There are some general rules for interpreting values of r_s, as follows:

- $0.9 \leq r_s \leq 1$ indicates a very strong correlation
- $0.7 \leq r_s \leq 0.9$ indicates a strong correlation
- $0.5 \leq r_s \leq 0.7$ indicates a moderate correlation

However, it is also possible to determine statistical significance by using a z test. In the case where the null hypothesis is that $r_s = 0$, the following z test can be carried out:

$$z = \frac{r_s - 0}{\sqrt{\dfrac{1}{n-1}}}$$
$$= r_s \sqrt{n - 1}$$

In this example:

$$z = 0.906 \sqrt{10 - 1}$$
$$= 2.718$$

$z = 2.718$ is statistically significant at the $p < 0.01$ significance level, and you can reject the null hypothesis.

Advanced Techniques

While Pearson's and Spearman's correlation techniques are the most commonly used, the point-biserial and phi correlation coefficients may also be used on some specialist settings, as described in this section.

Point-Biserial Correlation Coefficient

A psychologist interested in predicting emotional intelligence might administer a test battery of 200 questions that takes three hours to complete, in which the psychologist is concerned with determining the relationship between answers on a single categorical scale (loves animals/does not love animals) and overall emotional intelligence (EI). If 90% of the variability in EI could be accounted for by the answer to one question because of its high correlation, there wouldn't be much point in administering the whole test. However, neither Pearson's product-moment nor Spearman's rank correlation coefficient allows you to perform correlations with categorical variables: you need the *point-biserial correlation coefficient*. In this example, the single value of EI can be correlated with a 0 or 1 coded response from the "Do you love animals?" question.

The psychologist decides to test 10 participants to compute the point-biserial correlation coefficient (r_{pbi}), shown in Table 9-9.

Table 9-9. Love of animals and emotional intelligence

Love of animals (x)	Emotional intelligence (y)
0	67
0	77
1	98
1	95
1	85
0	68
0	71
1	89
1	82
1	79

The r_{pbi} can then be calculated as follows:

$$r_{pbi} = \frac{M_p - M_q}{S_t} \sqrt{pq}$$

Where M_p is the mean for the scores coded as 1, M_q is the mean for the scores coded as 0, S_t is the standard deviation for all scores, p is the proportion of scores coded as 1, and q is the proportion of scores coded as 0.

Given $s_t = 10.80$, $M_p = 88$, $M_q = 70.75$, $p = 0.6$, $q = 0.4$, then r_{pbi} is given by:

$$r_{pbi} = \frac{88 - 70.75}{10.8}\sqrt{0.24}$$

$$= 0.78$$

Thus, there is a high correlation between EI and the response to the question. The psychologist would no doubt want to ask participants more than one question to obtain a result closer to the full EI figure.

Phi Correlation Coefficient

A natural progression from investigating the relationship between one categorical (binary) variable and a variable measured on an interval or ratio scale is to measure the association between two categorical variables (r_φ). The logic is similar to the previous types of correlation coefficients you have learned about, but with some differences in the treatment of the categorical data.

Consider an example: a tropical diseases epidemiologist is interested in whether having visited a certain country (x) is associated with having a certain hemorrhagic fever (y). The data for 10 participants from a travel clinic has been collected over the past year, and is shown in Table 9-10.

Table 9-10. Country visit versus hemorrhagic fever

Country visit (x)	Hemorrhagic fever (y)
1	1
1	1
1	1
1	0
1	0
0	0
0	0
0	0
0	1
0	1

From simply observing the frequencies, you can see that 60% of people who visited the country had the fever, but a significant number of cases (40%) occurred even when the person had not visited the country. Perhaps we should consider whether participants had visited a neighboring country—but firstly, r_φ can be determined by coding up the occurrences into Cartesian coordinates, as shown in Table 9-11.

The formula for deriving $r\varphi$ is quite complex, but is based on the difference in the product of frequencies $f(0, 0)$ and $f(1, 1)$ and the product of frequencies $f(0, 1)$ and $f(1, 0)$, divided by an estimate of the products of the standard deviations for x and y. These are labeled C, B, A, and D, respectively. $r\varphi$ is then given by:

$$r_\varphi = \frac{AD - BC}{\sqrt{(A + B)(C + D)(A + C)(B + D)}}$$

$$= -0.2$$

Thus, there is no strong relationship between the two variables, and visiting a single country is clearly not the answer to the epidemiologist's problems.

Table 9-11. Country visit versus hemorrhagic fever (frequencies)

	Did not visit country (0)	Visited country (1)
Fever (1)	$(0, 1) f = 2 (A)$	$(1, 1) f = 3 (B)$
No fever (0)	$(0, 0) f = 3 (C)$	$(1, 0) f = 2 (D)$

10

Categorical Data

A categorical variable is one in which the responses consist of a set of categories rather than numbers that measure an amount or quantity of something on a continuous scale. For instance, a person may describe their gender in terms of "male" or "female" or a machine part may be classified as "acceptable" or "defective." More than two categories are also possible: for instance, a person might describe their political affiliation (in the United States) as "Republican," "Democrat," "Independent," or "Other."

Categorical variables may be inherently categorical, with no numeric scale underlying their measurement (such as political party affiliation) or may be created by categorizing a continuous or discrete variable. For instance, blood pressure is a measure of the pressure exerted on the walls of the blood vessels, measured in millimeters of mercury (Hg). Blood pressure is usually recorded with specific measurements such as 120/80 Hg, but it is often analyzed using categories such as low, normal, prehypertensive, and hypertensive. An example using a discrete variable is number of children in a household: while the data may be collected as the exact number of children, it may be analyzed in categories such as "0 children", "1–2 children," and "3 or more children."

Although the wisdom of classifying continuous or discrete measurements into categories is sometimes debatable (some researchers refer to it as "throwing away information" because it discards all the information about variance within the categories), it is a common practice in many fields. Categorizing is done for many reasons, from custom (for instance, if certain categorizations have become accepted in your professional field), to solving distribution problems with a particular data set.

Categorical data techniques may also be applied to ordinal variables, meaning those measured on a scale in which the categories may be ranked in order but without the assumptions that the distance between each category is equal. The well-known Likert scale, in which people choose their responses to questions from a set of ordered categories (such as Strongly Agree, Agree, Neutral, Disagree, and Strongly Disagree) is a classic example of an ordinal variable. There is a special set of analytic techniques, also discussed in this chapter, for ordinal data that takes advantage of the fact that ordered categories were used. Given a choice, specific ordinal techniques are preferred over categorical techniques for the analysis of ordinal data because they are more powerful.

A host of specific techniques have been developed to analyze categorical and ordinal data, and to integrate the analysis of categorical data into techniques such as linear regression. This chapter discusses the most common techniques used for exclusively categorical and ordinal data, with a few exceptions. Kappa is covered in Chapter 1; the Spearman correlation coefficient, point-biserial correlation coefficient and phi are discussed in Chapter 9; nonparametric methods are covered in Chapter 11; and the odds ratio, risk ratio, and the Mantel-Haenszel test are covered in Chapter 18.

The R × C Table

When an analysis concerns the relationship of two categorical variables, their distribution in the data set is often displayed in an *R×C table*, also referred to as a *contingency table*. The R in R×C refers to *row* and the C to *column*: a specific table is described by the number of rows and columns it contains. Rows and columns are always named in this order, a convention also followed in describing matrices and in subscript notation. Sometimes a distinction is made between 2×2 tables, which display the joint distribution of two binary variables, and tables of larger dimensions. This is not necessary because a 2×2 table can be thought of as an R×C table where R and C both equal 2. The phrase "R×C" is read as "R by C" and the same convention applies to specific tables sizes, so "2×2" is read as "2 by 2."

Suppose we are interested in studying the relationship between broad categories of age and health, the latter defined by the well-known five-point general health scale. We decide on the categories to be used for age and collect data from a sample of individuals, classifying them according to age (using our predefined categories) and health status (using the five-point scale). We then display this information in a contingency table, like in Table 10-1.

Table 10-1. Contingency table displaying health status by age category

	Excellent	Very good	Good	Fair	Poor
Under 18 years					
18–39 years					
40–64 years					
65 years and older					

This would be described as a 4×5 table because it contains four rows and five columns. Each cell would contain the count of people with the pair of characteristics described: the number of people under 18 years who were in excellent health, the number aged 18–39 years were in very good health, and so on.

The Chi-Square Distribution

When we do hypothesis testing with categorical variables, we need some way to evaluate if our results are significant. With R×C tables, the statistic of choice is often one of the *chi-square tests*, which draw on the known properties of the *chi-square distribution*. The chi-square distribution is a continuous theoretical probability distribution that is widely used in significance testing because many test statistics follow this distribution when the null hypothesis is true. The ability to relate a computed statistic to a known distribution makes it easy to determine the probability of a particular test result.

The chi-square distribution is a special case of the gamma distribution and has only one parameter, k, which specifies the degrees of freedom. The chi-square distribution has only positive values because it is based on the sum of squared quantities, as will be seen below, and is right-skewed. Its shape varies according to the value of k, most radically when k is a low value, as can be seen in Figure 10-1. As k approaches infinity, the chi-square distribution approaches (becomes very similar to) a normal distribution.

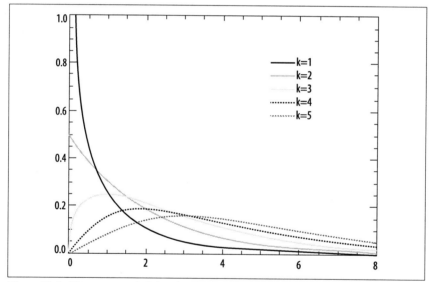

Figure 10-1. Chi-square probability distributions with different degrees of freedom

Statistics books sometimes include a table of critical values for the chi-square distribution, which can be useful if you don't have access to a computer. The tables define critical values for the different chi-square distributions, which can be used to compare the chi-square value from a particular study with the critical value. For instance, the critical value, assuming $\alpha = 0.05$, for the chi-square distribution with one degree of freedom is 3.84. Any test result above this value will be considered significant for a chi-square test of independence for a 2×2 table (described next).

Note that $3.84 = 1.96^2$ and that 1.96 is the critical value for the Z-distribution (standard normal distribution) for a two-tailed test when $\alpha = 0.05$. This result is not coincidental but is due to a mathematical relationship between the Z and chi-square distributions.

Stated formally: if X_i are independent, normally distributed variables with $\mu = 0$ and $\sigma = 1$, and the random variable Q is defined as:

$$Q = \sum_{i=1}^{k} X_i^2$$

Then Q will follow a chi-square distribution with k degrees of freedom.

The important points to remember are that you must know the degrees of freedom to evaluate a chi-square value, and that the critical values increase with the number of degrees of freedom. If $\alpha = 0.05$, the critical value for a chi-square distribution with one degree of freedom is 3.84, while for 10 degrees of freedom it is 18.31.

The Chi-Square Test

The chi-square test is one of the most common ways to examine relationships between two or more categorical variables. Not surprisingly, it involves calculating a number, called the chi-square statistic, which follows a chi-square distribution. For simplicity's sake I will explain the chi-square test first in terms of the 2×2 table, then look at some more complicated examples. In addition, there are several types of chi-square tests: this discussion will cover the most common, which is Pearson's chi-square test.

There are actually three ways of using the chi-square test, and while they are sometimes treated as identical we will differentiate among them here. The first is the *chi-square test for independence*. Taking the example of two variables, it tests the null hypothesis that the variables are independent of each other, i.e., that there is no relationship between them. The alternative hypothesis is that the variables are in fact related, so they are dependent rather than independent.

For instance, we might collect data on smoking status and diagnosis with lung cancer from a random sample of adults. Each of these variables is dichotomous: a person currently smokes or does not, and has a lung cancer diagnosis or does not. Our frequency table will look like Table 10-2.

Table 10-2. Smoking status and lung cancer diagnosis

	Lung cancer diagnosis	No lung cancer diagnosis
Currently smoke	60	300
Do not currently smoke	10	390

Just looking at this data, it seems that there is a relationship between smoking and lung cancer: 20% of the smokers have been diagnosed with lung cancer, while only about 2.5% of the nonsmokers have been. Appearances can be deceiving, so we will conduct a chi-square test for independence. Our hypotheses will be:

H_0: smoking status and lung cancer diagnosis are independent.
H_A: smoking status and lung cancer diagnosis are not independent.

Although chi-square tests are usually performed using a computer, particularly for larger tables, it is worthwhile to go through the steps of calculation for a simple example by hand. The chi-square test relies on the difference between *observed* and *expected* values in each of the cells of the 2×2 table. The observed values are simply what you found (observed) in your sample or data set, while the expected values are what you would have expected to find if the two variables were independent. To calculate the expected value for a given cell, we use this formula:

$$E_{ij} = \frac{i\text{th row total} \times j\text{th column total}}{\text{grand total}}$$

where E is the expected value for cell ij, and i and j designate the rows and columns of the cell. This subscript notation is used throughout statistics so it's worth reviewing here. Table 10-3 is a 2×2 table.

Table 10-3. Subscript notation for a 2×2 table

	$Cell_{11}$	$Cell_{12}$	Row 1 ($i = 1$)
	$Cell_{21}$	$Cell_{22}$	Row 2 ($i = 2$)
	Column 1 ($j = 1$)	Column 2 ($j = 2$)	

Table 10-4 adds row and column totals to our example.

Table 10-4. Smoking and lung cancer data with row and column totals

	Lung cancer diagnosis	No lung cancer diagnosis	Total
Currently smoke	60	300	360
Do not currently smoke	10	390	400
Total	70	690	760

The frequency for cell$_{11}$ is 60, the value for cell$_{12}$ is 300, the total for row 1 is 360, and the total for column 1 is 70.

The values for column and row totals are called *marginals* because they are on the margin of the table. They reflect the frequency of one variable in the study without regard to its relationship with the other variable, so the marginal frequency for lung cancer diagnosis in this table is 70. The numbers within the table (60, 300, 10, and 390 in this example) are called *joint* frequencies because they reflect the number of cases having specified values on both variables. For instance, the joint frequency for smokers with a lung cancer diagnosis is 60 in this table.

If the two variables are not related, we would expect that the frequency of each cell would be the product of its marginals, divided by the sample size. To put it another way, we would expect the joint frequencies to be affected only by the distribution of the marginals. For instance, if smoking and lung cancer were unrelated, we would expect the number of people who smoke and have lung cancer to be determined only by the number of smokers and the number of people with lung cancer in the sample. By this logic, the probability of lung cancer should be about the same in smokers and nonsmokers if it is true that smoking is not related to the development of lung cancer.

Using the formula above, we can calculate the expected values for each of the cells:

$$E_{11} = \frac{360 \times 70}{760} = 33.16$$

$$E_{12} = \frac{360 \times 690}{760} = 326.84$$

$$E_{21} = \frac{400 \times 70}{760} = 36.84$$

$$E_{22} = \frac{400 \times 690}{760} = 363.16$$

The observed and expected values for the lung cancer data are presented in Table 10-5; expected values for each cell are in parentheses. We need some way to determine if the discrepancies can be attributed to chance or if they represent a significant result. We can make this determination using the chi-square test.

Table 10-5. Observed and expected values for the lung cancer data

	Lung cancer diagnosis	No lung cancer diagnosis	Total
Currently smoke	60 (33.16)	300 (326.84)	360
Do not currently smoke	10 (36.84)	390 (363.16)	400
Total	70	690	760

The chi-square test is based on the squared difference between observed and expected values in each cell, using this formula:

$$\chi^2 = \sum_{i=1, j=1}^{rc} \frac{(O_{ij} - E_{ij})^2}{E_{ij}}$$

Which means:

1. Calculate the observed/expected values for cell 11.
2. Square the difference and divide by the expected value.
3. Do the same for the remaining cells.
4. Add the numbers calculated in steps 1–3 together.

Continuing with our example, for cell$_{11}$ this quantity is:

$$\frac{(O_{ij} - E_{ij})^2}{E_{ij}} = \frac{(60 - 33.16)^2}{33.16} = 21.7$$

Continuing with the other cells, we find values of 2.2 for cell$_{12}$, 19.6 for cell$_{21}$, and 2.0 for cell$_{22}$. The total is 45.5, which is within rounding error for the value calculated using SPSS (45.474).

In order to interpret a chi-square statistic, you need to know its degrees of freedom. Each chi-square distribution has a different number of degrees of freedom, and correspondingly different critical values. For a simple chi-square test, the degrees of freedom are $(r - 1)(c - 1)$, i.e., (the number of rows minus 1) times (the number of columns minus 1). For a 2×2 table, the degrees of freedom are $(2 - 1)(2 - 1)$ or 1; for a 3×5 table they are $(3 - 1)(5 - 1)$ or 8.

Having calculated the chi-square value and degrees of freedom by hand, we would consult a chi-square table to see if the chi-squared value calculated from our data exceeds the critical value for the relevant distribution. In this case the critical value for $\alpha = 0.05$ is 3.84, so we have sufficient evidence to reject the null hypothesis that the variables are independent. This process is further discussed in Chapter 7 on inferential statistics. More commonly, we would do the entire analysis by computer, which would give us a chi-squared value, its degrees of freedom, and the p-value. If the p-value is less than 0.05 we would normally reject the null hypothesis and conclude that the variables we studied are not independent. In this case, 21.7 is a highly significant value ($p < 0.001$) for a chi-square statistic with 1 degree of freedom, which leads us to the same conclusion, which is to reject the null hypothesis of independence between the variables.

The *chi-square test for equality of proportions* is computed exactly the same way as the chi-square test for independence, but the hypothesis tested is stated differently. The test for equality of proportions is used for data that is conceived as having been drawn from multiple independent populations: the hypothesis is that the distribution of some variable is the same in all populations. For instance, we could draw random samples from different ethnic groups and test whether the rates of lung cancer diagnosis were the same or different across the populations. The calculations would proceed as in the example above: people would be classified by ethnic group and lung cancer status, expected values would be computed, the value of the chi-square statistic and degrees of freedom calculated, and the statistic compared to a table of chi-square values for the appropriate degrees of freedom or the exact p-value obtained from a statistical software package.

The *chi-square test of goodness of fit* is used to test the hypothesis that the distribution of a categorical variable within a population follows a specific pattern of proportions, while the alternative hypothesis is that the distribution of the variable follows some other pattern. The test is calculated using expected values based on hypothesized proportions, and the different categories or groups are designated with the subscript i, from 1 to g:

$$\chi^2 = \sum_{i=1}^{g} \frac{(O_i - E_i)^2}{E_i}$$

For instance, suppose we believe that 10% of a particular population has low blood pressure (hypotension), 40% normal blood pressure, 30% prehypertension, and 20% hypertension. We can test this hypothesis by drawing a sample and comparing the observed proportions to those of our hypothesis (which are the expected values). Table 10-6 shows an example using hypothetical data.

Table 10-6. Expected and observed values for the distribution of blood pressure levels

	Hypotension	Normal	Prehypertension	Hypertension	Total
Expected proportion	0.10	0.40	0.30	0.20	1.00
Expected count	10	40	30	20	100
Observed count	12	25	50	13	100

The computed chi-square value is 21.7 with 3 degrees of freedom, and is highly significant (the critical value for $\alpha = 0.05$ is 9.49), so we would reject the null hypothesis that the blood pressure levels in the population followed the distribution we hypothesized. Degrees of freedom for the chi-square goodness of fit test is one less than the number of groups or proportions stated in the hypothesis. In this example, the degrees of freedom are $(4 - 1)$, or 3.

The Pearson's chi-square test is suitable for data in which all observations are independent (the same person is not measured twice, for instance) and the categories are mutually exclusive and exhaustive (so that no case may be classified into more than one cell, and all cases can be classified). It is also assumed that no cell has an expected value less than 1, and no more than 20% of the cells have an expected value less than 5. The reason for the last two requirements is that the chi-square is an asymptotic test and may not be valid for sparse data (data in which one or more cells have a low expected frequency).

Fisher's exact test, discussed below, does not require these assumptions and is a good substitute for the 2×2 chi-squared test with sparse data. Another solution, known as Yates' correction for continuity, is to subtract 0.5 from the absolute value of the difference between observed and expected values in each cell before squaring, as follows:

$$\chi^2 = \sum_{i = 1, j = 1}^{rc} \frac{(|O_{ij} - E_{ij}| - 0.5)^2}{E_{ij}}$$

Yates' correction reduces the chi-square value and thus the probability of a false positive result for data sets with sparse cells. Use of Yates' correction is not universally endorsed, however: some researchers feel it may be an over-correction leading to a loss of power and false negative results.

The chi-square test is often computed for tables larger than 2×2, although computer software is usually used for those analyses rather than hand calculations. There is no theoretical limit on the number of columns and rows that may be included, but two factors impose practical limits: the possibility of making a coherent interpretation of the results (try this with a 30×30 table!) and the necessity to avoid sparse cells, as noted above. Sometimes data is collected in a large number of categories but collapsed into a smaller number to get around the sparse cell problem. For instance, information about marital status may be collected using many categories (married, single never married, divorced, living with partner, widowed, etc.) but for a particular analysis the statistician may choose to reduce the categories (e.g., to married and unmarried) because of insufficient numbers in the smaller categories.

Fisher's Exact Test

Fisher's Exact Test (often called simply Fisher's) is a nonparametric test often substituted for the chi-square test with small or sparsely distributed data sets. Fisher's is based on the hypergeometric distribution and calculates the exact probability (p-value) of observing the distribution seen in the table, or a more extreme distribution: this is the meaning of the "exact" in the title. It is not an asymptotic test and therefore is not subject to the sparseness rules that apply to the chi-square tests (no cells with expected values less than 1, no more than 20% of cells with expected values less than 5). Computer software is usually used to calculate Fisher's, particularly for tables larger than 2×2, because of the repetitious nature of the calculations. A simple example with a 2×2 table is illustrated below.

Suppose we are interested in the relationship between use of a particular street drug and sudden cardiac failure in young adults. Because the drug is both illegal and new to our area, and because sudden cardiac death is rare in young adults, we were not able to collect enough data to allow us to conduct a chi-square test. Table 10-7 shows the data for analysis.

Table 10-7. Fisher's exact test: calculating the relationship between the use of novel street drug and sudden cardiac death in young adults

	Cardiac death	No cardiac death	Total
Used drug	7	2	9
Didn't use drug	5	6	11
Total	12	8	20

Our hypotheses are:

H_0: risk of sudden cardiac death is no more common among users of the new drug than in nonusers.

H_1: risk of sudden cardiac death is greater in people using the new drug.

Fisher's Exact Test calculates the probability of results at least as extreme as those found in the study. A "more extreme" result in this study would be one in which the difference in proportion of drug users versus nondrug users suffering sudden cardiac death was even greater than in the actual data (keeping the same sample size), as in Table 10-8.

Table 10-8. More extreme data distribution for drug use/cardiac death example

	Cardiac death	No cardiac death	Total
Used drug	8	1	9
Didn't use drug	4	7	11
Total	12	8	20

The formula to calculate the exact probability for a 2×2 table is:

$$p = \frac{r_1! r_2! c_1! c_2!}{n! a! b! c! d!}$$

where ! means factorial ($4! = 4 \times 3 \times 2 \times 1$) and cells and marginals are identified using the notation shown in Table 10-9.

Table 10-9. Table notation

a	b	r_1
c	d	r_2
c_1	c_2	n

In this case, $a = 8$, $b = 1$, $c = 4$, $d = 7$, $r_1 = 9$, $r_2 = 11$, $c_1 = 12$, $c_2 = 8$, and $n = 20$. The exact probability for Table 10-8 is:

$$p = \frac{9!\,11!\,12!\,8!}{20!\,7!\,2!\,5!\,6!} = 0.132$$

To find the *p*-value for this and all more extreme tables, we would have to repeat this calculation for each table and add them together. Since 0.132 is above the conventional *p*-value of 0.05, which is the usual standard for rejecting the null hypothesis, we don't need to continue with our labors, because the overall probability cannot get any lower by adding in probabilities from more tables. We can therefore conclude that our sample data does not provide sufficient evidence to reject the null hypothesis of no relationship between the two variables. Using SPSS, I found the exact significance for a one-sided hypothesis to be 0.16, which confirms the conclusion that this data does not present sufficient evidence to reject the null hypothesis.

McNemar's Test for Matched Pairs

McNemar's test is a type of chi-square test used when the data comes from paired samples. We might use McNemar's to examine the results of an opinion poll on some issue before and after a group of individuals viewed a political advertisement. In this example, each person would contribute two opinions, one before and one after viewing the advertisement. Another example would be the concordance of opinions on some issue among pairs of siblings. In this example, although different individuals are involved, they are so closely related or affiliated that we would expect them to be more similar than an independent sample of individuals from the population. Analogous to the sibling example, McNemar's can also be used to analyze data collected from two groups of individuals who have been closely matched on important characteristics such that they can no longer be considered independent. For instance, medical studies sometimes look at the occurrence of a particular disease, related to a risk factor, among two groups of individuals matched on characteristics such as age, gender, and ethnicity.

Suppose we want to measure the effectiveness of a political advertisement in changing people's opinions. One way to do this would be to ask people if they are for or against capital punishment, both before and after viewing a 30-second television commercial advocating that capital punishment should be abolished. Consider the hypothetical data set in Table 10-10.

Table 10-10. McNemar's test of opinions on capital punishment, before and after viewing a television commercial

		After viewing the commercial		
Before viewing the commercial		For capital punishment	Against capital punishment	
	For capital punishment	15	25	40
	Against capital punishment	10	20	30
		25	45	70

More people were against capital punishment after viewing the commercial, but is this difference significant? We can test this using McNemar's chi-square test, calculated using the following formula:

$$\chi^2 = \frac{(b-c)^2}{b+c}$$

This formula uses a method of referring to cells by letters, using the plan shown in Table 10-11.

Table 10-11. Method of referring to cells in a 2×2 table by letters

a	b
c	d

Note that this formula is based exclusively on the distribution of discordant pairs (*b* and *c*), in this case those in which a person changed their opinion after viewing the commercial. McNemar's has a chi-squared distribution with one degree of freedom. In this example:

$$\chi^2 = \frac{(25-10)^2}{25+10} = \frac{225}{35} = 6.43$$

This is sufficient evidence to reject the null hypothesis (the critical value[*] is 3.84 for a chi-square distribution with one degree of freedom) and conclude that people's opinions do change after viewing the commercial. I also determined from a computer analysis that the exact probability of getting a chi-square statistic with one degree of freedom at least as extreme as 6.43, if people's opinions did not change before and after viewing the commercial, is 0.017, reinforcing the fact that the change observed in this study is significant.

Correlation Statistics for Categorical Data

The most common correlation statistic, Pearson's correlation coefficient, requires variables measured on at least the interval level. A number of different measures of correlation have been developed for categorical and ordinal data. These are often produced using a statistical software package, although most can also be calculated by hand. These measures of correlation share most characteristics with Pearson's correlation. One is that their range is from −1 to +1 (although some have range restrictions, as noted below), with 0 indicating no relationship. Another is *symmetry*, meaning that either variable can be considered independent or dependent.

As with Pearson's correlation, the correlation statistics discussed in this section are measures of association only, and statements about causality cannot be supported by a correlation coefficient alone. There are a plethora of these measures, some of which are known under several names: a few of the most common are discussed here. A good approach if you're using a new statistical software package is to see which measures are supported by that package, and then investigate which are appropriate for your data, because there are so many different correlation statistics.

Binary Variables

Phi is a measure of the degree of association between two binary variables, i.e., two categorical variables, each of which can take on only two values. Phi is the same as r (Pearson's correlation) when variables are scored as 0 and 1, and is further discussed in Chapter 9. *Cramer's V* is analogous to phi for tables larger than 2×2; it is usually calculated using statistical software, as are the other measures discussed in this paragraph. If two binary variables are thought to represent underlying continuous measurement scales (for instance, if test scores on a scale of 0–100 are dichotomized for analytical purposes as pass/fail), the *tetrachoric correlation coefficient* is an appropriate statistic to use.

Ordinal Variables

The most common correlation statistic for ordinal data (in which data is ordered but cannot be assumed to have equal distance between values) is *Spearman's rank-order coefficient*. It is based on the ranks of data points (first, second, third, and so on) rather than their values, and is sometimes used in favor of Pearson's correlation to lessen the influence of outliers (extreme values) even for variables

---

[*] Tables of critical values for the chi-square and other statistics are available online from the Web site of the U.S. National Institute of Standards and Technology *http://www.itl.nist.gov/div898/ handbook/eda/section3/eda3674.htm*.

measured at the interval or ratio scale. Spearman's rank-order coefficient is discussed in detail in Chapter 9.

Goodman and Kruskal's gamma, often called simply *gamma*, is a measure of association for ordinal variables computed by calculating the number of concordant and discordant pairs among two variables. It is sometimes called a measure of *monotonicity* because it tells you how often the variables have values in the order expected. For instance, if I tell you that two variables in a data set have a positive relationship, and that case 2 has a higher value on the first variable than does case 1, you would expect that case 2 also has a higher value on the second variable. This would be a *concordant pair*. If case 2 had a lower value on the second variable, it would be a *discordant pair*. To calculate gamma by hand, we would first create a frequency distribution on the two variables, retaining their natural order.

Consider a hypothetical data set relating BMI (body mass index, a measure of weight relative to height) and blood pressure levels. In general, high BMI is associated with high blood pressure, but this is not the case for every individual. Some overweight people have normal blood pressure, and some normal-weight people have high blood pressure. Is there a significant relationship between weight and blood pressure in the data set shown in Table 10-12?

Table 10-12. Example data to calculate gamma

			Blood pressure		
			Normal	Prehypertensive	Hypertensive
BMI	Normal		15 (*a*)	15 (*b*)	5 (*c*)
	Overweight		10 (*d*)	15 (*e*)	20 (*f*)

The equations to calculate gamma rely on the cell designations shown in Table 10-13.

Table 10-13. Cell designations to compute gamma

a	*b*	*c*
d	*e*	*f*

First we have to find the number of concordant pairs (P) and discordant pairs (Q), as follows:

$$P = a(e + f) + bf = 15(15 + 20) + 15(20) = 525 + 300 = 825$$
$$Q = c(d + e) + bd = 5(10 + 15) + 15(10) = 125 + 150 = 275$$

Gamma is then calculated as:

$$\gamma = \frac{P - Q}{P + Q} = \frac{825 - 275}{825 + 275} = 0.5$$

Gamma is a symmetrical measure because it does not matter which variable is considered the predictor and which the outcome: the value of gamma will be the same in either case. It does not correct for tied ranks within the data. For large samples gamma has an approximately normal distribution, making it possible to calculate standard errors and *p*-values, and these numbers are provided when gamma is calculated using a computer program. I calculated gamma for this data using SPSS, which gave me a standard error of 0.145 and a *p*-value of 0.002, so I concludes that the data shows a relationship between BMI and blood pressure.

Maurice Kendall developed three slightly different types of ordinal correlation as alternatives to gamma. Statistical programs sometimes use slightly different formulas to calculate these statistics, so the exact formula used by any particular package should be confirmed with the software manual. All Kendall's tau statistics, like gamma, are symmetrical measures.

Kendall's tau-a is based on the number of concordant versus discordant pairs, divided by a measure based on the total number of pairs (*n* = the sample size):

$$\tau_a = \frac{P - Q}{\left(\frac{n(n-1)}{2}\right)}$$

Kendall's tau-b is a similar measure of association based on concordant and discordant pairs, adjusted for the number of ties in ranks. It is calculated as (*P* − *Q*) divided by the geometric mean of the number of pairs not tied on X (X_0) and the number of pairs not tied on Y (Y_0). Kendall's tau-b has a known sampling distribution and statistical packages usually report its standard error and significance. Tau-b can approach 1.0 or −1.0 only for square tables (tables with the same number of rows and columns). The formula for Kendall's tau-b is:

$$\tau_b = \frac{P - Q}{\sqrt{(P + Q + Y_0)(P + Q + X_0)}}$$

where X_0 = the number of pairs not tied on X, and Y_0 = the number of pairs not tied on Y.

Kendall's tau-c is used for nonsquare tables and, like tau-b, has a known sampling distribution. Tau-c is calculated as:

$$\tau_c = (P - Q)\left[\frac{2m}{n^2(m-1)}\right]$$

where *m* is the number of rows or columns, whichever is smaller, and *n* is the sample size.

Somers's d is an asymmetrical version of gamma, so calculation of the statistic varies depending on which variable is considered the predictor and which the outcome. Somers's d also differs from gamma because it is corrected for the number of pairs tied on the predictor variable. So if the hypothesis is that X predicts Y, Somers's d is corrected for the number of pairs tied on X. If the hypothesis is that Y predicts X, it is corrected for the number of pairs tied on Y. This is often phrased as saying that Somers's d "penalizes" the data for ties: all this really means is that, as in tau-b, tied pairs are removed from the denominator. Using the notation that X_0 = the number of pairs not tied on X, and Y_0 = the number of pairs not tied on Y, Somers's d is calculated as:

$$d(\text{predicting } Y \text{ from } X) \; = \; \frac{P - Q}{P + Q + X_0}$$

$$d(\text{predicting } X \text{ from } Y) \; = \; \frac{P - Q}{Q + P + Y_0}$$

A symmetric value of Somers's d may be calculated by averaging the two asymmetric values calculated with these formulas.

The Likert and Semantic Differential Scales

Several types of scales have been developed to measure qualities that have no natural metric, such as opinions, attitudes, and perceptions. The best known of these scales is the Likert scale, introduced by Rensis Likert in 1932, and widely used today in fields ranging from education to health care to business management. In a typical Likert scale question, a statement is presented and the respondent is asked to choose from an ordered list of responses. For instance:

My classes at Lincoln East High School prepared me for university studies.
1. Strongly agree
2. Agree
3. Neutral
4. Disagree
5. Strongly disagree

This is a classic ordinal scale: we can be reasonably sure that "strongly agree" represents more agreement than "agree," and "agree" represents more agreement than "neutral," but we can't be sure if the increment of agreement between "agree" and "strongly agree" is the same as the increment between "neutral" and "agree," or if these increments are the same for each respondent.

Categorical and ordinal methods, as described in this chapter, are always appropriate for the analysis of Likert scale data, and so are some of the nonparametric methods described in Chapter 11. The fact that Likert scale responses are often identified with numbers has sometimes led researchers to analyze the data as if it were collected on an interval scale. For instance, you can find published articles that report the mean and variance for data collected using a Likert scale. A researcher choosing to follow this path (treating Likert data as interval) should be aware that this is a controversial approach that will be rejected by many editors (including myself), and that the burden is on the researcher to justify any departure from ordinal or categorical methods of analysis for Likert scale data.

Five levels of response are commonly used with Likert scales, because three is thought to not allow sufficient variation of response, while seven is believed to offer too many choices. There is also some evidence that people are reluctant to select the extreme values of a scale when a large number of choices are offered. Some researchers prefer to use an even number of responses, usually four or six, in order to avoid a middle category that may be chosen by default by some respondents.

The *semantic differential scale* is similar to the Likert scale, except that individual data points are not labeled, merely the extreme values. The Likert question above could be rewritten as a semantic differential question as follows:

> Please rate your academic preparation at Lincoln East High School in relation to the demands of university study.
> Excellent preparation   1 2 3 4 5   Inadequate preparation

Because individual data points do not have to be labeled, semantic differential items often offer more data points to the respondent. Ten data points is a popular choice because people are familiar with a 10-point judging scale (hence the popular phrase "a perfect 10"). Like Likert scales, semantic differential scales are by nature ordinal, although when a larger number of data points are offered, some researchers argue that they can be analyzed as interval data.

Rensis Likert (1903–1981)

Rensis Likert (pronounced Lick-urt, with the accent on the first syllable) was an American social scientist who specialized in research on organizational behavior and management theory. Likert received his B.A. in sociology from the University of Michigan in 1926 and his Ph.D. in psychology from Columbia in 1932: he developed the Likert scale as part of his dissertation research. Likert was a founder of the University of Michigan Institute for Social Research and served as its director from 1946 to 1970; he spent his later years consulting for corporations and writing books on management theory. A central aspect of his work will endear him to self-motivated students and employees around the world: Likert introduced the concepts of participation management and the human-centered organization, based on his findings that there was an inverse relationship between coercive management supervision and employee productivity.

Exercises

Here are some review questions on the topics covered in this chapter.

Question

What are the dimensions of these tables? What are the degrees of freedom for an independent samples chi-square test calculated from data of these dimensions?

Answer

2×4 and 4×3: remember, tables are described as $R \times C$, i.e., (# of rows) by (# of columns). The degrees of freedom are 3 for the first table $[(2-1)(4-1)]$ and 6 for the second $[(4-1)(3-1)]$, because degrees of freedom for chi-squares are calculated as $[(r-1)(c-1)]$.

Question

What is the null hypothesis for the chi-square test of independence?

Answer

The variables are independent, which also means that the joint probabilities may be predicted using only the marginal probabilities.

Question

What is an appropriate statistic to measure the relationship between the two independent variables displayed in the following 2×2 table? What is the value of that statistic and what conclusion would you draw from it?

	D+	D−
E+	25	10
E−	2	5

Answer

Because two cells have expected values of less than five (cells c and d), Fisher's Exact Test should be used. The value is 0.077 (obtained using computer software), which does not provide sufficient evidence to reject the null hypothesis of no relationship between E and D.

Question

What are the expected values for the cells in this table? What is the value of the chi-square statistic? What conclusion would you draw about the relationship between exposure and disease, given this data?

	D+	D−
E+	25	30
E−	15	5

Answer

Here are the expected values.

	D+	D−
E+	29.3	25.7
E−	10.7	9.3

Chi-square(1) = 5.144, p = 0.023. We can therefore conclude that we have sufficient evidence to reject the null hypothesis that exposure and disease are unrelated.

Question

The following table represents political affiliations of married couples. Compute
the appropriate statistic to see if the affiliations of husbands and wives are independent of those of their spouses.

		Wife	
		Republican	Democrat
Husband	Republican	20	30
	Democrat	20	20

Answer

McNemar's test is appropriate because the data comes from correlated pairs. The
value of McNemar's chi-square is 2.00, which is not sufficient to reject the null
hypothesis that the political affiliations of spouses are independent of the affiliation of the other spouse.

Question

Which of Kendall's tau statistics would be appropriate for the following data?

		Satisfaction with job		
		Dissatisfied	Neutral	Satisfied
Educational Level	< HS	45	20	10
	HS grad	15	15	20
	Some college	30	10	25
	College grad	10	15	30

Answer

Kendall's tau-c should be used because the table is not square (it has four rows
and three columns).

Question

What is the argument against analyzing Likert and similar attitude scales as
interval data?

Answer

There is no natural metric for constructs such as attitudes and opinions. We can
devise scales that are ordinal (the responses can be ranked in order of strength of
agreement, for instance) to measure such constructs, but it is impossible to determine if the intervals among points on such scales are equally spaced. Therefore,
data collected using Likert and similar types of scales should be analyzed at the
ordinal or categorical level rather than the interval or ratio level.

Question

In what circumstance would you compute the Cramer's V statistic?

Answer

Cramer's V is an extension of the phi statistic and should be calculated to determine the strength of association between two categorical variables that have more than two levels. For binary variables, Cramer's V is equivalent to phi.

Simpson's Paradox

Simpson's paradox is a circumstance in which the direction of an association reverses when data from several groups is combined. This paradox is well known among baseball fans, for instance: it is possible for player A to have a higher batting average (proportion of hits) than player B in each of two years, yet have a lower batting average when data from the two years is combined. Consider the example in Table 10-14.

Player A had a higher batting average each year, yet over both years combined, a lower average. This phenomenon occurs due to the different number of cases observed for each player in each year. Simpson's paradox was also at the root of a controversy about gender discrimination in university admissions a few years ago. A lawsuit filed against the University of California was denied when it was shown that apparent gender discrimination (a lower percentage of women than men admitted overall to the university) could be explained by the fact that admissions were determined on a department-by-department basis, and that most women applied to departments where the percentage of applicants accepted was low, while most men applied to departments where the percentage of applicants accepted was higher. In fact, in most departments a slightly lower percentage of men than women were accepted, but this distinction was reversed when admissions data from all departments was combined.

Simpson's paradox is also seen in the evaluation of medical treatments, where treatment A may be superior to treatment B in each of two samples, yet inferior when the samples are combined. Some statisticians argue that circumstances such as this should not be labeled a "paradox" at all, since to do so implies that there is a causal relationship between the two variables.

Table 10-14. Simpson's paradox in baseball

Player	2000			2001			Combined		
	Hits	At-bats	Average	Hits	At-bats	Average	Hits	At-bats	Average
A	10	50	0.200	200	600	0.333	210	650	323
B	85	400	0.213	50	145	0.345	135	545	0.248

11

Nonparametric Statistics

The basis of statistics is parameter estimation, i.e., when an attempt is made to estimate the parameters (mean and standard deviation) of a population from a random sample. However, most statistical techniques rely on the underlying distribution being of a particular type, such as the normal distribution, for inferences made from the relevant statistical tests to be valid. What about scenarios where the underlying data is known to be nonnormal? In these cases, a different set of statistical techniques, known as nonparametric statistics, can be fruitfully applied to understand data. These techniques are often known as distribution-free since they make no assumptions about the underlying distribution of the data.

Nonparametric statistics are often applied to data sets where ranks rather than raw scores are used. For example, scholastic testing often involves some ranking of students from highest to lowest scores, and the ranks rather than the scores are often used in analysis. Taking the mean of the ranks of these scores is not a useful measure of central tendency in this scenario. Alternatively, Likert scales asking participants to rate their satisfaction with a product on a scale of values from 1–10, where 1 is very dissatisfied and 10 is extremely satisfied, the appropriate measure of central tendency would be the median rather than the mean, since the scores are ordinal rather than interval or ratio—that is, a score of 10 does not indicate 10 times the satisfaction of a corresponding score of 1. This is precisely the type of scenario where inferential tests that do not rely on parameterization are most useful.

In this chapter, you will learn about the most commonly used nonparametric procedures, including the median test, the Mann-Whitney U test, the Wilcoxon matched pairs signed rank test, the Kruskal-Wallis test, and the Friedman test.

Note that while nonparametric techniques are more *robust** than their parametric counterparts, they typically have lower power (i.e., they are less sensitive), and are most appropriately used for smaller rather than larger samples, since some nonnormal distributions (such as Student's t distribution) approximate a normal distribution for large N (from the Central Limit Theorem).

Nonnormal Data

A common practice of some researchers in certain disciplines is to assume that all data is normally distributed, and to apply the relevant parametric test to "prove" differences between groups. Such "proof" has no validity if the underlying assumptions about distributions are not met. However, these researchers could just as easily use nonparametrics, and immediately enhance the credibility of their work. Many naturally occurring variables approximate a normal distribution for large samples, such as height. Yet other physical variables, such as the distribution of weight in a population, are subject to variation, even within the same population over time, as the current obesity epidemic in first-world countries demonstrates.

Another issue is the scale of data being used, a topic discussed in detail in Chapter 1. For a ratio or interval scale, such as height or height differences, a 7-foot tall tree is twice as tall as a 3.5-foot tree. But for ordinal data, such as test scores that are ordinal, it's not clear that in a set of grades [a, b, c, d] that an "a" grade is twice as good as a "c" grade. This is because grades are ultimately associated with ranks and not interval or ratio data.

Denoting ranks using alphabetical rather than numerical categories is very useful in one respect: there is no temptation, then, to treat ordinal data as if it were interval or ratio, and computing mean values as a true representation of central tendency. Unfortunately, this is common practice in some types of survey research. For example, undergraduate students may be asked to rate their lecturers at the end of each term on 10 measures such as clarity of speaking, feedback on assignments, etc., on an ordinal scale of 1 to 5 as follows:

- 1—Needs improvement
- 2—Below expectation
- 3—Meets expectation
- 4—Exceeds expectation
- 5—Outstanding

To obtain an average performance measure, the human resources department decides to calculate the mean for each academic across all 10 measures. If Professor Smith obtains a mean score of 2.3, what does this really mean? Of course, it is a meaningless calculation, since there is no constant interval between

---

* In statistics, robustness is a quality of tests whose validity is not unduly violated by departures from underlying assumptions—without having to rely on the Central Limit Theorem. As an example, many multivariate statistical tests are very sensitive to assumption violations, but nonparametrics are generally the most robust.

measures on the scale. In other words, below expectation is not half as bad as needs improvement, and exceeds expectation is not twice as good as below expectation. Rather, a better measure of central tendency would be the median rank (i.e., what rank lies in the middle of the distribution) or the mode rank (i.e., what rank did most students give to this academic). Using numerical labels in this example has presented a temptation to treat ordinal ranks as interval data, which is completely inappropriate.

The more appropriate approach is to start analyzing nonnormal and/or ranked data using descriptive statistics that are designed to make sensible characterizations of the data, and then to use these statistics as the basis for inferential testing. For ordinal data, the typical way to begin analysis is to order the scores, usually from lowest to highest, and assign them a rank to show the order in which each score appears. Consider a football competition at the end of each season; every team has won a number of games, and this score is used to order and rank each team in terms of their performance. The two top ranked teams will compete in a "final match," irrespective of whether the difference in the number of games won by each team is 1 or 10—the two teams are still ranked 1 and 2, and it's possible for the rank 2 team to win the competition by winning the "final match."

Nonparametric tests are suitable for both between-subjects and within-subjects designs, described in Chapter 5, as well as tests of association. In Chapter 9, for example, you learned about Spearman's R, which is a correlation coefficient computed from ranks rather than data measured using interval or ratio scales. In this chapter, the focus is on exploring nonparametrics for between- and within-subjects comparisons.

The most commonly used nonparametrics for between-subjects comparisons are the Mann-Whitney U test, the median test, and the Kruskal-Wallis test. For within-subjects designs, the Wilcoxon matched pairs signed rank test and the Friedman test are generally used.

Between Subjects Designs

This section reviews some commonly used nonparametric tests for between-subjects designs, generally based on the rank sum and mean rank measures.

Wilcoxon's Rank Sum Test and the Mann-Whitney U Test

Two main descriptive statistics are used to characterize ordinal data: the *rank sum* and the *mean rank*. To illustrate how these statistics can be used, consider an example; an Olympic Games selection committee must choose a champion tae kwon do team from two states (California and Nevada) to represent the U.S. Since there are both individual and group events for which the members have trained together, the teams can't be combined to produce a composite team of the most highly performing individuals. Each team member has been given an overall performance score, based on the number of bricks that they managed to break during a five-minute testing session. The results are shown in Table 11-1.

Table 11-1. Performance scores for tae kwon do teams from two states

California	Nevada
4	2
5	3
6	3
6	4
7	4
8	5
9	10
9	10
9	11

Trying to interpret the results in this case is difficult just by visual inspection; the scores for the California team are more consistent and clustered in a smaller range, while the Nevada results are bimodal and have a greater range. Since the top three performers are from Nevada, you might be tempted to select this team, but the median score for Nevada is just 4, compared to California, with 7. This is not particularly helpful, since the three top scores for Nevada are beyond inclusion in the median range, and yet these athletes would most likely win the individual competition on their own merits. Similarly for the mean, the California team can break 7 bricks on average, and the Nevada team only 5.78, but since brick-breaking is a discrete event, the fractional of the score doesn't have a physical meaning.

The most appropriate way to describe the data is to assign a rank to each case, and then add all of the ranks together for each team. This usually gives an accurate indication of where the values are likely to be grouped on the scale. To assign ranks, every team member from both teams is ranked from top to bottom. This process is shown in Table 11-2.

Table 11-2. Team rankings

California	Nevada	Rank
	2	1
	3	2
	3	3
4		4
	4	5
	4	6
5		7
	5	8
6		9
6		10
7		11
8		12

Table 11-2. Team rankings (continued)

California	Nevada	Rank
9		13
9		14
9		15
	10	16
	10	17
	11	18

Note that, where tied ranks occur—i.e., the same rank occurs for more than one case in either team—the average rank is instead computed from the sum of the ranks concerned and divided by their number. The new ranking, including tied ranks, is shown in Table 11-3.

Table 11-3. Ranks for individual tae kwon do performance scores, including ties

California	Nevada	Rank
	2	1
	3	2.5
	3	2.5
4		5
	4	5
	4	5
5		7.5
	5	7.5
6		9.5
6		9.5
7		11
8		12
9		14
9		14
9		14
	10	16.5
	10	16.5
	11	18

The rank sum is then calculated for each group by simply adding together their respective ranks:

$$\Sigma_R(\text{California}) = 5 + 7.5 + 9.5 + 9.5 + 11 + 12 + 14 + 14 + 14 = 96.5$$

$$\Sigma_R(\text{Nevada}) = 1 + 2.5 + 2.5 + 5 + 5 + 7.5 + 16.5 + 16.15 + 18 = 74.5$$

The magnitude of the rank sum indicates how close together the ranks are for each group. Thus, the California group overall tends to cluster at high values, where the Nevada group tends to be grouped toward lower values, even though Nevada has the three highest scorers. Going a step further, and to compensate for unequal group sizes to make a fair comparison, the mean rank $\bar{R}$ can be computed by dividing each sum by the respective N. The results are shown below:

$$\bar{R}(\text{California}) = \frac{96.5}{9} = 10.72$$

$$\bar{R}(\text{Nevada}) = \frac{74.5}{9} = 8.28$$

Thus, on the whole, the California team ranks higher than the Nevada team. Thus, using rank-based methods, the selectors should select the California team, since the mean rank is higher. However, it is also possible to calculate a z-test for the rank sum to determine whether the difference between the two groups is statistically significant. If the null hypothesis is that the two groups would have equal mean ranks, we can compute the expected sum as follows:

$$\mu_W = \frac{n_1(n_1 + n_2 + 1)}{2}$$

$$= \frac{9(9 + 9 + 1)}{2}$$

$$= 85.5$$

In the example above, you can see that one group (California) has a rank sum above the expected mean, and the other group (Nevada) has a rank sum below the expected mean. The z-test can be computed from the mean and standard deviation of W, which is the difference between the smallest observed rank sum and the expected rank sum, as shown below:

$$z = \frac{W - \mu_W}{\sigma_W}$$

The estimate for σ_W is given by:

$$\sigma_W = \sqrt{\frac{n_1 n_2(n_1 + n_2 + 1)}{12}}$$

In this example:

$$\sigma_W = \sqrt{\frac{9 \times 9(9 + 9 + 1)}{12}}$$

$$= 11.32$$

Thus:

$$z = \frac{74.5 - 85.5}{11.32}$$

$$= -0.97$$

This result is not statistically significant at either $p < 0.01$ or $p < 0.05$. Thus, you would fail to reject the null hypothesis.

The Wilcoxon rank sum test is considered mathematically equivalent to the Mann-Whitney U test since the same z score will be produced, and can be used as a substitute for the two-sample t-test, where the normality assumption of the underlying data is questionable. In summary, the Mann-Whitney U test can be used to test the null hypothesis that two groups have identical distributions and/or identical medians (rather than means). The Mann-Whitney U test can be applied to data measured on ordinal, integer, or ratio scales.

Median Test

The *median test* makes use of ranks and the binomial distribution to test hypotheses, i.e., where there are only two possible outcomes. The test is based on the distribution of dichotomous variables using either a known median rank, or comparing the differences between two groups with median ranks estimated from a sample (i.e., one-sample and two-sample median tests, respectively). In epidemiology, the median test is often used to generalize findings between studies. For example, a researcher may be interested in testing a hypothesis whether a new metabolic disorder, provisionally termed Type X diabetes, is a disease of older rather than younger people. Having previously studied Type II diabetes, the median age of onset was found to be 35.5 years. After studying Type X diabetes, the researcher believes the age of onset is greater. Thus, the null hypothesis is that $\pi = 0.50$, while the hypothesis is that $\pi > 0.50$. After examining the age of onset in a clinical sample of 40, 36 had an age greater than 35.5 years, and thus the null hypothesis was rejected at $p < 0.05$.

Further metabolic research in the lab subsequently suggests that there may be two subtypes of Type X diabetes—Type X1 and X2—and raises the question of whether subtypes are associated with age: that is, do younger patients tend to have Type X1 and older patients Type X2? The research decides to look at another sample of 40 cases, 20 of which have been provisionally given Type X1 and 20 classified as Type X2. The median age was 36 years. For Type X1, 12 cases were above the median age and 8 cases were below the median age. For Type X2, 9 cases were above the median age and 11 cases were below the median age. The null hypothesis is that π(Type X1) = π(Type X2) = 0.50, whereas the alternative hypothesis is that π(Type X1) < π(Type X2). The tabulated frequencies are shown in Table 11-4, and would omit exact median scores if they occurred.

Table 11-4. Frequencies of age of occurrence for Type X1 and Type X2 diabetes

	Above median	Below median	Total
Type X1	12	8	20
Type X2	9	11	20
Total	20	20	40

A chi-square test (discussed in Chapter 10) can be used to test the significance. You can use the fast computational formula for χ^2 analysis, where the fields are described as in Table 11-5.

Table 11-5. Chi-square test for significance

Type	Above median	Below median	Row sums
X1	a	b	$a+b=n1$
X2	c	d	$c+d=n2$
Column sums	$a+c$	$b+d$	N

Thus:

$$\chi^2 = \frac{n(ad - bc)^2}{(a + b)(c + d)(a + c)(b + d)}$$

$$= \frac{40((12 \times 11) - (8 \times 9))^2}{(12 + 8)(9 + 11)(12 + 9)(8 + 11)}$$

$$= 0.902$$

The results indicate that $\chi^2 = 0.902$, which is not statistically significant at $p < 0.05$, $df = 1$. Thus, there is no clear difference between the two subtypes in terms of age, and the original one-way test is clearly most useful in terms of understanding the relationship between age and diabetes subtypes.

Kruskal-Wallis H Test

The Kruskal-Wallis H test extends the Mann-Whitney U test by allowing for multiple groups to be compared, in order to test the null hypothesis that there is no median difference for at least two of the groups. You may be wondering how this relates to the *t*-test; in summary, the *t*-test can only test comparisons between two groups, just like a correlation is a measure of association between two variables. However, as you will learn in Chapter 12, the general linear model allows more general comparisons to be made between groups and variables, without being limited to a single comparison within one test.

In the case of comparisons between more than two groups, using the Mann-Whitney U test requires you to undertake all possible pair-wise comparisons between each pair of groups. For example, in the Wilcoxon rank sum test example, a comparison was made between the California and Nevada groups. However, if the selectors were required to choose between California, Nevada, and Utah, then three comparisons would be required to establish if any two groups differed significantly from each other:

- California versus Nevada
- California versus Utah
- Utah versus Nevada

The benefit of using the Kruskal-Wallis H test is that all possible combinations can be tested at once using a generalized version of the Wilcoxon rank sum test. However, statistical significance is not established using a *z* score; rather, a chi-square test is used to determine whether all of the populations being examined have the same ordinal distribution.

H can be calculated as follows:

$$H = \frac{n\left(\sum_i \frac{(\bar{R}_i - (n+1))^2}{2}\right)}{\frac{n(n+1)}{12}}$$

Returning to the tae kwon do example, Table 11-6 shows the scores from the teams in three states (now including Utah), and the rankings are shown in Table 11-7.

Table 11-6. Performance scores for tae kwon do teams from three states

California	Nevada	Utah
4	2	10
5	3	9
6	3	10
6	4	6
7	4	6
8	5	7
9	10	8
9	10	7
9	11	6

Table 11-7. Ranks for individual tae kwon do performance scores from three states, including ties

California	Nevada	Utah	Rank
	2		1
	3		2.5
	3		2.5
4			5
	4		5
	4		5
5			7.5
	5		7.5
6			11
6			11
		6	11
		6	11
		6	11
7			15
		7	15
		7	15
8			17.5

Table 11-7. *Ranks for individual tae kwon do performance scores from three states, including ties (continued)*

California	Nevada	Utah	Rank
		8	17.5
9			20.5
9			20.5
9			20.5
		9	20.5
	10		24.5
	10		24.5
		10	24.5
		10	24.5
	11		27

$\Sigma_R(\text{California}) = 5 + 7.5 + 11 + 11 + 15 + 17.5 + 20.5 + 20.5 + 20.5 = 128.5$

$\Sigma_R(\text{Nevada}) = 1 + 2.5 + 2.5 + 5 + 5 + 7.5 + 24.5 + 24.5 + 27 = 99.5$

$\Sigma_R(\text{Utah}) = 11 + 11 + 11 + 15 + 15 + 17.5 + 20.5 + 24.5 + 24.5 = 150$

$\bar{R}(\text{California}) = \dfrac{128.5}{9} = 14.28$

$\bar{R}(\text{Nevada}) = \dfrac{99.5}{9} = 11.06$

$\bar{R}(\text{Utah}) = \dfrac{150}{9} = 16.66$

You can determine the expected value for all of the mean ranks using the following formula:

$$\bar{R} = \dfrac{N + 1}{2}$$

$$= 14.0$$

Against this expected value, you will be interested in verifying whether any of the groups deviate from this expected mean, or whether only one or more groups deviate from the expected value. We then compute the H statistic as shown below:

$$H = \dfrac{9[(14.28 - 14.0)^2 + (11.06 - 14.0)^2 + (16.66 - 14.0)^2]}{\dfrac{27(27 + 1)}{12}}$$

$$= \dfrac{9[0.08 + 8.64 + 7.08]}{63}$$

$$= \dfrac{142.18}{63}$$

$$= 2.26$$

To calculate the significance of H, you compare it to χ^2 at the $p = 0.05$ level, for df = 8, which is 15.51; the null hypothesis can be rejected, since at least one team outperforms the others. However, the selectors actually want to know which teams are significantly better performing in this case. Thus, an *a posteriori* procedure, such as Fisher's least significant difference, can be used to determine a value that represents the least significant difference between two groups, in terms of their mean ranks. In this example, we can use the following:

$$z = \sqrt{\frac{2[N(N+1)]}{12n}}$$

$$= \sqrt{\frac{2[27(27+1)]}{108}}$$

$$= 3.74$$

Table 11-8 shows the differences between mean ranks for all three groups, and the significant differences are marked with a * ($p < 0.05$). As you can see, the only significant difference is between Utah and Nevada, so the selectors should have confidence in selecting Utah over Nevada, but not Utah over California.

Table 11-8. Significance for mean rank differences assessed a posteriori

	California	Nevada	Utah
California	–	3.22	2.38
Nevada	3.22	–	5.6*
Utah	2.38	5.6*	–

Within-Subjects Designs

This section reviews some commonly used nonparametric tests for within-subjects designs.

Wilcoxon Matched Pairs Signed Rank Test

The Wilcoxon matched pairs signed rank test can be used as a nonparametric replacement or substitute for the one-sample *t*-test, in the situation where a pre-treatment measure is compared with a post-treatment measure, and the null hypothesis is that the difference is zero (i.e., the treatment has no effect). The Wilcoxon can also be considered the within-subjects equivalent of the Wilcoxon rank sum test, as reviewed in the previous section. The Wilcoxon does not assume normality, but does assume at least a symmetric distribution.

In the case of the Wilcoxon matched pairs versus the rank sum test, the former assumes that pairs of scores can be matched in a meaningful way, e.g., repeated measures from the same participant, while the latter can cater for comparisons between groups of different sizes.

Following from the previous tae kwon do example, where the Wilcoxon rank sum test was used to select the California team over the Nevada team, the Olympic Games selectors are now faced with the question of choosing which event to enter the team: patterns or sparring. The selectors have decided to only enter one event and devote all their energies to training for that event, since the skills required and training techniques involved are quite different (patterns focuses on style, sparring requires form and fitness). The selectors want to answer the following research question: when compared with their own performance on the two events, do the team members consistently perform better in one event or the other?

The selectors randomly allocate each California team member to perform five minutes of patterns or five minutes of sparring, and then alternate events, after a 10-minute rest break. A panel of experts judges the performance of each team member, and the scores are averaged to arrive at a final score out of 10. The differences between the two measures are then calculated and tabulated, as shown in Table 11-9.

Table 11-9. Ranks for individual tae kwon do members on sparring and patterns

Member	Sparring	Patterns	Difference
1	6	8	+2
2	7	5	−2
3	8	7	−1
4	8	8	0
5	10	9	−1
6	9	8	−1
7	9	10	+1
8	8	5	−3
9	10	8	−2

In this example, member 4 has an identical score, and is therefore excluded from the analysis, since his performance could count equally toward either activity (sparring or patterns). The next stage is to rank the differences in performance for all nontied scores, as shown in Table 11-10. The average rank is computed if more than one score ties for the same rank.

Table 11-10. Rank differences for individual tae kwon do members between sparring and patterns

Member	1	7	2	3	5	6	8	9
Difference	+2	+1	−1	−1	−1	−1	−1	−2
Rank	1	2	5	5	5	5	5	8

If there were a genuine difference between the groups in either direction (i.e., in favor of sparring or patterns), then you would expect to see a clustering among the rankings and their differences (which, by visual inspection, is clearly the case here). The rank sum of the differences can be used to calculate whether the positive differences are sufficient to cancel out the effect of the negative differences, thereby supporting the null hypothesis of no difference between the two groups:

$$\mu_T = \frac{n(n+1)}{4}$$

$$= \frac{8(8+1)}{4}$$

$$= 18$$

To calculate the statistical significance of any difference between the two groups, Wilcoxon's T is calculated by adding the rank sums for both the positive and negative cases separately. The rank sum of smallest magnitude is then taken as T, and the mean difference calculated between T and the estimated μ_T. In this example, the rank sums can be derived as follows:

$$\Sigma_n(\text{Positive}) = 2 + 1 = 3$$

$$\Sigma_n(\text{Negative}) = 1 + 1 + 1 + 1 + 1 + 2 = 7$$

To calculate the z-test, you can use the following formula:

$$z = \frac{T - \mu_T}{\sigma_T}$$

where:

$$\sigma_T = \sqrt{\frac{n(n+1)(2n+1)}{24}}$$

$$= \sqrt{\frac{8(8+1)(16+1)}{24}}$$

$$= 7.141$$

thus:

$$z = \frac{3 - 18}{7.141} = -2.10$$

Friedman Test

The *Friedman test* is an extension of the matched pairs signed rank test for multiple related samples. Recalling the tae kwon do example, an Olympic team will need to perform at consistently high levels over five hours of competition, so repeated measures of performance can be taken during each hour, to determine if there are any differences between the means. Each hourly score is ranked in order of performance. The null hypothesis is that there will be no differences between the means; however, you can imagine that for many teams, there will be significant differences over time between the first and last measurement, or even a similar performance level at the first and last stages, but with tiredness intervening during the intermediate periods to reduce performance.

The use of the Friedman test is not limited to measures made over time, but could also be used to evaluate the effect of drug treatments or any other experimental situation where a nonparametric approach may be most appropriate.

The test is used as follows: consider b members of a tae kwon do team whose performance is measured at t time points. Let x_{ij} represent the performance score at time i for team member j, where $i = 1, 2, .., t$ and $j = 1, 2, ..., b$. Scores are ranked and replaced with their respective ranks. The rank sum at each time interval i is given by s_i, where $I = 1, 2, ..., t$, and the Friedman statistic T can then be calculated as:

$$T = \frac{12 \sum s_i^2}{bt(t+1)} - 3b(t+1)$$

The sparring performance scores for the eight members of the Texas tae kwon do team are shown in Table 11-11, at three different hourly time periods, and their respective ranks are shown in Table 11-12.

Table 11-11. Sparring performance scores at three different hourly time periods

	1 Hour	2 Hours	3 Hours
Member 1	9	8	7
Member 2	9	7	8
Member 3	6	8	7
Member 4	8	7	6
Member 5	8	7	6
Member 6	9	8	7
Member 7	9	8	7
Member 8	7	5	6

Table 11-12. Sparring performance ranks at three different hourly time periods

	1 Hour	2 Hours	3 Hours
Member 1	3	2	1
Member 2	3	1	2
Member 3	1	3	2
Member 4	3	2	1
Member 5	3	2	1
Member 6	3	2	1
Member 7	3	2	1
Member 8	3	1	2
Total	22	15	11

The squared sum of rank sums for each time period can be calculated as follows:

$$\sum s_i^2 = 22^2 + 15^2 + 11^2 = 830$$

Therefore, given $b = 8$ team members and $t = 3$ time periods, T can be calculated as follows:

$$T = \frac{12 \times 830}{8 \times 3 \times (3 + 1)} - 3 \times 8 \times (3 + 1)$$

$$= \frac{9960}{96} - 96$$

$$= 7.75$$

Using critical values from a chi-square distribution, and $df = 2$, there are statistically significances at $p < 0.05$, but not at $p < 0.01$. Thus, you would reject the null hypothesis, and be confident in selecting a team that had significantly higher performance than their counterparts.

Exercises

Here are some exercises to review the topics covered in this chapter.

Question

There is a different rank procedure to use when there are tied ranks, based on mid-ranks.[*] Imagine that the testing procedure over three time intervals is then performed separately for the Alaskan tae kwon do team. The performance scores are shown in Table 11-13.

Table 11-13. Sparring performance scores at three different hourly time periods (with ties)

	1 Hour	2 Hours	3 Hours
Member 1	8	8	6
Member 2	6	6	7
Member 3	6	8	7
Member 4	8	7	6
Member 5	9	9	7
Member 6	9	8	7
Member 7	8	7	6
Member 8	8	7	7

Answer

The scores are firstly ranked scores, as shown in Table 11-14.

---

[*] Some sources also recommend using a correction factor with tied ranks, which may be available in the statistical package of your choice.

Table 11-14. Sparring performance ranks at three different hourly time periods (with ties)

	1 Hour	2 Hours	3 Hours
Member 1	2.5	2.5	1
Member 2	1.5	1.5	3
Member 3	1	3	2
Member 4	3	2	1
Member 5	2.5	2.5	1
Member 6	3	2	1
Member 7	3	2	1
Member 8	3	1.5	1.5
Rank Sum	19.5	17	11.5

The squared sum of rank sums for each time period can be calculated as follows:

$$\sum s_i^2 = 19.5^2 + 17^2 + 11.5^2 = 801.5$$

Therefore, given $b = 8$ team members and $t = 3$ time periods, T can be calculated as follows:

$$T = \frac{12 \times 801.5}{8 \times 3 \times (3+1)} - 3 \times 8 \times (3+1)$$

$$= \frac{9618}{96} - 96$$

$$= 4.188$$

In this case, after consulting a table of critical values for the chi-square distribution for $df = 2$, you would fail to reject the null hypothesis at both $p < 0.05$ and $p < 0.01$.

Question

A marketing professional is interested in determining whether there are differences in soccer supporters and soft drink preferences, to guide appropriate advertisement placement in a future campaign. 100 randomly selected supporters from each team of the two top-ranking soccer teams (W, X) are asked if they prefer one soft drink (A or B) over another, when exiting a football match. The results are shown in Table 11-15, based on whether the values are above or below the median. Does the data support the hypothesis that different team supporters have different soft drink preferences?

Table 11-15. Median results of drink preferences (brand A, B) between two soccer teams' fans (W, X)

Team	Above median	Below median	Row sums
W	30	70	100
X	60	40	100
Column sums	90	110	200

Answer

You can use the fast computational formula for χ^2 analysis, where the fields are described as shown in the table below.

Team	Above median	Below median	Row sums
W	a	b	$a + b = n_1$
X	c	d	$c + d = n_2$
Column sums	$a + c$	$b + d$	n

Thus:

$$\chi^2 = \frac{n(ad - bc)^2}{(a + b)(c + d)(a + c)(b + d)}$$

$$= \frac{200((30 \times 40) - (70 \times 60))^2}{(30 + 70)(60 + 40)(30 + 60)(70 + 40)}$$

$$= 18.18$$

The result is $\chi^2 = 18.18$, $p < 0.01$. Thus, you can conclude that Team X supporters are more likely than Team W supporters to score above the median.

12

Introduction to the General Linear Model

In Chapter 9, you learned to describe the association between two variables by using a simple graphical technique in a two-dimensional (x, y) plane. You also learned to quantify the *bivariate relationship* by computing a correlation coefficient. You may have been surprised by how easy it was to relate the mathematical relationship between two variables, especially for simple cases such as $r = 1.00$, where a perfect correlation can be graphically described by a straight line, with a specific slope and intercept.

It is possible to take the relationship one step further and use characteristics, such as the *slope* and *intercept*, to build a functional *mathematical model*, and determine the precise deviation from the model for observed data. In this approach, the correlation coefficient and the *coefficient of determination* still have an important role to play; however, the use of *linear regression* to test the *goodness of fit* of observed data to a theoretical model goes one step further in being able to characterize existing data, and predict values of dependent variables from independent variables. This process occurs literally by simple algebraic operations, such as substitution.

Linear regression is an extremely valuable technique, which is often used for prediction in models where no experimental control has been applied to the collection of data. For example, you may want to determine the relationship between training and performance in athletics. However, where appropriate experimental design is in place, then it may be better to use the Analysis of Variance (ANOVA) technique, which—like the *t*-test—can be used to determine the likelihood of different samples being drawn from the same population (i.e., using a hypothesis testing framework, where there is a hypothesized difference between two groups). For example, you may want to determine whether athletes who have taken a mineral supplement perform better than athletes who have taken a placebo.

Linear regression and ANOVA are both based on the *general linear model*, which, in simple terms, is the geometry and algebra of straight lines. The applications of the general linear model are widespread, as linear models have often been found to be useful approximations to more difficult real-world relationships. Chapter 15 describes some of the more esoteric types of regression (including fitting of nonlinear models).

This chapter introduces linear regression and ANOVA through the concept of the general linear model, which encompasses both techniques. Bivariate regression will be used to introduce the basic assumptions of linear regression, such as *homoscedasticity*, and basic concepts such as the least-squares method of line fitting and the meaning of slope and intercept.

The General Linear Model

What is meant by the term "functional mathematical relationship" between an independent (or explanatory) variable x and a dependent (or response) variable y? Simply put, $y = f(x)$ means that you can calculate any value of y if you know the value of x. The function $f()$ can be any valid mathematical function:

- $y = x$ means that the value of y is always the same as the value of x, e.g., (x, y) = (1, 1), (2, 2), (3, 3), etc.
- $y = ax$ means that the value of y is always some multiple of the value of x, determined by the constant a, e.g., where $a = 2$, $(x,y) = (1, 2)$, (2, 4), (3, 6), etc.
- $y = ax + b$ means that the value of y is always some multiple of the value of x, determined by the constant a, plus the value of a constant b, e.g., where $a = 2$, $b = 1$, (1, 3), (2, 5), (3, 7), etc.
- $y = x^2$ means that the value of y is always the value of x multiplied by itself, e.g., (x,y)= (1, 1), (2, 4), (3, 9), etc.

In its bivariate form, the general linear model can always be described by $y = ax + b$. In the multiple case, i.e., where there are n independent variables $(x_1, x_2, x_3, ...x_n)$, each is assigned a separate slope $(a_1, a_2, a_3 ..., a_n)$, and the form of the model is $y = a_1x_1 + a_2x_2 + a_3x_3 + ... + a_nx_n + b$. But you'll learn more about the multiple case in Chapter 14; the point is that a really simple model of straight lines on a plane is extremely powerful and can be expanded in a number of different and very useful ways.

In general, the best way to start looking for relationships between dependent and independent variables is to use a graph in the two-dimensional plane, along with a correlation. However, if you are dealing with large numbers of possibly related variables, it may be less time-consuming to start with a correlation table, and only visually investigate relationships between variables that are moderately or highly correlated. The risk with this strategy is that the underlying model that relates the two variables may be nonlinear, so relying on correlation alone may not suggest any relationship.

Linear Regression

Imagine you have a relationship between two variables (x, y) that is described by the model $y = ax + b$. If the observed data perfectly matches the model, then a correlation of $r = 1.00$ will be observed, and also the coefficient of determination $r^2 = 1.00$. However, there may be any number of reasons why the observed data may not actually match that predicted; for instance:

- There may be other independent variables that independently account for variation in y, other than x. For example, athletic performance may be influenced by a range of factors, including coaching strategies, number of hours and their impact, etc.

- There may be some element of random variation in the observations, such as bias or systematic error. For example, two clinicians measuring tumor growth may use slightly different procedures to arrive at a different volume measurement from the same observation.

- The variables may represent phenomena that are not fixed in value over time. For example, in a study of attitudes, participants' attitudes may change during the experimental period.

However, accepting that error will always be present, how do we actually "fit" the observed data to a hypothetical model? In the following example, we review the process for carrying out linear regression, and examine some of the pitfalls associated with "line fitting."

Imagine a psychologist interested in the relationship between height and intelligence, operationalized as IQ. The psychologist believes that there is good evidence for a causal relationship between height and IQ, and has investigated numerous biological pathways that suggest that IQ is a dependent variable, and height is an independent variable. Some example measures relating IQ to height are shown in Figure 12-1: the correlation between the two variables is high, at $r = 0.83$; thus, using the coefficient of determination, more than 70% of the variability in IQ can be attributed to height. Looking at the plot of the data, there is a clear linear relationship between IQ and height: as height rises, IQ also rises correspondingly. However, notice that there are also many cases where IQ ≈ 110, for a range of different heights (1.4m–1.65m). There is obviously a strong effect of four observations of large heights on the quantified association; if the psychologist had good reasons for considering these cases as outliers, and they were removed from the analysis, then a weak correlation would be observed $r = 0.33$, with $r^2 = 0.11$, as shown in Figure 12-2. In this situation, the psychologist must examine whether the first or second relationship really represent, the true state of nature.

Linear regression can be used to go one step further than correlation. In this case, an imaginary straight line can be drawn on the graph, representing the hypothesized linear relationship between the two variables. The deviations from the imaginary line can be used to calculate the correlation coefficient, but most importantly, the linear model can be used to predict all dependent variable values along the straight line, based on values of the independent variable. This type of

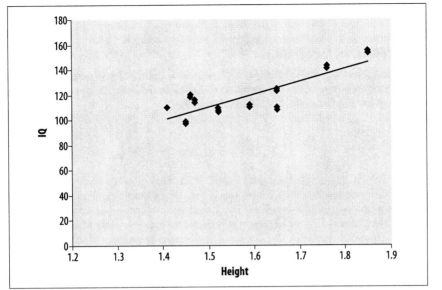

Figure 12-1. Association between two variables height (IV) and IQ (DV)

prediction is extremely useful, as you know—with a certain level of accuracy, from the coefficient of determination—how accurate your prediction is going to be. Using a regression line also means that you don't need to experimentally measure every possible value of an independent variable to predict the corresponding value of the dependent variable; however, as you can see from the example in Figure 12-1, extrapolation and prediction in ranges where you haven't measured may be dangerous to rely upon. For example, if the psychologist had never measured any heights greater than 165cm, she would never have realized how strongly correlated the two variables were across the range of measurable height.

Compare Figures 12-1 and 12-2 side-by-side (Figure 12-2 is at the end of this section): what is the main characteristic that differentiates the two lines, and their vastly different correlation coefficients? Clearly, the slope of each line is different; when there is a very shallow vertical rise in the slope compared to the horizontal run, as in Figure 12-2, the correlation is low. However, when there is an equivalent vertical rise to the horizontal run, then the correlation is at its highest. Conversely, if the vertical rise is much greater than the horizontal run, then the correlation would also be weak. The relationship between rises and runs can be described more formally using calculus, but the important message for regression is to understand how the slope of the regression directly influences correlations.

So, given the two possible relationships between IQ and height, as shown in Figures 12-1 and 12-2, how do you go about deciding which regression line is the "best fit" to the observed data? Regression involves picking the "best" line that fits the observed data, where "best" means that the deviations between each data point of the linear model and the observed data are minimized. To ensure that the best overall line is selected as the "best fit," you need to have some way of averaging across all of the observations to ensure that no single deviation results in an inappropriate line being selected.

In linear regression, the deviations are measured against the dependent variable for each observation of the independent variable. In addition, it is usual to utilize the squared deviations, and sum them all together:

$$SSE = \sum_{i=1}^{N} (y_i - \hat{y}_i)^2$$

Where SSE is the *sum of squared errors*, N is the number of cases, y_i is the observed dependent variable value, and $\hat{y}_i$ is the expected value of the observation, as predicted from the model. The literal goal of *least-squares* is to find the line that minimizes SSE for all observations. Note that error here means the quantitative difference between the expected and observed values, whether the error arises from measurement error, bias, or a genuine deviation from the prediction made by the underlying model.

Why are the deviations squared instead of their raw values? There are several justifications that can be made, but one important one is that large differences are weighted more highly than smaller differences—so, to most effectively minimize the SSE, in terms of a model fit, it is more desirable to reduce larger than smaller differences.

The formula for SSE can be rewritten in terms of the general linear model in the following way:

$$SSE = \sum_{i=1}^{N} (y_i - b - ax_i)^2$$

It is then possible to use calculus to minimize this function. A set of computational formulae is given below:

$$Sxx = \sum x^2 - \frac{(\sum x)^2}{N}$$

$$Sxy = \sum xy - \frac{(\sum x)(\sum y)}{N}$$

$$a = \frac{Sxy}{Sxx}$$

$$b = \frac{\sum y}{N} - a\frac{\sum x}{N}$$

For the IQ and height example:

$$\sum x = 33.25$$

$$\sum y = 2{,}486$$

$$\sum x^2 = 53.01$$

$$\sum y^2 = 299{,}676$$

$$\bar{x} = \frac{33.25}{21} = 1.58$$

$$\bar{y} = \frac{2{,}486}{21} = 118.38$$

$$Sxx = 53.01 - \frac{(33.25 \times 33.25)}{21} = 0.36$$

$$Sxy = 3{,}973.04 - \frac{(33.25 \times 2{,}486)}{21} = 36.87$$

$$a = \frac{36.87}{0.36} = 102.35$$

$$b = 118.38 - (102.35 \times 1.58) = -43.67$$

Therefore, the model that best fits the line in this example is:

$$y = ax + b$$

$$= 102.35x - 43.67$$

Or, to predict IQ on the basis of height:

$$IQ = 102.35 \times Height - 43.67$$

Therefore, the IQ of any person can be estimated using the model, based on their height, and the value of slope and intercept estimated using this procedure. Note that the accuracy of any individual prediction is determined by the coefficient of determination: if this is close to $r^2 = 1.00$, then we would feel very confident in the strength of predictions made for new cases, especially where the value of the dependent variable has previously been observed and entered into the model. Also, as more cases with the same ordered pairs are observed through replicated observations, further independently and randomly obtained samples etc., the overall confidence in the model will increase. However, in omitting the four tallest cases from Figure 12-1—as described in Figure 12-2—a very different model would be produced, suggesting a strong lack of relationship between IQ and height. Therefore, you need to be careful in extrapolating a known model to regions on the plane where observations have never been made.

In some cases, single parameter estimation may be used. For example, where the intercept is zero, ratio estimation may be used to make predictions in the model. Thus, if the model was $IQ = 102.35 \times Height$, simply multiply a specific height by 102.35. Alternatively, if the slope was 1, then difference estimation could be used.

Thus, if the model was *IQ = Height* – 43.67, then adding a value to the righthand side of the equation would predict IQ. An example would be an intervention to make people taller (and by inference, increase their IQ); in both cases (ratio estimation and difference estimation), a single parameter could be used to predict the experimental effect, if a causal relationship existed.

Most importantly in the height and IQ example, since there have been no experimental controls or manipulations, the relationship between IQ and height is purely predictive and not causative, i.e., simply developing a mathematical model and using least-squares regression and correlation analysis to determine the strength of relationship between two variables cannot be used to infer causation. Indeed, the equation could easily be rewritten to show that height can be just as easily predicted from IQ as IQ can be predicted from height. The only way that causation could be inferred would be from using the appropriate experimental design, as discussed in Chapter 5.

You may be wondering which value is more useful—the correlation coefficient or the slope of a regression line. The correlation coefficient is useful in determining the strength of association between two variables—once established, the slope of the regression can be used to estimate by *how much* one variable rises (or falls) as a function of another. If the relationship is weak, then the slope of a regression line is meaningless—so both measures are useful, and have an important role in characterizing and predicting relationships between two variables respectively.

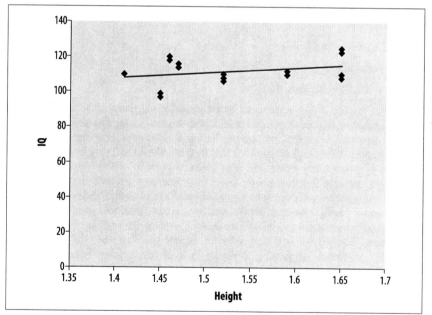

Figure 12-2. Association between two variables height (IV) and IQ (DV) in a restricted range

Assumptions

As with most statistical procedures, there are a number of assumptions that underlie the validity of linear regression, which—if violated—limit any inferences that may be drawn. Key assumptions include:

- That the most relevant independent and dependent variables were selected for inclusion in the model. This is particularly important when moving from characterization and prediction to hypothesis testing, and especially when there are multiple variables that are correlated with each other.
- That the variables have been measured in such a way to minimize measurement error and bias.
- That the independent variables are independent and, if any inferences are intended to be drawn, that an experimental manipulation has been applied. Correlated independent variables in multiple linear regressions (*multicollinearity*) are discussed further in Chapter 14.
- That the data are randomly sampled, and/or that the inferences drawn from the analysis are qualified accordingly.
- That the mathematical function fitted is an accurate model of the phenomenon under study.
- That the variance of the independent and dependent variables is the same (*homogeneity of variance*), i.e., the conditional variance of the independent variable given the dependent variable is the same for all values of the dependent variable.

If there is any doubt about the quality or validity of the variables being used or the data collected, it makes sense to replicate the analysis, where possible, with a different sample, and/or to use an alternative but related variable or approach that may be less prone to measurement error or bias.

In terms of fitting models, the least squares procedure can also be used to fit nonlinear models directly, or a linear model may be fit over a certain range where the nonlinear function is approximately linear (such as a threshold function). The difficulty in using a linear model to approximate a nonlinear model is the potential bias introduced in terms of the functional relationship between the dependent and independent variables.

One way to forensically examine whether a nonlinear model is more appropriate is to use an explicit test for linearity, or to examine the residuals; for a stepwise function, for example, low residuals may be observed for the middle section of the function, but greater residuals at either end. In this case, a nonlinear model may be better.

Examining the residuals for various sorts of issues is an important activity to undertake, once a regression analysis has been performed. Two key issues can be investigated: determining autocorrelation, and verifying the assumption of homoscedasticity. The former is usually associated with longitudinal data, while the latter is often associated with cross-sectional studies.

Autocorrelation arises when there is a large correlation between different observations, especially where observations are recorded over a period of time. After fitting a linear model, autocorrelation will be readily apparent in the residuals if it is present in the data. Where the residuals are rising over time they are said to be positively correlated, or negatively correlated where they are decreasing over time. The risk is that a bias will be introduced into the estimate of standard error. In the case of simple linear regression, autocorrelation may indicate that a more sophisticated model is required to account for the relationship, thus, a search for a second and additional explanatory variable would normally occur.

In the case of *homoscedasticity*, where residuals are assumed to be unrelated to any variables in the model, a violation is known as *heteroscedasticity*. This can be determined by examining the residuals; if there is an overall linear relationship, but there are large variations in the residuals observed for some levels of the independent variable, then heteroscedasticity may exist. Like autocorrelation, the risk is that a bias will be introduced into the estimate of standard error.

In both cases, tests to detect biases and appropriate corrections can be found in most statistical packages.

Analysis of Variance (ANOVA)

Simply put, the Analysis of Variance (ANOVA) is a technique commonly used to test whether there are statistically significant differences between two or more independent groups. These groupings are made on the basis of levels of independent variables, as described in Chapter 7. The simplest form of ANOVA is a one-way analysis, where the intention is to determine whether there is an overall main effect of different levels of an independent variable on a dependent variable. The outcome of performing an ANOVA is an *F ratio*, which can be used to determine whether statistically significant differences exist between the groups. For example, in a simple experiment to determine the effect of caffeine on learning, a psychologist might have two levels of the independent variables (experimental: caffeine 100mg; control: placebo). The F ratio can then be tested for significance at different levels of α, typically $p < 0.01$ or $p < 0.05$, for a dependent variable, such as number of correct responses in a math quiz.

Like linear regression, ANOVA can be extended to account for differences across multiple independent variables, as well as *interactions* between the variables. For example, a psychologist may wish to determine whether caffeine or nicotine has an effect on learning (both known as main effects), as well as any interaction between caffeine and nicotine.

ANOVA can also be used to test the significance of the coefficient of determination, arising from regression analysis.

One-Way ANOVA

The simplest form of ANOVA is a one-way ANOVA, where the simple question of differences between two or more treatment levels can be tested in terms of a main effect. This is equivalent to performing a *t*-test. The hypothesis is that one or more populations differ significantly in their means. Figure 12-3 shows an example of four sample groups, each with a slightly different mean, but with significant overlap in their distributions; in this case, there is a large amount of variation within each sample. Thus, it is unlikely that you would find any statistically significant differences between the means, and indeed, the directly testable null hypothesis would be that the populations' means were equal, i.e.:

$$\overline{X}_1 = \overline{X}_2 = \overline{X}_3 = \overline{X}_4$$

No difference would imply that the samples were drawn from the same population, and that the division into groups was not appropriate.

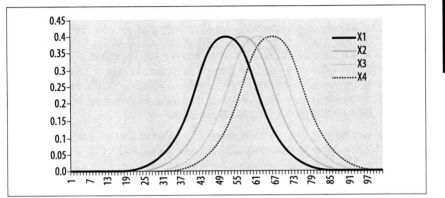

Figure 12-3. Four normally distributed groups with large variation within samples

However, consider the alternate case where there is relatively little variation within two samples; here, the distributions have very little overlap, and you can make a reasonable prediction that the means are statistically significant, as shown in Figure 12-4. That is, you may predict that the two samples have been drawn from different populations. Note that in the case of more than two samples, a difference between at least two of the groups would lead you to reject the null hypothesis. The hypothesized variation between means is the basis for the analysis of *variance*. The analysis is based on the total mean—as measured from all of the observations across all samples—and the mean of each individual sample. The total variation across all samples is known as the Total Sum of Squares (SS_{Total}), i.e., the total sum of squared deviations; the variation between samples is known as the Sum of Squares Between ($SS_{Between}$); and the variation within samples is known as the Sum of Squares Within (SS_{Within}). In simple terms, and taking into account degrees of freedom for each variation term, the F test for ANOVA is simply the ratio between the Sum of Squares Between and the Sum of Squares Within. Thus, if the variation between samples becomes relatively large compared to the variation within samples, you would expect the groups to be drawn from different populations, and a significant mean difference to exist.

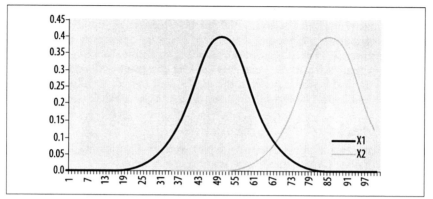

Figure 12-4. Two normally distributed groups with little variation within samples

Let's look at an example. An economist is interested in the effect of interest rates on house prices, and uses a series of *Monte Carlo*[*] simulations to arrive at median house prices for four different levels of interest: 2.5%, 5%, 7.5%, and 10%. The output of the models, after three different iterations, is shown in Table 12-1 and plotted in Figure 12-5, shown later in this chapter.

From the graph, you can see that there is a general rise in house prices; the medians for the different rates are $164,331, $199,628, $214,402, and $265,521, respectively. This appears to indicate that interest rate rises have a positive effect on house prices. But is the apparent difference "real," in the sense that the apparent rise is not caused by confounding factors, such as biases in measurement? If there is a hypothesized "real" difference in median house prices between the four interest rate levels, then the null hypothesis would be no difference between the groups. This is exactly the main effect test that ANOVA is suitable for.

In the analysis that follows, you will see how ANOVA (and its F ratio) can be implemented by hand; however, you would normally use a statistical package to calculate the F ratio, especially where multiple independent variables are included in the analysis. Of course, you may feel that to test the null hypothesis, you could simply perform:

$$\binom{4}{2} = 6$$

t-tests, but the overall α level would be greatly increased because of the multiple comparisons.

Each observed median house price is denoted y_{ij}, where i is the group, and j is the observation. In this example, the number of groups is $a = 4$, the number of observations per group is $n = 3$, and the interest rate levels are denoted $j = \{1, 2, 3, 4\}$

---

[*] The Monte Carlo method involves the use of random numbers as inputs to a deterministic function to solve a problem using repeated trials. Intuitively, you can imagine how aggregating the results from numerous simulation trials using random numbers can be used to minimize the effect of various forms of bias. Monte Carlo methods are often used when mathematical problems are too complicated to solve analytically.

for {2.50%, 5.00%, 7.50%, 10.00%} respectively. To keep things simple, there are equal numbers of observations for each group, but in practice, you may have some missing observations to deal with.

Table 12-1. Simulation results for effect of different interest rates on median house prices

2.50%	5.00%	7.50%	10.00%
12,655	19,877	21,033	25,023
17,877	20,122	21,188	27,877
18,766	19,888	22,099	26,755

Table 12-2 shows the assignment to treatment groups using the algebraic notation.

Table 12-2. Simulation results for different groups with algebraic notation

	y_{1j}	y_{2j}	y_{3j}	y_{4j}
	$y_{11} = 12,655$	$y_{21} = 19,877$	$y_{31} = 21,033$	$y_{41} = 25,023$
	$y_{12} = 17,877$	$y_{22} = 20,122$	$y_{32} = 21,188$	$y_{42} = 27,877$
	$y_{13} = 18,766$	$y_{23} = 19,888$	$y_{33} = 22,099$	$y_{43} = 26,755$
Total	49,298	59,887	64,320	79,655

$$\sum\sum y_{ij} = 253,160$$

As the name suggests, ANOVA is literally an analysis of variance; if the null hypothesis is accepted, then it is possible to compare variances between groups and variances within groups. Consider the different averages associated with the groups:

Grand Mean: $\dfrac{(\sum\sum y_{ij})}{12} = \dfrac{253,160}{12} = 21,096$

Group Means:

$\bar{y}_1 = \dfrac{49,293}{3} = 16,432$

$\bar{y}_2 = \dfrac{59,887}{3} = 19,962$

$\bar{y}_3 = \dfrac{64,320}{3} = 21,440$

$\bar{y}_4 = \dfrac{79,655}{3} = 26,551$

Assuming the model:

$$y_{ij} = \mu + \alpha_i + \varepsilon$$

where μ is the population mean and ε is error (or variation ascribed to each house price), the α_i becomes the effect of the independent variable (interest rate). α_i can be estimated by:

$$\overline{y}_i - y_i$$

thus, the data can be rewritten in *additive model* terms, as shown in Table 12-3.

Table 12-3. Simulation results rewritten using additive model

2.50%
$12,655 = 21,096 + (16,432 - 21,096) - 3,777$
$17,877 = 21,096 + (16,432 - 21,096) + 1,455$
$18,766 = 21,096 + (16,432 - 21,096) + 2,334$
5.00%
$19,877 = 21,096 + (19,962 - 21,096) - 85$
$20,122 = 21,096 + (19,962 - 21,096) + 160$
$19,888 = 21,096 + (19,962 - 21,096) - 74$
7.50%
$21,033 = 21,096 + (21,440 - 21,096) - 407$
$21,188 = 21,096 + (21,440 - 21,096) - 252$
$22,099 = 21,096 + (21,440 - 21,096) + 659$
10.00%
$25,023 = 21,096 + (26,551 - 21,096) - 1,528$
$27,877 = 21,096 + (26,551 - 21,096) + 1,326$
$26,755 = 21,096 + (26,551 - 21,096) + 204$

If the null hypothesis is true, then all $\alpha_i = 0$, and also:

$$\sum_i \alpha_i = 0$$

The assumption is that the error term is *independently normally distributed* (IND) with $\mu = 0$, and the same variance for all α_i. As you will see in many research publications, violations of these assumptions are commonplace and often uncorrected, bringing into question the validity of the analysis and the inferences drawn.

Using ANOVA for hypothesis testing requires that you compare three types of variation: total, within-group, and between-group, where the total variance can be partitioned into either within-group or between-group variation:

Total variance
   The mean sum of squared deviations from the grand mean:

$$\frac{\sum_i \sum_j (y_{ij} - \bar{y})^2}{na - 1}$$

Within-group variance

The mean sum of squared deviations from the group average:

$$\frac{\sum_i \sum_j (y_{ij} - y_i)^2}{a(n-1)}$$

Between-group variance

The mean sum of squared deviations from the grand mean but multiplied by n:

$$n\left[\frac{\sum_i (\bar{y}_i - \bar{y})^2}{a-1}\right]$$

Why do you need to calculate so many variances? Simply put, the within-group variance will be greater than the mean when the null hypothesis is rejected, while the between-group variance will be equal to the mean when the null hypothesis is accepted. The F statistic, calculated below, is the ratio of the among-group variance to the within-group variance. Thus, a test of whether a group of means is significantly different from each other becomes a test of whether their variances are significantly different.

To calculate the numerators of the different variances, you can use the following formulae:

$$SS_{Total} = \sum_i \sum_j (y_{ij} - \bar{y})^2$$

$$SS_{Within} = \sum_i \sum_j (y_{ij} - y_i)^2$$

$$SS_{Between} = n \sum_i (\bar{y}_i - \bar{y})^2$$

Recalling that:

$SS_{Total} = SS_{Within} + SS_{Between}$
$SS_{Between} = 3 \times (16{,}432 - 21{,}096)^2 +$
$\quad 3 \times (19{,}962 - 21{,}096)^2 +$
$\quad 3 \times (21{,}440 - 21{,}096)^2 +$
$\quad 3 \times (26{,}551 - 21{,}096)^2$
$\quad = 158{,}743{,}533$
$SS_{Total} = (831{,}898{,}910 + 1{,}195{,}522{,}557 + 1{,}379{,}684{,}234 + 2{,}119{,}107{,}683) -$
$\quad 12\,(21{,}096)^2$
$\quad = 185{,}381{,}251$
$SS_{Within} = 185{,}381{,}251 - 158{,}743{,}533$
$\quad = 26{,}637{,}718$

$$F = \frac{\dfrac{SS_{Between}}{a-1}}{\dfrac{SS_{Within}}{n-a}} = 15.89$$

Here, $a - 1$ is the df for $SS_{Between}$ and $n - a$ is the df for SS_{Within}. Thus, the critical value for $F_{3,8}$ at $p = 0.05$ is 4.066; you would reject the null hypothesis on this occasion. Figure 12-5 illustrates.

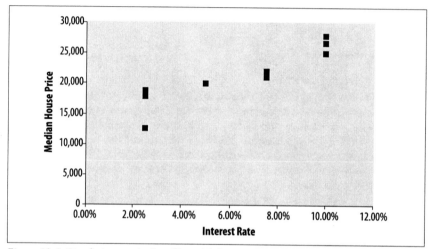

Figure 12-5. Simulation results for effect of different interest rates on median house prices

Post Hoc Tests

In the one-way ANOVA example, the F-test was statistically significant, and the null hypothesis rejected. However, you still need a way to determine which of the means were significantly different from each other, i.e., the result of the F-test does not imply that *all* samples were drawn from different populations, merely that at least *one* was drawn from a different population from the others.

To determine which groups were actually different, you need to undertake *post hoc tests*. In most cases, you would use the Scheffe test, but there may be justification for using one of the many others, e.g., depending on whether the homoscedasticity assumptions can be met, or how liberal you are prepared to be in terms of committing Type I errors. The Scheffe test is the most conservative test, and should be used unless there is a good reason not to.[*]

The Scheffe test will produce a significance value for each pair-wise comparison, i.e., $(n - 1) \times a$ distinct combinations:

---

[*] Post hoc testing should not be confused with post hoc hypothesis generation, in which, having discovered an unexpected difference or failing to find one using a Scheffe test, a shopping expedition is undertaken, where you simply pick a test until it gives you the result you are looking for.

$$\left(\frac{a}{n-1}\right)$$

Other post hoc tests, in order of conservatism, include:

- Fisher's least significant difference
- Bonferroni correction
- Duncan's new multiple-range test
- Student-Newman-Keuls' test
- Tukey's honestly significant difference test

Fisher's least significant difference is based on a two-sample t-test, as described in Chapter 9, where the estimate of pooled variance s^2_p is used, since the variances are assumed to be equal.

The pooled variance can be estimated by:

$$s^2_p = \frac{\sum(y_1 - \bar{y}_1)^2 + \sum(y_2 - \bar{y}_2)^2}{(n_1 - 1)(n_2 - 1)}$$

$$= \frac{(n_1 - 1)s^2_1 + (n_2 - 1)s^2_2}{n_1 + n_2 - 2}$$

where y_1 and y_2 are the two groups for whom the pair-wise comparison is made. The t-test then has the form:

$$t = \frac{\bar{y}_1 - \bar{y}_2}{\sqrt{\dfrac{2s^2_p}{n}}}$$

Fisher's least significant difference should only be used after an overall F-test has established a significant difference. Otherwise, the risk of Type I errors greatly increases, given the number of comparisons being made. Thus, at $p = 0.05$, if 20 comparisons are made, you would expect at least one erroneous result. It has been shown mathematically that the error rate for least significant difference is no greater than a t-test under these circumstances.

Exercises

In the following exercises, you will explore the relationship between linear regression and the one-way ANOVA. The technique that you use to either relate two variables, build a model, or test for mean differences will always make use of the underlying general linear model.

Question

Some nutritionists have suggested a link between coffee consumption and IQ, so you decide to investigate the strength of association by using linear regression. Table 12-4 shows a set of data collected from a randomly selected sample of people in a railway carriage. A short IQ inventory is administered, and participants are asked how many cups of coffee they have drunk in the past 24 hours. Fortunately, everybody catching the train only drinks the same coffee drink (size and strength) made by the same café.

Your role is to determine the linear model and strength of relationship between the two variables.

Table 12-4. Relationship between IQ and coffee consumption

Coffee	IQ
2	123
1	112
1	102
1	98
0	79
0	87
1	102
2	120
2	120
3	145

Answer

You plot the coordinates contained in Table 12-4, and conclude that there is a positive association between IQ and coffee consumption. You then calculate the correlation coefficient using the techniques described in Chapter 10, giving $r = 0.98$, and the coefficient of determination $r^2 = 0.95$. This indicates that 95% of the variability in IQ can be accounted for by coffee consumption, so you decide to proceed with the linear regression. Your next task is to build the model:

$$\sum x = 13.0$$

$$\sum y = 1,088$$

$$\sum x^2 = 25.0$$

$$\sum y^2 = 121,720$$

$$\bar{x} = \frac{13}{10} = 1.3$$

$$\bar{y} = \frac{1,088}{10} = 108.8$$

$$Sxx = 25.0 - \frac{(13.0 \times 13.0)}{10} = 8.1$$

$$Sxy = 1,575 - \frac{(13.0 \times 1,088)}{10} = 160.6$$

$$a = \frac{160.6}{8.1} = 19.82$$

$$b = 108.8 - (19.82 \times 1.3) = 83.02$$

Therefore, the model that best fits the line in this example is:

$$y = ax + b$$
$$= 19.82x + 83.02$$

Or, to predict IQ on the basis of coffee consumption:

$$IQ = 19.82 \times Coffee + 83.02$$

Question

In the previous question, you established a strong linear relationship between IQ and coffee consumption. However, since the design was not experimental, you are unable to determine whether drinking coffee causes a corresponding change in IQ, or the equally likely possibility that clever people simply like to drink more coffee (e.g., because of a personality trait associated with high IQ).

How would you determine a causal relationship?

Answer

To really determine if there is a causal effect, you set up a randomized experiment in which participants are randomly selected from the telephone book, and asked to not drink any coffee or caffeine-containing substances in the prior to the experiment 24 hours. The participants are then assigned randomly to a treatment (100mg caffeine tablet) or placebo (a tablet with equivalent size and taste to the treatment), after being matched on age and sex. The treatment is always administered at nine in the morning by the same experimenter in the same room, with the same temperature. Neither the experimenter nor the participant knows whether they are receiving the treatment or the placebo. In the treatment group, after waiting 15 minutes for the caffeine to establish a pharmacological effect for the treatment, a small battery of verbal tasks—highly correlated with a full IQ test—is administered, and the responses noted by the experimenter. The placebo condition is run in exactly the same way, except that (obviously) the placebo tablet is administered in place of the caffeine.

The results from the experiment are shown in Table 12-5. In this example, $a = 2$ and $n = 10$.

Table 12-5. Treatment and control groups for coffee (IV) and placebo consumption, and the effect on IQ (DV)

Treatment	Control
110	100
100	95
120	100
125	122
120	115
120	88
115	97
90	87

Table 12-5. Treatment and control groups for coffee (IV) and placebo consumption, and the effect on IQ (DV) (continued)

Treatment	Control
95	92
88	76

Recalling that:

$$SS_{Total} = SS_{Within} + SS_{Between}$$

$$SS_{Between} = 10 \times (108.3 - 102.75)^2 + 10 \times (97.2 - 102.75)^2 = 616.05$$

$$SS_{Total} = (119{,}019 + 96{,}096) - 20\,(102.75)^2 = 3{,}963.75$$

$$SS_{Within} = 200{,}594 - 616.05 = 199{,}977.6$$

$$F = \frac{\dfrac{SS_{Between}}{a-1}}{\dfrac{SS_{Within}}{n-a}} = 3.31$$

Here, $a - 1$ is the *df* for $SS_{Between}$ and $n - a$ is the *df* for SS_{Within}. Thus, the critical value for $F_{1,18}$ at $p = 0.05$ is 3.16; as a result, you would reject the null hypothesis on this occasion. Note that the critical value for $F_{1,18}$ at $p = 0.01$ is 5.092.

13

Extensions of Analysis of Variance

In Chapter 12, you learned about the general linear model and its applications in linear regression and one-way Analysis of Variance (ANOVA). In the algebraic derivation of the general linear model, from an analysis of the two-dimensional number plane, the possible extension to the multidimensional case was alluded to. *Factorial ANOVA* involves the use of models that include more than one independent variable, while *Multivariate ANOVA* (or MANOVA) uses models with multiple dependent variables.

In this chapter, you will learn about more of these complex ANOVA designs, including two-way and three-way factorial ANOVA, and MANOVA for at least two dependent variables. Issues surrounding the use of factorial and nonfactorial designs, and the *Analysis of Covariance* (ANCOVA), will be covered, while in Chapter 14, corresponding multidimensional extensions to linear regression will be covered.

Realistically, most ANOVA designs are typically factorial, and at least two-way, depending on your field of interest. In addition, during model building based around groups, it may become apparent that there is a confound influencing variation in the dependent variable as observed. For example, a test of athletic performance between two schools is confounded if one happens to be a specialist athletics school, with twice as many contact sport hours as the second school. The number of contact sport hours is thus considered a covariate, and ANCOVA allows an adjustment on the dependent variable to be made to cancel out its effect, and consider (in this example) whether there may be differences between the two schools independent of the covariate.

This chapter will also present ANOVA for repeated measures, and unique assumptions for these designs, such as *sphericity*, as well as considering the use of mixed designs, combining between-subjects and within-subjects comparisons.

As the designs and examples become more complicated, the use of SPSS output and syntax as examples will be used. Please refer to Appendix B for more details on how to use this statistical package. However, the ANOVA tables generated by SPSS will be comparable with your package of choice.

Factorial ANOVA

Factorial designs are often used in experiments to understand the combined effect of at least two different factors on a dependent variable. These include the main effects that could be tested individually for each factor, using t-tests, as well as an additional source of variation—the interaction. Typically, participants are randomly allocated to a control or treatment group, and a placebo or experimental treatment applied. In this case, it is possible to study the main effect of each treatment, as well as their interaction (i.e., based on a significant effect at a specific combination of levels). Factors must lend themselves to categorization in terms of different levels (either naturally or artificially constructed); otherwise, multiple linear regression (Chapter 14) may be more appropriate.

The major assumptions for factorial ANOVA are the independence of observations and the homogeneity of variance, as per other techniques based on the general linear model. Fortunately, statistical packages generally provide methods to test homogeneity of variance (e.g., *Levene's test*, used below), while the independence of observations is an issue that is generally dealt with at the experimental design stage.

Two common types of factorial designs are $a \times b$ (two-way) and $a \times b \times c$ (three-way). Main effects and interactions can be tested for statistical significance by using the appropriate factorial ANOVA test. In this section, you will learn when and how to apply factorial ANOVA to determine whether population means differ between groups based on a number of different independent variables.

Two-Way ANOVA

Physical performance measures often vary in populations, and declines in grip strength, for instance, may be correlated with a number of different clinical conditions. Your research team is interested in studying how two factors, gender and alcohol consumption, are related to grip strength. That is, you want to answer three research questions:

- Does gender influence grip strength?
- Does alcohol consumption influence grip strength?
- Do gender and alcohol consumption interact to influence grip strength?

In this case, the *interaction* expresses a hypothesis that gender modifies or qualifies the relationship between alcohol consumption and grip strength. Since the independent variables (gender and alcohol consumption) divide neatly into distinct categories, hypotheses can be drawn that specify the direction of any effect; for example:

- Does the male population generally have greater grip strength than the female population?

- Does the population of alcohol consumers have lower grip strength than the teetotaler population?
- Does the effect of low alcohol consumption on grip strength depend on gender? In other words, is grip strength significantly lower for female alcohol consumers compared to male teetotalers?

Geometrically, it is easy to determine if an interaction exists, since the dependent variable means across all levels will be parallel for different IVs if an interaction does not exist, or intersecting if there is an interaction.

Figure 13-1 demonstrates a clear difference between the DV (grip strength) for two IVs (gender and number of alcoholic drinks), but no interaction between the two DVs. In other words, women tend to have lower grip strength than men, and drinking alcohol tends to reduce grip strength.

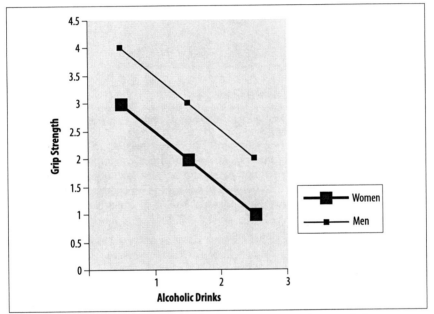

Figure 13-1. Main effects for gender and drink consumption on grip strength, but no interaction

Conversely, if there were an interaction between the two IVs, as shown in Figure 13-2, then the two lines would cross (or at least not be parallel). In this case, there is an interaction because differences appear in grip strength; low grip strength is associated with being female and having moderate alcohol consumption, and with being male and having high alcohol consumption, while high grip strength is associated with being male and having moderate alcohol consumption, and being female and having high alcohol consumption.

In experimental disciplines, it would be normal practice to define groups only where there has been an experimental manipulation. For example, alcohol consumption, as a category, would be based on two randomly assigned groups,

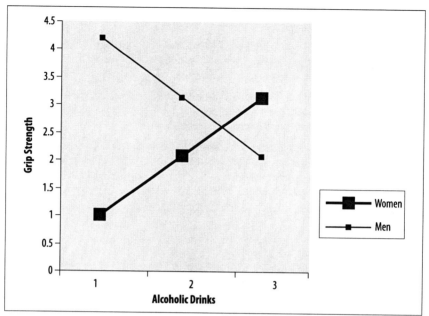

Figure 13-2. Interaction between gender and drink consumption on grip strength

one receiving alcohol, and one a control. But in prospective, life course studies, for example, such variables cannot be ethically artificially manipulated over time, and the quality of experimental control would be very low indeed. Other useful groups, such as gender, are also difficult to reassign in an experimental paradigm. The message is to always treat the results of ANOVA in a manner consistent with the data collection.

Table 13-1 shows sample data for the first 12 cases collected in the grip strength lab. Six women and six men had their grip strength measured, and each gender group also had three drinkers and three nondrinkers (defined as drinking at least weekly, or never drinking).

Table 13-1. Relationship between grip strength (DV) and gender and alcohol consumption (IVs)

Gender	Alcohol	Grip strength (psi)
Female	Weekly	19
Female	Weekly	20
Female	Weekly	21
Female	Never	30
Female	Never	25
Female	Never	28
Male	Weekly	31
Male	Weekly	30
Male	Weekly	35

Table 13-1. Relationship between grip strength (DV) and gender and alcohol consumption (IVs) (continued)

Gender	Alcohol	Grip strength (psi)
Male	Never	32
Male	Never	35
Male	Never	32

The two main effects are testing two mean population differences, based on the null hypothesis:

- $\overline{\text{Female}} - \overline{\text{Male}} = 0$, where the sample means are 23.83 and 32.5, respectively
- $\overline{\text{Weekly}} - \overline{\text{Never}} = 0$, where the sample means are 26.00 and 30.33, respectively

To test the interaction, the sample means are compared as follows:

- $\overline{\text{Female, Weekly}} = 20$
- $\overline{\text{Female, Never}} = 27.67$
- $\overline{\text{Male, Weekly}} = 32$
- $\overline{\text{Male, Never}} = 33$

If a significant effect is found, one way to determine the direction is to plot the means. After a two-way ANOVA is conducted using a statistical package, a number of tables are generally produced, but the key results are shown in Table 13-2. Normally, you can ignore the F-tests for the intercept and the corrected model, and focus on the results for each main effect and interaction. The coefficient of determination, as applied to the underlying linear model, shows $R^2 = 0.901$ (or adjusted $R^2 = 0.864$). This represents the amount of variation in the DV accounted for by the two significant main effects and the interaction, i.e., 90.1% of variation is accounted for. A more conservative R^2 is provided by the multiple correlation coefficient, which is discussed in detail in Chapter 14.

R can also be used as an indicator of *effect size*, with values of $R = 0.10, 0.36, 0.51$ indicating small, medium, or large effects respectively. Other statistics (such as eta) can also be computed to indicate effect size, and many disciplines now require effect sizes to be reported in academic journals. The important point, in terms of inference, is not just pursuing ANOVA as a means to determining whether there are "statistically significant differences" between groups, but whether these differences are meaningful.

All three effects tested in the design are significant:

Gender main effect: $F(1, 8) = 52.00$, $p < 0.001$
> The direction of the effect shows that women generally have lower grip strength than men.

Alcohol main effect: $F(1, 8) = 13.00$, $p = 0.007$
> The direction of the effect shows that alcohol reduces grip strength.

Gender × alcohol interaction: F(1, 8) = 7.692, p = 0.024

The interaction shows that gender and alcohol interact, with female alcohol drinkers having much worse grip strength than male nondrinkers.

Levene's test, which is a test of the assumption of homogeneity of error variances, was also performed. The test result was $F(3, 8) = 1.273$, $p = 0.35$, which is not significant. Therefore, the error variance of the DV across all groups is equal, and the ANOVA has not violated this assumption (Table 13-2).

Table 13-2. ANOVA testing differences in grip strength (DV) for gender and alcohol consumption (IVs)

Source	Sum of squares	df	Mean square	F	Sig.
Corrected model	315.000	3	105.000	24.231	0.000
Intercept	9520.333	1	9520.333	2197.000	0.000
Gender	225.333	1	225.333	52.000	0.000*
Alcohol	56.333	1	56.333	13.000	0.007*
Gender * alcohol	33.333	1	33.333	7.692	0.024*
Error	34.667	8	4.333		
Total	9870.000	12			
Corrected total	349.667	11			

Three-Way ANOVA

The two-way factorial model can easily be extended to three factors. After demonstrating significant main effects for gender and alcohol consumption on grip strength, your research team investigates other possible factors that may influence grip strength. In the literature, there appears to be a lot of discussion about the influence of age on grip strength, with a marked decline appearing after the age of 40. So, you decide to add an age category (below 40 or above 40) to determine if age has any influence or as much influence on grip strength, as the other factors.

Table 13-3 shows the raw data for the study.

Table 13-3. Relationship between grip strength (DV) and gender, alcohol consumption, and age (IVs)

Gender	Alcohol	Grip strength (psi)	Age
Female	Weekly	19	Below 40
Female	Weekly	20	Above 40
Female	Weekly	21	Below 40
Female	Never	30	Above 40
Female	Never	25	Below 40
Female	Never	28	Above 40

Table 13-3. Relationship between grip strength (DV) and gender, alcohol consumption, and age (IVs) (continued)

Gender	Alcohol	Grip strength (psi)	Age
Male	Weekly	31	Below 40
Male	Weekly	30	Above 40
Male	Weekly	35	Below 40
Male	Never	32	Above 40
Male	Never	35	Below 40
Male	Never	32	Above 40

Table 13-4 shows the three-factor version of the ANOVA table that corresponds to the two-factor case shown in Table 13-2. The results can be summarized as follows:

- A significant main effect was found for gender, $F(1, 4) = 72.00$, $p = 0.001$.
- A significant main effect was found for alcohol consumption, $F(1, 4) = 18.00$, $p = 0.013$.
- No significant main effect was found for age, $F(1,4) = 0.22$, $p = 0.662$.
- No significant interaction was found for gender×alcohol, $F(1,4) = 5.56$, $p = 0.078$.
- No significant interaction was found for gender×age, $F(1, 4) = 5.56$, $p = 0.078$.
- No significant interaction was found for age×alcohol, $F(1, 4) = 0.89$, $p = 0.34$.
- No significant interaction was found for age×alcohol×gender, $F(1, 4) = 0.89$, $p = 0.34$.

Note that those are two-way interactions for each pair of factors as well as a three-way interaction. In this case, the main effects of gender and alcohol were replicated, as per the previous example, but no interaction was observed (in the previous example, there was an interaction observed between alcohol consumption and gender). In this example, $R^2 = 0.966$ represents the amount of variation in the DV accounted for by the two significant main effects, and even the more conservative adjusted $R^2 = 0.906$. This represents a very large effect size.

Table 13-4. ANOVA testing differences in grip strength (DV) based on gender, alcohol consumption, and age (IVs)

Source	Type III Sum of squares	df	Mean square	F	Sig.
Corrected model	337.667	7	48.238	16.079	0.009
Intercept	8362.667	1	8362.667	2787.556	0.000
Gender	216.000	1	216.000	72.000	0.001*
Alcohol	54.000	1	54.000	18.000	0.013*
Age	0.667	1	0.667	0.222	0.662
Gender * alcohol	16.667	1	16.667	5.556	0.078
Gender * age	16.667	1	16.667	5.556	0.078

Table 13-4. ANOVA testing differences in grip strength (DV) based on gender, alcohol consumption, and age (IVs) (continued)

Source	Type III Sum of squares	df	Mean square	F	Sig.
Alcohol * age	2.667	1	2.667	0.889	0.399
Gender * alcohol * age	2.667	1	2.667	00.889	0.399
Error	12.000	4	3.000		
Total	9870.000	12			
Corrected total	349.667	11			

MANOVA

Factorial ANOVA deals with cases in which there are multiple IVs, while MANOVA allows designs with multiple DVs to be analyzed. Continuing with the physical performance example, grip strength is often analyzed along with other measures, such as the number of steps that can be mounted and demounted in a given timeframe (usually 30-second intervals). While all measures of physical performance are expected to be correlated to some extent, there may be good reasons (clinical or otherwise) for treating the effects separately. Otherwise, if the measures are highly correlated (known as *multicollinearity*), and make no significant contribution to the model, then the best measure of performance may be selected, and the other measure not collected—this can reduce time, cost, and inconvenience to study participants.

MANOVA has many assumptions, such as independent observations, multivariate normality, homogeneity of variance, equality of group sizes (or at most, a 1.5 ratio between N of the smallest group and the largest group), and homogeneity of covariance for dependent variable pairs across all groups (which can be tested by Box's test of equality of covariance matrices).

Table 13-5 shows sample data in which there are two independent variables (gender and alcohol consumption) and two dependent variables (grip strength and number of steps mounted in 30s). The MANOVA model can be used to test the following effects:

- Main effects of the independent variables (i.e., do alcohol and/or gender independently influence physical performance)
- Interactions between the independent variables (i.e., do alcohol and/or gender modulate the effect of each other to influence physical performance)
- The degree of relationship between the dependent variables
- How significant are the dependent variables, in terms of being affected by gender or alcohol consumption

Again, Levene's test could be used to determine whether the assumption of the equality of error variances has been violated. If the assumption is violated, a *transformation* may be used to meet the assumption.

Table 13-5. Relationship between grip strength and steps (DVs) and gender and alcohol consumption (IVs)

Gender	Alcohol	Grip strength (psi)	Steps
Female	Weekly	19	Below 40
Female	Weekly	20	Above 40
Female	Weekly	21	Below 40
Female	Never	30	Above 40
Female	Never	25	Below 40
Female	Never	28	Above 40
Male	Weekly	31	Below 40
Male	Weekly	30	Above 40
Male	Weekly	35	Below 40
Male	Never	32	Above 40
Male	Never	35	Below 40
Male	Never	32	Above 40

To test the multivariate case, a multivariate F-test is used to evaluate the hypotheses. The results of the multivariate F-test—shown in Table 13-6—indicate a main effect for each of the independent variables, but no interaction:

- For gender, $F(2.00, 7.00) = 52.23$, $p = 0.000$
- For alcohol, $F(2.00, 7.00) = 10.94$, $p = 0.007$
- For gender $\times$ alcohol, $F(2.00, 7.00) = 4.70$, $p = 0.051$

Notice that there are multiple F-tests computed for several different statistics: *Wilks's Lambda, Pillai's Trace, Hotelling's Trace,* and *Roy's Largest Root.* In the case of two dependent variables, these statistics will produce the same F values. Normally, Pillai's Trace is used, but only if the result of *Box's test* is not significant.

Tests of between-subjects effects are shown in Tables 13-6 and 13-7, to highlight the importance of each dependent variable. Interpreting these results is similar to analyzing two factorial designs with a separate DV; an R^2 is computed for each model, based on each dependent variable ($R^2 = 0.901$ for grip strength and $R^2 = 0.909$ for steps). The results are quite interesting:

- There is a significant main effect for gender on grip strength and steps.
- There is a significant main effect for alcohol on grip strength and steps.
- There is a significant interaction between gender and alcohol on grip strength.
- There is no significant interaction between gender and alcohol on steps.

By examining the univariate F-tests, you can determine which dependent variable made the greatest contribution to the significant multivariate result; in this case, grip strength made a more significant contribution than steps. Generally, unless the multivariate F-test is significant, the results of the univariate tests are disregarded.

In the ideal case, dependent variables are completely uncorrelated (orthogonal). If this is not possible, orthogonal factors can be produced using advanced techniques like orthogonal decomposition, which can be implemented by using principal components analysis (presented in Chapter 17).

Note that MANOVA is a complex procedure, including variations incorporating analysis of covariates (MANCOVA) and mixed MANOVA or doubly multivariate MANOVA where multiple DVs are collected on different occasions. These techniques are extremely powerful and commonly used in experimental disciplines. You should refer to a specialist multivariate analysis text for further details.

Table 13-6. ANOVA testing differences in grip strength and steps (DVs) based on gender and alcohol consumption (IVs) (multivariate tests)

Effect		Value	F	Hypothesis df	Error df	Sig.
Intercept	Pillai's Trace	0.999	3556.956	2.000	7.000	0.000
	Wilks's Lambda	0.001	3556.956	2.000	7.000	0.000
	Hotelling's Trace	1016.273	3556.956	2.000	7.000	0.000
	Roy's Largest Root	1016.273	3556.956	2.000	7.000	0.000
Gender	Pillai's Trace	0.937	52.233	2.000	7.000	0.000*
	Wilks's Lambda	0.063	52.233	2.000	7.000	0.000*
	Hotelling's Trace	14.924	52.233	2.000	7.000	0.000*
	Roy's Largest Root	14.924	52.233	2.000	7.000	0.000*
Alcohol	Pillai's Trace	0.758	10.940	2.000	7.000	0.007*
	Wilks's Lambda	0.242	10.940	2.000	7.000	0.007*
	Hotelling's Trace	3.126	10.940	2.000	7.000	0.007*
	Roy's Largest Root	3.126	10.940	2.000	7.000	0.007*
Gender * alcohol	Pillai's Trace	0.573	4.700	2.000	7.000	0.051
	Wilks's Lambda	0.427	4.700	2.000	7.000	0.051
	Hotelling's Trace	1.343	4.700	2.000	7.000	0.051
	Roy's Largest Root	1.343	4.700	2.000	7.000	0.051

Table 13-7. Tests of between-subjects effects

Source	Dependent variable	Type III Sum of squares	df	Mean square	F	Sig.
Corrected model	Grip	315.000(a)	3	105.000	24.231	0.000
	Steps	147.000(b)	3	49.000	26.727	0.000
Intercept	Grip	9520.333	1	9520.333	2197.000	0.000
	Steps	10680.333	1	10680.333	5825.636	0.000
Gender	Grip	225.333	1	225.333	52.000	0.000*
	Steps	120.333	1	120.333	65.636	0.000*
Alcohol	Grip	56.333	1	56.333	13.000	0.007*
	Steps	21.333	1	21.333	11.636	0.009*
Gender * alcohol	Grip	33.333	1	33.333	7.692	0.024*
	Steps	5.333	1	5.333	2.909	0.126
Error	Grip	34.667	8	4.333		
	Steps	14.667	8	1.833		
Total	Grip	9870.000	12			
	Steps	10842.000	12			
Corrected total	Grip	349.667	11			
	Steps	161.667	11			

ANCOVA

Analysis of Covariance (ANCOVA) is a variation of factorial ANOVA that allows the potentially confounding effect of a covariate to be canceled out in ANOVA. You may wonder why, if there is a factor that influences variability in the DV, it is not just included in the model as a normal factor. The answer is that the DV can be adjusted for changes that would not have occurred if the confounding effect of the covariate had been in place.

Continuing with the grip strength example, the research team becomes concerned that frequency of attending the gym may be influencing the observed relationship between gender and alcohol consumption (IVs) and grip strength (DV). An ANCOVA is carried out using a categorical gym attendance variable with two levels (never attends, has attended at least once in the previous week), with the data shown in Table 13-8. To test the assumption of homogeneity of error variance, Levene's test is computed, with $F(3,8) = 1.290$, $p = 0.342$, which is not significant, so the assumption holds. Proceeding with the ANCOVA, you find that the previously established relationship between gender, alcohol consumption, and grip strength holds true, and that the covariate (gym attendance) is not significant, as shown in Table 13-9:

- For gender, $F(1, 7) = 45.55$, $p = 0.000$
- For alcohol, $F(1, 7) = 10.319$, $p = 0.015$
- For gender × alcohol, $F(1, 7) = 6.739$, $p = 0.036$
- For gym atttendance, $F(1, 7) = 0.008$, $p = 0.929$

So, gym attendance, in itself, does not have a significant effect on the DV, and the previously established significant main effects and interaction still remain significant after ANCOVA. $R^2 = 0.901$ and adjusted $R^2 = 0.844$, indicating a large effect size, and a very significant proportion of variance in the DV being accounted for by the main effects of gender and alcohol consumption, and their interaction.

Table 13-8. Relationship between grip strength (DV) and gender and alcohol consumption (IVs) and covariate (frequency of gym attendance)

Gender	Alcohol	Grip strength (psi)	Gym
Female	Weekly	19	Never
Female	Weekly	20	Never
Female	Weekly	21	Weekly
Female	Never	30	Never
Female	Never	25	Weekly
Female	Never	28	Weekly
Male	Weekly	31	Never
Male	Weekly	30	Never
Male	Weekly	35	Weekly
Male	Never	32	Weekly
Male	Never	35	Never
Male	Never	32	Weekly

Table 13-9. ANOVA testing differences in grip strength (DV) between gender and alcohol consumption (IVs) with a covariate (frequency of gym attendance) (between-subjects effects)

Source	Type III Sum of squares	df	Mean square	F	Sig.
Corrected model	315.042	4	78.760	15.923	0.001
Intercept	867.191	1	867.191	175.317	0.000
Gym	0.042	1	0.042	0.008	0.929
Gender	225.333	1	225.333	45.555	0.000*
Alcohol	51.042	1	51.042	10.319	0.015*
Gender * alcohol	33.333	1	33.333	6.739	0.036*
Error	34.625	7	4.946		
Total	9870.000	12			
Corrected total	349.667	11			

Repeated Measures ANOVA

As described in Chapter 5, a repeated measures design involves measuring the same dependent variables across two or more intervals (usually time intervals). In medical and behavioral research, pre- and post-treatment designs are very common, where the effect of a drug or some other intervention needs to be assessed. In these designs, participants act as their own controls, minimizing error due to variation between individuals.

In this example, imagine you are a psychiatrist interested in the effects of a new anti-anxiety drug on cognitive performance, since clinical observations indicate that it has some detrimental effect. The two research questions are:

- Does taking the drug reduce cognitive performance?
- Are there some doses that are relatively safer than others?

To answer these questions, different doses of the drug can be randomly administered to a sample, with four different treatment levels (100, 200, 300, and 400mg) and a baseline (placebo) condition, as shown in Table 13-10. If any of the treatments are significantly different to the baseline, the first research question can be answered in the affirmative. The second research question can then be investigated to determine which doses are significantly different to the baseline. Regression analysis is also suited to these designs by allowing for different models (linear and nonlinear) to be fitted to determine the dose-response relationship.

Table 13-10. Effect of anti-anxiety drug (IV) on cognitive performance (DV)

Baseline	100mg	200mg	300mg	400mg
50	48	45	44	44
42	40	36	35	34
56	50	48	47	47
49	48	45	43	42
50	47	48	43	43
43	41	37	36	35
56	53	49	48	47
59	58	57	56	53
48	46	43	42	39
43	42	36	35	34

While the assumptions for repeated measures ANOVA are similar to other ANOVA designs, an additional complication is *sphericity*, which is the assumption of homogeneity of variances and covariances within each level of the within-subjects factor. In this example, sphericity means that if the cognitive performance scores for 100mg were subtracted from the 200mg scores, the variances would be equal. For many real-world data sets, this assumption is often violated; you can test whether a violation has occurred by performing *Mauchly's Test of Sphericity*. Alternatively, the degrees of freedom can be adjusted by using the

Greenhouse-Geisser, Huynh-Feldt, or *lower bound* corrections. The effects can be tested using univariate or multivariate techniques, with the latter being appropriate if the assumptions for repeated measures ANOVA are not met.

The results of the analysis are shown in Tables 13-11 through 13-14. Firstly, Mauchly's Test of Sphericity shows that no violation of the assumption of homogeneity of variance and covariance has occurred, $W = 0.353$, $p = 0.572$ with $df = 9$. Secondly, if the assumption had been violated, the Greenhouse-Geisser, Huynh-Feldt, or lower-bound corrections shown could be applied to the degrees of freedom for the design. Thirdly, multivariate F-tests show that there is a significant effect of administering the drug on cognitive performance, $F(4, 6) = 92.568$, $p = 0.000$. By plotting the means, you can easily see that the direction is negative, i.e., cognitive performance is reduced by taking the drug. Finally, tests of within-subjects contrasts show significant linear and quadratic components, with $F(1, 9) = 367.347$, $p = 0.000$, and $F(1, 9) = 6.446$, $p = 0.032$ respectively. Thus, at the $p < 0.01$ significance level, you would say that the relationship between the DV and IV was strictly linear, but at the $p < 0.05$ level, there was both a linear and a quadratic component. This indicates that the effect is not just a simple reduction in cognitive performance as a function of drug dosage, but that there is a significant nonlinear effect as well. Further modeling may indicate why this additional effect is observed, and whether the quadratic component is concave or convex.

Table 13-11. ANOVA testing effect on cognitive performance (DV) after administration of anti-anxiety drug (IV) (Mauchly's Test of Sphericity)

Within-subjects effect	Mauchly's W	Approx. chi-square	df	Sig.
Drug	0.353	7.729	9	0.572

Table 13-12. Epsilon

Epsilon		
Greenhouse-Geisser	Huynh-Feldt	Lower-Bound
0.677	0.997	0.250

Table 13-13. Multivariate tests

Effect		Value	F	Hypothesis df	Error df	Sig.
Drug	Pillai's Trace	0.984	92.568	4.00	6.000	0.000*
	Wilks's Lambda	0.016	92.568	4.000	6.000	0.000*
	Hotelling's Trace	61.712	92.568	4.000	6.000	0.000*
	Roy's Largest Root	61.712	92.568	4.000	6.000	0.000*

Table 13-14. Tests of within-subjects contrasts

Source	Drug	Type III Sum of squares	df	Mean square	F	Sig.
Drug	Linear	400.000	1	400.000	367.347	0.000*
	Quadratic	10.314	1	10.314	6.446	0.032*
	Cubic	1.000	1	1.000	1.731	0.221
	Order 4	1.286	1	1.286	0.964	0.352
Error (drug)	Linear	9.800	9	1.089		
	Quadratic	14.400	9	1.600		
	Cubic	5.200	9	0.578		
	Order 4	12.000	9	1.333		

Mixed Designs

In the previous section, you saw how useful a repeated measures design was in a study to investigate the effects of a drug on cognitive performance, where participants acted as their own controls. However, in many cases, you will also want to perform an experimental manipulation between subjects, or you may want to examine the effect across different natural groups. In this case, you can use a mixed design, with both between- and within-subjects factors. Additionally, you may need to control for the effect of a covariate; in the case of dose-response relationships in pharmacology or clinical studies, you may need to control for the confounding effects of the co-administration of other drugs. In this section, you will see how these scenarios can be implemented using a mixed design.

Within-Subjects and Between-Subjects

In this example, we extend the previous repeated measures design to examine whether the effect of the anti-anxiety drug (the within-subjects factor) has the same effect on both men and women (the between-subjects factor), or whether there is no difference. The data is shown in Table 13-15.

Table 13-15. Effect on cognitive performance (DV) after administration of anti-anxiety drug (WS IV) across two gender groups (BS IV)

Dose 1	Dose 2	Dose 3	Dose 4	Dose 5	Gender
50	48	45	44	44	Male
42	40	36	35	34	Male
56	50	48	47	47	Male
49	48	45	43	42	Male
50	47	48	43	43	Male
43	41	37	36	35	Female
56	53	49	48	47	Female
59	58	57	56	53	Female
48	46	43	42	39	Female
43	42	36	35	34	Female

The results for the multivariate tests are shown in Tables 13-16 through 13-20. Once again, Mauchly's Test of Sphericity was not significant, so the sphericity assumption was not violated. You can see that there is a significant effect for the drug factor, $F(4, 5) = 101.118$, $p = 0.000$, but no interaction with gender, $F(4, 5) = 1.946$, $p = 0.241$. This indicates that the within-subjects effect is present, as it was in the previous example, but that no significant difference was observed between men and women. These results were confirmed by the univariate tests of within-subjects effects, with $F(4, 32) = 91.486$, $p = 0.000$ for drug, and $F(4,32) = 1.180$, $p = 0.241$ for gender. The univariate test for the between-subjects factor (gender) was also not significant, $F(1, 8) = 0.13$, $p = 0.910$.

Table 13-16. ANOVA testing effect on cognitive performance (DV) after administration of anti-anxiety drug (WS IV) across two gender groups (BS IV) (Mauchly's Test of Sphericity)

Within-subjects effect	Mauchly's W	Approx. chi-square	df	Sig.	Epsilon(a)		
					Greenhouse-Geisser	Huynh-Feldt	
Drug	.239	9.181	9	0.436	0.607	1.000	0.250

Table 13-17. Multivariate tests

Effect		Value	F	Hypothesis df	Error df	Sig
Drug	Pillai's Trace	0.988	101.118	4.000	5.000	0.000*
	Wilks's Lambda	0.012	101.118	4.000	5.000	0.000*
	Hotelling's Trace	80.894	101.118	4.000	5.000	0.000*
	Roy's Largest Root	80.894	101.118	4.000	5.000	0.000*
Drug * gender	Pillai's Trace	0.609	1.946	4.000	5.000	0.241
	Wilks's Lambda	0.391	1.946	4.000	5.000	0.241
	Hotelling's Trace	1.557	1.946	4.000	5.000	0.241
	Roy's Largest Root	1.557	1.946	4.000	5.000	0.241

Table 13-18. Test of within-subjects effects

Source		Type III Sum of squares	df	Mean square	F	Sig.
Drug	Sphericity assumed	412.600	4	103.150	91.486	0.000*
	Greenhouse-Geisser	412.600	2.429	169.844	91.486	0.000*
	Huynh-Feldt	412.600	4.000	103.150	91.486	0.000*
	Lower-bound	412.600	1.000	412.600	91.486	0.000*

Table 13-18. Test of within-subjects effects (continued)

Source		Type III Sum of squares	df	Mean square	F	Sig.
Drug * gender	Sphericity assumed	5.320	4	1.330	1.180	0.338
	Greenhouse-Geisser	5.320	2.429	2.190	1.180	0.336
	Huynh-Feldt	5.320	4.000	1.330	1.180	0.338
	Lower-bound	5.320	1.000	5.320	1.180	0.309
Error (drug)	Sphericity assumed	36.080	32	1.128		
	Greenhouse-Geisser	36.080	19.434	1.857		
	Huynh-Feldt	36.080	32.000	1.128		
	Lower-bound	36.080	8.000	4.510		

Table 13-19. Tests of within-subjects contrasts

Source	Drug	Type III Sum of squares	df	Mean square	F	Sig.
Drug	Linear	400.000	1	400.000	363.636	0.000*
	Quadratic	10.314	1	10.314	6.171	0.038*
	Cubic	1.000	1	1.000	1.538	0.250
	Order 4	1.286	1	1.286	1.181	0.309
Drug * gender	Linear	1.000	1	1.000	0.909	0.368
	Quadratic	1.029	1	1.029	0.615	0.455
	Cubic	0.000	1	.000	0.000	1.000
	Order 4	3.291	1	3.291	3.024	0.120
Error (drug)	Linear	8.800	8	1.100		
	Quadratic	13.371	8	1.671		
	Cubic	5.200	8	0.650		
	Order 4	8.709	8	1.089		

Table 13-20. Tests of between-subjects effects

Source	Type III Sum of squares	df	Mean square	F	Sig.
Intercept	102152.000	1	102152.000	478.710	0.000
Gender	2.880	1	2.880	0.013	0.910
Error	1707.120	8	213.390		

Within-Subjects and Between-Subjects and Covariates

If there is a potential covariate that may be confounding an ANOVA result, it's always best to include this in the analysis so that its effect can be accounted for and adjustments made. Following with the mixed design presented previously, you suspect that prior administration of another drug (Drug B) may be giving rise to the observed effect of the anti-anxiety medication on cognitive performance. So, you decide to explicitly enter the prior administration of Drug B into the model as a covariate. Table 13-21 shows the new data.

Table 13-21. *Effect on cognitive performance (DV) after administration of anti-anxiety drug (WS IV) across two gender groups (BS IV) taking into account covariate (prior administration of Drug B)*

Dose 1	Dose 2	Dose 3	Dose 4	Dose 5	Gender	Drug B
50	48	45	44	44	Male	Yes
42	40	36	35	34	Male	No
56	50	48	47	47	Male	Yes
49	48	45	43	42	Male	No
50	47	48	43	43	Male	Yes
43	41	37	36	35	Female	No
56	53	49	48	47	Female	Yes
59	58	57	56	53	Female	No
48	46	43	42	39	Female	Yes
43	42	36	35	34	Female	No

The results of the analysis are shown in Tables 13-22 through 13-26. Firstly, you perform Mauchly's Test of Sphericity and conclude that no violation has occurred, $W = 0.204$, $p = 0.496$ for $df = 9$.

Next, multivariate tests are performed. In addition to the previous tests for a main effect of drug (within-subjects), and a drug (within-subjects) $\times$ gender (between-subjects) interaction, a drug (within-subjects) $\times$ Drug B (covariate) is also tested. The interaction is not significant, $F(4, 4) = 0.791$, $p = 0.587$.

Univariate test results for within-subjects effects indicate no drug $\times$ gender or Drug $\times$ Drug B interactions, with $F(4, 28) = 0.787$, $p = 0.352$ and $F(4, 28) = 0.873$, $p = 0.587$ respectively, but a significant main effect for drug, $F(4, 32) = 9.913$, $p = 0.000$.

In terms of the direction of effect, tests of within-subjects contrasts indicate a significant linear component for drug, $F(1, 7) = 32.770$, $p = 0.001$, but no other component was significant for the drug main effect, nor was any other component significant for any of the interactions (Drug $\times$ Drug B and drug $\times$ gender).

Since you are testing a between-subjects factor, and a covariate, between-subjects are also performed. No significant effects were observed for either Drug B or gender, $F(1, 7) = 1.074$, $p = 0.335$ and $F(1, 7) = 0.103$, $p = 0.757$ respectively.

Table 13-22. ANOVA testing effect on cognitive performance (DV) after administration of anti-anxiety drug (WS IV) across two gender groups (BS IV) taking into account covariate (prior administration of Drug B) (Mauchly's Test of Sphericity)

Within-subjects effect	Mauchly's W	Approx. chi-square	df	Sig.	Epsilon		
					Greenhouse-Geisser	Huynh-Feldt	
Drug	0.204	8.599	9	0.496	0.539	1.000	0.250

Table 13-23. Multivariate test

Effect		Value	F	Hypothesis df	Error df	Sig.
Drug	Pillai's Trace	0.905	9.498	4.000	4.000	0.025*
	Wilks's Lambda	0.095	9.498	4.000	4.000	0.025*
	Hotelling's Trace	9.498	9.498	4.000	4.000	0.025*
	Roy's Largest Root	9.498	9.498	4.000	4.000	0.025*
Drug * Drug_B	Pillai's Trace	0.442	.791	4.000	4.000	0.587
	Wilks's Lambda	0.558	.791	4.000	4.000	0.587
	Hotelling's Trace	0.791	.791	4.000	4.000	0.587
	Roy's Largest Root	0.791	.791	4.000	4.000	0.587
Drug * gender	Pillai's Trace	0.600	1.502	4.000	4.000	0.352
	Wilks's Lambda	0.400	1.502	4.000	4.000	0.352
	Hotelling's Trace	1.502	1.502	4.000	4.000	0.352
	Roy's Largest Root	1.502	1.502	4.000	4.000	0.352

Table 13-24. Tests of within-subjects effects

Source		Type III Sum of squares	df	Mean square	F	Sig.
Drug	Sphericity assumed	45.931	4	11.483	9.913	0.000*
	Greenhouse-Geisser	45.931	2.155	21.318	9.913	0.002*

Table 13-24. Tests of within-subjects effects (continued)

Source		Type III Sum of squares	df	Mean square	F	Sig.
	Huynh-Feldt	45.931	4.000	11.483	9.913	0.000*
	Lower-bound	45.931	1.000	45.931	9.913	0.016*
Drug * Drug_B	Sphericity assumed	3.647	4	0.912	0.787	0.543
	Greenhouse-Geisser	3.647	2.155	1.693	0.787	0.482
	Huynh-Feldt	3.647	4.000	0.912	0.787	0.543
	Lower-bound	3.647	1.000	3.647	0.787	0.404
Drug * gender	Sphericity assumed	4.047	4	1.012	0.873	0.492
	Greenhouse-Geisser	4.047	2.155	1.878	0.873	0.445
	Huynh-Feldt	4.047	4.000	1.012	0.873	0.492
	Lower-bound	4.047	1.000	4.047	0.873	0.381
Error (drug)	Sphericity assumed	32.433	28	1.158		
	Greenhouse-Geisser	32.433	15.082	2.150		
	Huynh-Feldt	32.433	28.000	1.158		
	Lower-bound	32.433	7.000	4.633		

Table 13-25. Tests of within-subjects contrasts

Source	Drug	Type III Sum of squares	df	Mean square	F	Sig.
Drug	Linear	41.002	1	41.002	32.770	0.001*
	Quadratic	3.796	1	3.796	2.148	0.186
	Cubic	0.435	1	0.435	0.733	0.420
	Order 4	0.699	1	0.699	0.685	0.435
Drug * Drug_B	Linear	0.042	1	0.042	0.033	0.860
	Quadratic	1.001	1	1.001	0.567	0.476
	Cubic	1.042	1	1.042	1.754	0.227
	Order 4	1.562	1	1.562	1.530	0.256
Drug * gender	Linear	1.042	1	1.042	0.833	0.392
	Quadratic	0.630	1	0.630	0.356	0.569
	Cubic	0.042	1	0.042	0.070	0.799
	Order 4	2.334	1	2.334	2.286	0.174

Table 13-25. Tests of within-subjects contrasts (continued)

Source	Drug	Type III Sum of squares	df	Mean square	F	Sig.
Error (drug)	Linear	8.758	7	1.251		
	Quadratic	12.370	7	1.767		
	Cubic	4.158	7	0.594		
	Order 4	7.146	7	1.021		

Table 13-26. Tests of between-subjects effects

Source	Type III Sum of squares	df	Mean square	F	Sig.
Intercept	12893.858	1	12893.858	60.982	0.000
Drug_B	227.070	1	227.070	1.074	0.335
Gender	21.870	1	21.870	0.103	0.757
Error	1480.050	7	211.436		

14

Multiple Linear Regression

In Chapter 12, you learned how to use bivariate linear regression to build simple linear models suitable for characterization of the relationship between two variables (typically one dependent variable and one independent variable). Clearly, many variables in the physical world can have multiple IVs independently, accounting for some portion of variance in the DV. Note the difference between "single linear regression" and "multiple linear regression"—the former refers to a setting in which there are multiple *responses* in a response vector emphasizing that we are in the setting with a *single* outcome but multiple predictors. This chapter discusses multiple linear regression as an extension of simple linear regression. Assumptions specific to multivariate regression, such as *multicollinearity* among predictor variables, are discussed and methods for model-building are presented.

Multiple Regression Models

The use of simple linear regression models and the bivariate correlation coefficient and its square (the coefficient of determination) are useful for illustrating simple examples; in reality, very few physical systems or fields of interest rely on a single independent and dependent variable pair. Consider models used to study climate change, such as General Circulation Models (GCMs) and even more sophisticated Atmosphere-Ocean General Circulation Models (AOGCMs). These models have been developed over the past 30 years to allow the increasingly accurate forecast of weather patterns. The models used involve understanding and quantifying relationships between potentially hundreds and thousands of variables in many different qualitative categories. For example, in the mid-1970s, models focused on variables derived from atmospheric conditions, while in the near future, models will be available that are based on atmospheric data combined with land surface, ocean and sea ice, sulphate and nonsulphate aerosol, carbon cycle, dynamic vegetation, and atmospheric chemistry data. By combining these additional sources of variation into a large-scale statistical model, predictions of weather activity of qualitatively different types has been made possible at different spatial and temporal scales.

There are two things to note about models of this type that are extremely important for the development of multiple linear regression models. Firstly, each new set of variables added to the model should uniquely and significantly account for some of the variation observed in the dependent variable; secondly, since you expect that some (many?) of the variables will be correlated, you need a technique that selects the set of variables that provide the best fit to the dependent variable. As you shall see, multiple linear regression algorithms provide these tools (and more!), allowing you to add new variables into a model to see if they make any difference to the variation accounted for. For example, if you look at the timescale of developments of the GCM and AOGCM models, each qualitative set of variables has been added over time and evaluated to determine if any new and useful information could be extracted from them vis-à-vis predictability of dependent variables. In cases where no additional information is gained, the variables concerned can be withdrawn from the model, and time and money saved by having to only collect the smallest number of variables possible to predict weather phenomena.

Formally, multiple linear regression models take the form:

$$Y = \beta + \alpha_1 X_1 + \alpha_2 X_2 + ... \alpha_n X_n$$

where Y is the dependent variable, β is the intercept, X_1, X_2, ... X_n are the independent variables, and α_1, α_2, ... α_n are the slopes. Note that the dependent variable (Y) and independent variables (X_1, X_2, ... X_n) are observed data, while the intercept (β) and slopes (α_1, α_2, ... α_n) are computed by the multiple linear regression algorithm. In algebraic terms, the goal of the algorithm is to *minimize* the following sum:

$$\sum_{i=1}^{N} (Y - \beta - \alpha_1 X_1 - \alpha_2 X_2 - ... - \alpha_n X_n)^2$$

For bivariate relationships, visualizing the relationship between two variables can be easily accomplished using a two-dimensional number plane. This is also possible for three-dimensional models (i.e., one dependent variable and two independent variables), but becomes much harder for higher-dimensional models.

In the following example, a three-dimensional model is used to illustrate how minimizing the expression above can be achieved by minimizing the residuals between the observed dependent variable value and its predicted value on the three-dimensional number plane $Y = \beta + \alpha_1 X_1 + \alpha_2 X_2$. Imagine that you are an atmospheric researcher interested in characterizing and predicting the relationship between average temperature at ground level, and the atmospheric levels of carbon dioxide (CO_2) and methane (CH_4), measured in parts per million (ppm) and parts per billion (ppb), respectively. The model assumes that changes in the surface temperature have increased linearly over the past 100 years, i.e., there is a correlation between year and surface temperature increase of 0.01 degrees Celsius per year. Thus, for characterization, the model would use surface temperature as the dependent variable, but you may use year as the dependent variable if you wanted to predict the concentrations of CO_2 and CH_4 in future years. Sample data is shown in Table 14-1.

Table 14-1. Sample data showing the relationship between temperature, CH_4 concentration, and CO_2 concentration

Temperature	CO_2	CH_4
20.1	288.8803	1475.822
20.2	292.5981	1490.691
20.3	297.8734	1492.851
20.4	291.7772	1507.829
20.5	305.7114	1531.721
20.6	313.7092	1518.571
20.7	324.1619	1539.442
20.8	304.6294	1547.516
20.9	325.738	1548.636
21	308.9492	1525.255

Figure 14-1 shows the relationship between temperature, CO_2, and CH_4. You can see that all three variables tend to rise with each other in a positive manner, which corresponds with generally held theories that rises in CO_2 and CH_4 are associated with rising surface temperatures.

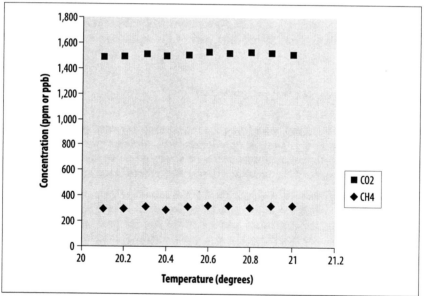

Figure 14-1. Relationship between CO_2, CH_4, and temperature

What you hope to achieve by performing multiple linear regression is to account for more variation in the DV by using multiple IVs. However, the variation accounted for by each factor will be additive only if the two factors are orthogonal, so you can't simply take the coefficient of determination from the results of two simple linear regression analyses and add them together to arrive at the combined variation accounted for in the multivariate case. However, if you have used one of the orthogonal decomposition techniques described in Chapter 16, such as principal component analysis, then the factors would indeed be *orthogonal* and the variance additive.

To illustrate, construct two simple linear regression models and perform linear regression on them, and compare the result with the multivariate case, as shown in Examples 14-1 through 14-3, which show the bivariate and multivariate regression models for temperature, CO_2, and CH_4.

Example 14-1. Bivariate: temperature $= \beta + \alpha CO_2$

Source	SS	df	MS		Number of obs =	10
					F(1, 8) =	25.77
Model	0.62954855	1	0.62954855		Prob > F =	0.0010
Residual	0.195450687	8	0.024431336		R-squared =	0.7631
					Adj R-squared =	0.7335
Total	0.824999237	9	0.091666582		Root MSE =	0.15631

temperature	Coef.	Std. Err.	t	P>|t|	[95% Conf. Interval]	
co2	0.0104444	0.0020575	5.08	0.001	0.0056998	0.0151891
_cons	4.697114	3.123358	1.50	0.171	-2.505362	11.89959

Example 14-2. Bivariate: temperature $= \beta + \alpha CH_4$

Source	SS	df	MS		Number of obs =	10
					F(1, 8) =	12.18
Model	0.498017047	1	0.498017047		Prob > F =	0.0082
Residual	0.32698219	8	0.040872774		R-squared =	0.6037
					Adj R-squared =	0.5541
Total	0.824999237	9	0.091666582		Root MSE =	0.20217

temperature	Coef.	Std. Err.	t	P>|t|	[95% Conf. Interval]	
ch4	0.0180594	0.0051737	3.49	0.008	0.0061289	0.0299899
_cons	15.03461	1.581345	9.51	0.000	11.38802	18.6812

Example 14-3. Multivariate: temperature $= \beta + \alpha_1 CO_2 + \alpha_2 CH_4$

Source	SS	df	MS		Number of obs =	10
					F(2, 7) =	12.15
Model	0.640537185	2	0.320268593		Prob > F =	0.0053
Residual	0.184462052	7	0.026351722		R-squared =	0.7764
					Adj R-squared =	0.7125
Total	0.824999237	9	0.091666582		Root MSE =	0.16233

temperature	Coef.	Std. Err.	t	P>|t|	[95% Conf. Interval]	
co2	0.0046003	0.007124	0.65	0.539	-0.0122452	0.0214458

```
 ch4 |  0.008522   0.0036645   2.33   0.053   -0.000143   0.0171871
_cons |  6.209997   4.001369    1.55   0.165   -3.251736   15.67173
```

The results contain two important inferential tests: univariate *t*-tests on each coefficient, as well as an overall F-test. These can be used for testing hypotheses that each IV makes a significant contribution to the model, or that the model as a whole is statistically significant in terms of the relationships it specifies between the DVs and IVs. The results of these tests should always be interpreted by using the appropriate estimate of effect size.

From the results of the analysis, you can observe that:

1. The regression equation resulting from the analysis of the temperature $= \beta + \alpha\, CO_2$ model can be expressed as follows:

   temperature $= 4.70 + 0.10\, CO_2$ with $R^2 = 0.76$, $F(1, 8) = 12.18$, $p = 0.008$

   t-tests indicate that both CO_2 concentration and the intercept make a significant contribution to the model, with $t = 3.49$, $p = 0.008$ and $t = 9.51$, $p = 0.000$, respectively.

2. The regression equation resulting from this analysis of the temperature $= \beta + \alpha\, CH_4$ model can be expressed as follows:

   temperature $= 15.03 + 0.18\, CH_4$ with $R^2 = 0.60$, $F(1, 8) = 25.77$, $p = 0.001$

   t-tests indicate that only CH_4 concentration makes a significant contribution to the model, with $t = 5.08$, $p = 0.001$ and $t = 1.50$, $p = 0.171$, respectively.

3. The regression equation resulting from the analysis can be expressed as follows:

   temperature $= 6.210 + 0.009\, CH_4 + 0.005\, CO_2$ with $R^2 = 0.77$, $F(2, 7)=12.15$, $p = 0.005$

   None of the variables makes a significant individual contribution to the model, with $t = 2.33$, $p = 0.053$ coming closest for CH_4 concentration. This can be attributed to the multicollinearity between the two IVs.

Thus, CH_4 concentration is the best univariate predictor of temperature ($R^2 = 0.76$), although CO_2 is also quite high ($R^2 = 0.60$). However, by adding CO_2 into the bivariate model to form a multivariate model, only 1% more variation in temperature can be uniquely explained. Given this result, you can conclude that using the bivariate model with CH_4 as the IV explains almost as much variation in the DV as the multivariate model, and thus you would only use the bivariate model for characterization and prediction.

Recall in this example that there was a perfect correlation between increases in year and temperature, so temperature was selected as the DV, and the model was treated as if it were a cross-sectional design. However, in any realistic assessment of global warming, it's likely that—to some extent—temperature can be predicted from the previous year's temperature. In this case, temperature can be treated as a *lagged dependent variable*. Thus, historical values of temperature can (and should) be explicitly included in the model. For example, the model:

temperature$_y = \beta + \alpha_1 CO_2 + \alpha_2 CH_4 + \alpha_3$temperature$_{y-1}$

explicitly acknowledges the predictive power of this year's temperature (year y) by the previous year's temperature (year $y - 1$). Removing autocorrelated sources of variation is an important part of *time series analysis*, which is discussed in more detail in Chapter 17. The important message is not to feel constrained to use standard textbook models for understand the relationships that you can observe, especially if you are trying to predict the impact of changing the levels of one or more factors.

Standardized Coefficients

A central question in multiple regression is which of two IVs may have the greatest impact. In this example, and in the current climate of global warming treaties, carbon trading schemes, etc., is it more important to reduce CH_4 or CO_2? In other words, which chemical actually has the greatest effect on temperature? If we simply compared the size of the estimated parameters, α_2 is twice as large as α_1, suggesting that CH_4 has a greater effect on temperature or is more important than CO_2. You can see this is not true by the t-tests, but in any case, it is a spurious comparison, since CH_4 and CO_2 are measured in different units (ppb and ppm, respectively). In simple volumetric terms, for example, a one ppm change is much more significant than a one ppb change.

As CH_4 and CO_2 are both chemicals, it is possible to compare like-with-like on an atomic basis, but in many cases of linear regression, the IV measures are not directly comparable. In this case, the coefficients for each IV can be standardized by using the standard deviation. Such standardized coefficients are known as *beta coefficients*. Beta coefficients are usually reported as part of the regression analysis, but it is necessary to also compute the standard deviations to interpret the results. In Examples 14-4 and 14-5, the descriptive statistics for the model variables are shown, along with the beta coefficients. Thus, a one standard deviation change in CO_2 would result in a temperature change of 0.198 standard deviations, and a one standard deviation change in CH_4 would result in a temperature change of 0.713 standard deviations. Thus, a change in CH_4 has a greater relative impact on temperature than CO_2. As a result, it may be argued that countering the effects of CH_4 may have a greater impact than reducing CO_2 on global warming, although in reality, it is unlikely that there is one "silver bullet" that can solve global warming! Examples 14-4 and 14-5 show the standardized coefficients in a regression model for temperature, CO_2, and CH_4.

Example 14-4. Descriptive statistics

Variable	Obs	Mean	Std. Dev.	Min	Max
temperature	10	20.55	0.3027649	20.1	21
co2	10	305.4028	13.02559	288.8803	325.738
ch4	10	1517.833	25.32265	1475.822	1548.636

Example 14-5. Standardized coefficients

temperature	Coef.	Beta

Example 14-5. Standardized coefficients (continued)

```
co2 |  0.0046003    0.1979162
ch4 |  0.008522     0.7127669
```

Proxy and Dummy Variables

Although regression is often presented as a technique that can only be used with real-valued, continuous variables, categorical variables can also be used. Indeed, in many fields like psychology and the social sciences, using categorical variables is critical to understanding the relationships between different group members in which the categories are mutually exclusive. Where categorical variables are used in regression analysis, they may also be referred to as *dummy variables*, since they nominally code for a category rather than lying on an interval or ratio scale. For example, if you are interested in gender as a DV, it may be normal to code $S = 0$ (for males) and $S = 1$ (for females), but these are nominal values. In the social sciences, building up regression models around many sets of IVs is very useful for understanding what causes changes in a DV; for example, models with wealth as a DV may take into account parental income (low, high), social class (low, middle), education (high school, college), and so on. If a model is statistically significant, and if the factors each make a distinct contribution, such models can be very useful in predicting the DV when all of the IV dummy variable values are known.

However, dummy variables are often criticized for lack of *face validity* when a direct measurement is not actually made to determine category membership. "Sex" as a dummy variable is a good example: sex is determined at face value by primary and secondary sex characteristics; at the genetic level, most people have either XX chromosomes (female) or XY (male). Individuals with Kleinfelter's syndrome, on the other hand, have three chromosomes (XXY); consequently, the mapping between sexes "at face value" with the possible permutations at the chromosomal level is not clear. This is why many social scientists prefer to use the dummy variable "gender," which is clearly defined in terms of social roles that have face validity. Similar problems occur with the use of proxy variables like race, where there is little genetic basis for the definition of the category.

In situations where there are more than two possible values for a category, dummy variables can also be used. For example, class often has three possible groupings: working, middle, and ruling. In this case, one grouping acts as a *reference group* for two dummy variables that are created for the other two groupings. Any grouping could be chosen as the reference group. For example, a case would have the "working" variable set to 1 only where the individual was classified as being "working class"; otherwise, the value would be set to 0.

The following model would be used to regress a DV against the three IV categories:

$$Y = \beta + \alpha_1 X + \alpha_2 \text{middle} + \alpha_3 \text{ruling}$$

After running the analysis, coefficients (α_1 and α_2) would be computed for each of the dummy variables (middle, ruling), while the intercept would be for the reference group, since $\alpha_1 = 0$ and $\alpha_2 = 0$ for them. α_1 and α_2 provide an estimate of the relative difference between the reference group and the grouping identified by the dummy variable. For instance, the coefficient for middle would indicate the difference between the working and middle groups, and the associated univariate t-test would indicate whether the difference was statistically significant. However, the result does not indicate an effect size, and hence whether the differences are particularly meaningful.

Unfortunately, models of this kind are often used to make all sorts of wild claims about members of different groups who have been coded up using different dummy variables; as with the case of gender/sex described above, you need to be very clear about how to establish the validity of the particular groupings that you are interested in, noting the potential biases in both self-reported and simplistic observational measures.

Table 14-2 shows some sample data, with the IV for working class used as the reference variable, middle and ruling class membership the dummy variables, and IQ as the DV.

Table 14-2. Sample data for the reference IV X, dummy IVs middle and ruling class, and DV (IQ)

IQ	X	Middle	Ruling
70.0	1	0	0
75.0	1	0	0
80.0	1	0	0
85.0	1	0	0
90.0	0	1	0
95.0	0	1	0
100.0	0	1	0
120.0	0	1	0
105.0	0	0	1
110.0	0	0	1
115.0	0	0	1

The model is significant, with $F(2, 8) = 11.86$, $p = 0.004$. The coefficient of determination $R^2 = 0.748$, but the adjusted $R^2 = 0.685$, with the ANOVA results shown in Table 14-3. The univariate tests for the dummy variables show that both are significant, with $t = 3.607$, $p = 0.007$ and $t = 4.569$, $p = 0.002$ for middle and upper, respectively, as shown in Table 14-4. These results indicate that the difference between the working and middle and working and ruling groups are statistically significant. Since the standardized coefficient for $\beta_{Ruling} = 0.915$ is greater than $\beta_{Middle} = 0.723$, the relative gap in IQ is greater between working and ruling groups than working and middle groups.

Table 14-3. ANOVA results for the regression model based on working, middle, and ruling class membership, and DV (IQ)

Model		Sum of squares	df	Mean square	F	Sig.	
1	Regression	2056.250	2	1028.125	11.856	0.004a	
	Residual	693.750	8	86.719			
	Total	2750.000	10				

Table 14-4. Univariate significance tests for the regression model based on working, middle, and ruling class membership, and DV (IQ)

Model	Unstandardized coefficients		Standardized coefficients	t	Sig.
	B	Std. error	Beta		
Middle	23.750	6.585	0.723	3.607	0.007
Upper	32.500	7.112	0.915	4.569	0.002

Regression Algorithms

In the previous examples, there were cases in which adding or removing an independent variable resulted in the model reaching statistical significance or not. While some statistical packages just enter all IVs specified into the model and report the overall statistical significance and the univariate significance for each IV, other algorithms are much more sophisticated in identifying the best model, based on including or excluding specific IVs. The analysis can often be provided with default probabilities to include or exclude variables, while attempting to minimize error.

There are two different types of algorithm: *stepwise* and *blocking*. Stepwise algorithms allow factors to be entered in three different ways:

Backward entry
> All independent variables are removed sequentially until the model is no longer significant, after they have initially been entered as a block using the enter method.

Forward entry
> Independent variables are added until statistical significance is reached.

Stepwise
> Begins with backward entry, and once significance has been reached, adds IVs in again—based on the size of *F*—and determines overall significance again. The goal is to avoid local minima in the global optimization of the model fit. The algorithm then evaluates the model again, to determine if another IV should be removed. Since this process could end up in an infinite loop, the maximum number of steps is often specified. Note that forward and backward selection and forward and backward stepwise are different techniques.

Blocking methods can be described as follows:

Enter
> All variables are entered in nonincreasing order of tolerance, or in a specific order.

Remove
> Removes a group of variables in a single block.

Forced entry
> A list of IVs is specified, and all are entered.

Depending on your method of entry, it is possible that you may receive different and/or potentially inaccurate results. Let's examine the effect of using the different stepwise techniques. Imagine you are an educator interested in the relationship between IQ and traditional measures of general ability, such as performance on numerical, reading, verbal, and reasoning skills, as well as nontraditional measures, such as musical and physical performance. Sample data is shown in Table 14-5.

Table 14-5. Data showing the relationship between traditional measures and nontraditional measures of general ability, and IQ

IQ	Numerical	Reading	Verbal	Physical	Musical	Reasoning
85.0	3.0	5.0	7.0	10.0	6.0	10.0
90.0	3.0	6.0	7.0	10.0	6.0	10.0
95.0	4.0	6.0	7.0	9.0	7.0	8.0
100.0	4.0	7.0	8.0	9.0	7.0	5.0
100.0	5.0	7.0	8.0	8.0	8.0	6.0
100.0	5.0	8.0	8.0	7.0	9.0	5.0
105.0	6.0	8.0	8.0	6.0	8.0	4.0
105.0	6.0	8.0	8.0	5.0	7.0	5.0
110.0	7.0	9.0	8.0	4.0	6.0	6.0
110.0	7.0	9.0	8.0	3.0	6.0	9.0
115.0	8.0	10.0	9.0	3.0	5.0	10.0
120.0	9.0	10.0	9.0	1.0	4.0	9.0

Linear Regression

You decide to explore the relationships between the variables, calculating all pairwise correlations and their statistical significance, as shown in Table 14-6. Unsurprisingly, the most traditional measures (numerical, reading, and verbal) are highly positively correlated with IQ.

However, reasoning was almost completely uncorrelated, and there was a strong negative relationship between IQ and physical performance. A nonsignificant relationship was observed between IQ and musical ability.

Table 14-6. Pairwise relationships between traditional measures and nontraditional measures of general ability, and IQ

	IQ	Numerical	Reading	Verbal	Physical	Musical	Reasoning
IQ	1.000	0.978**	0.976**	0.914**	−0.955**	−0.427	−0.073
Numerical	0.978**	1.000	0.963**	.887**	−0.986**	−0.481	0.026
Reading	0.976**	0.963**	1.000	.912**	−0.954**	−0.381	−0.055
Verbal	0.914**	0.887**	0.912**	1.000	−0.836**	−0.337	−0.103
Physical	−0.955**	−0.986**	−0.954**	−.836**	1.000	0.503	−0.062
Musical	−0.427	−0.481	−.381	−0.337	0.503	1.000	−0.738**
Reasoning	−0.073	0.026	−.055	−0.103	−0.062	−0.738**	1.000

At this point, you may well decide to use a blocking method and enter the significantly correlated variables into a model. However, you are really interested in teasing out what factors will produce the overall best model, given the fair amount of multicollinearity, so let's explore what happens when you use two of the most popular stepwise methods: forward and backward.

Forward

With the strongest pairwise correlation with IQ (r = 0.978), numerical is entered as the first factor, with the inclusion p-value set to $p <= 0.05$. For this model, R^2 = 0.956, adjusted R^2 = 0.952. The overall model was significant, $F(1, 10)$ = 217.36, p = 0.000, as shown in Table 14-7.

Table 14-7. ANOVA testing significance of forward (stepwise) multiple linear regression

Model		Sum of squares	df	Mean square	F	Sig.
1	Regression	1073.528	1	1073.528	217.362	0.000
	Residual	49.389	10	4.939		
	Total	1122.917	11			

The numerical IV made a significant contribution to the model, t = 14.743, p = 0.000, as shown in Tables 14-8 and 14-9, which also shows that all of the other IVs were excluded, and did not make a significant contribution to the model. The benefit of the forward stepwise algorithm is that you quickly reach the model that captures the greatest amount of variance. However, since many IVs are significantly correlated, you cannot be sure from a theoretical and model building perspective which IV is causal. This is where the other stepwise methods come into their own. Tables 14-8 and 14-9 show the univariate significance tests for the regression model.

Table 14-8. Included variables

Model	Unstandardized coefficients		Standardized coefficients	t	Sig.
	B	Std. error	Beta		
1 (Constant)	74.318	2.043		36.374	0.000
Numerical	5.122	0.347	0.978	14.743	0.000

Table 14-9. Excluded variables

Model	Beta in	t	Sig.	Partial correlation	Collinearity statistics
					Tolerance
1 Reading	.467	2.239	0.052	0.598	0.072
Verbal	.219	1.648	0.134	0.482	0.213
Physical	.288	.716	0.492	0.232	0.029
Musical	.057	.737	0.480	0.239	0.768
Reasoning	−.098	−1.594	0.146	−0.469	0.999

Backward

Forward stepwise models work by adding variables until some criterion is met (such as the t value indicating univariate significance), while backward stepwise models work in the opposite way: all variables are included, and then excluded in the reverse order of contribution made to the model, i.e., the IVs making the smallest contribution are removed, and the analysis is rerun after each removal.

Table 14-10 shows that five models were considered; after each iteration, one IV is removed, starting with verbal and proceeding to physical, musical, and reasoning. Recall that the forward stepwise method resulted in only one IV—numerical—being included, so the backward method has provided a more insightful result, in the sense that both numerical and reading measures of general ability are thought to reflect quite different underlying information processing skills.

Table 14-10. Backward stepwise model for linear regression

Model	Variables entered	Variables removed	Method
1	Reasoning, numerical, musical, verbal, reading, physical	.	Enter
2	.	Verbal	Backward (criterion: Probability of F-to-remove >= 0.100).
3	.	Physical	Backward (criterion: Probability of F-to-remove >= 0.100).
4	.	Musical	Backward (criterion: Probability of F-to-remove >= 0.100).
5	.	Reasoning	Backward (criterion: Probability of F-to-remove >= 0.100).

Table 14-11 shows the standardized coefficients for each model iteration, as well as the corresponding t values and their significance. At each stage, you can see how the variable that has the least univariate significance is excluded, and the model is then rerun. In each iteration, also notice how the standardized coefficients converge toward a final value; in the final iteration, $\beta_{Numerical}$ has been reduced from 0.731 to 0.528, while the other included IV ($\beta_{Reading}$) has stayed relatively constant, from 0.487 to 0.467. From a model-building perspective, based on a theoretical understanding of the abilities that may contribute to IQ, the relatively higher contribution of numerical over reading skills is more satisfactory than the model arising from the forward stepwise model, which excluded any significant contribution from reading.

Table 14-11. Standardized coefficients for each model iteration

Model	Unstandardized coefficients	Standardized coefficients	t	Sig.		
1	(Constant)	64.480	20.702		3.115	0.026
	Numerical	3.827	2.369	0.731	1.616	0.167
	Reading	3.070	1.749	0.487	1.755	0.140
	Verbal	.048	2.628	0.003	0.018	0.986
	Physical	1.011	1.423	0.305	0.710	0.509
	Musical	−1.222	0.864	−0.167	−1.414	0.216
	Reasoning	−.742	0.445	−0.169	−1.668	0.156
2	(Constant)	64.514	18.819		3.428	0.014
	Numerical	3.851	1.822	0.735	2.114	0.079
	Reading	3.088	1.301	0.490	2.373	0.055
	Physical	1.026	1.040	0.310	0.986	0.362
	Musical	−1.224	0.777	−0.167	−1.575	0.166
	Reasoning	−.743	0.402	−0.169	−1.848	0.114
3	(Constant)	80.511	9.530		8.448	0.000
	Numerical	2.449	1.137	0.467	2.153	0.068
	Reading	2.863	1.279	0.454	2.239	0.060
	Musical	−1.179	0.775	−0.161	−1.522	0.172
	Reasoning	−0.785	0.399	−0.179	−1.968	0.090
4	(Constant)	68.274	5.524		12.360	0.000
	Numerical	3.149	1.122	0.601	2.806	0.023
	Reading	2.476	1.352	0.393	1.831	0.105
	Reasoning	−0.294	0.253	−0.067	−1.161	0.279
5	(Constant)	64.655	4.649		13.908	0.000
	Numerical	2.765	1.093	0.528	2.529	0.032
	Reading	2.945	1.316	0.467	2.239	0.052

Table 14-12 shows the ANOVA table for all models. Note the increase in F at each iteration, even though fewer IVs are in the model; caution should be exercised since each F value is computed using different degrees of freedom.

Table 14-12. ANOVA table for each model iteration

Model		Sum of squares	df	Mean square	F	Sig.
1	Regression	1105.368	6	184.228	52.491	0.000
	Residual	17.548	5	3.510		
	Total	1122.917	11			
2	Regression	1105.367	5	221.073	75.582	0.000
	Residual	17.550	6	2.925		
	Total	1122.917	11			
3	Regression	1102.521	4	275.630	94.600	0.000
	Residual	20.395	7	2.914		
	Total	1122.917	11			
4	Regression	1095.770	3	365.257	107.638	0.000
	Residual	27.147	8	3.393		
	Total	1122.917	11			
5	Regression	1091.192	2	545.596	154.779	0.000
	Residual	31.725	9	3.525		
	Total	1122.917	11			

In summary, you can see that different stepwise methods can give different results. You shouldn't use stepwise methods as some kind of post-hoc hypothesis generating tool; instead, build models that actually make sense in terms of the underlying theory. In these two examples, you have seen how including a term for both "left brain" and "right brain" contributions to IQ produced a better outcome than a single IV.

Common Problems with Multiple Regression

As with all techniques based on the general linear model, violating assumptions of multiple linear regression reduces the validity of any results. For example, in many of the examples presented in this chapter, the IVs are significantly correlated. The best way to deal with this problem is to perform some form of orthogonal decomposition on the IVs prior to model building, so that you are guaranteed to have a set of independent IVs in the model. Techniques to create orthogonal IVs are discussed in Chapter 16.

As you saw from the presentation of stepwise methods, adding IVs in or leaving them out can significantly alter the model's significance. Wherever relevant variables are excluded, the coefficients calculated in the analysis must be regarded as suspect, since they will be biased at the expense of a true population estimate. This is why regression shouldn't be considered an endpoint; rather, it is one step in a journey that should include alternative model specifications and/or measures that are shown to act in the same way, and can therefore be considered robust.

Similarly, including irrelevant variables is also problematic; while it may seem that an irrelevant variable should have no impact on the model, unless the variable is completely independent of the other IVs (which would be rare indeed, especially in a large model), it should not be included. This will eventually lead to an underestimate of the true relationship between an IV and a DV, and may result in a Type II error, since significant IVs may be treated as insignificant in the analysis.

Another key problem in multiple linear regression is assuming linearity in the face of *nonlinearity*. In many disciplines, it's common to use a (simple) linear approximation to a (more complex) nonlinear model. This may be perfectly acceptable for functions like a sigmoid, which are linear in specific ranges; if the coordinates of your IV and DV only fall within the linear range, a linear model is entirely suitable. However, you should never assume linearity, and if you are dealing with a genuinely nonlinear phenomenon, it's best to create and test against a model that is nonlinear.

Several examples of multicollinearity have been presented in the chapter. Multicollinearity arises when IVs are correlated with each other. A distinct phenomenon arises when correlation is observed in the residuals, i.e., the differences between observed and predicted values for the DV. This phenomenon is known as *autocorrelation* or *serial correlation*, and can invalidate models analyzed using the *Ordinary Least Squares* (OLS) algorithm. The assumptions for OLS include:

- The mean of all residuals is zero, even though there is variation in the pool of residual values
- Residuals are uncorrelated
- The residuals are independent of and uncorrelated with the IVs (*homoscedasticity*)

Where a cross-sectional design is used, the chance of (say) residuals being correlated with each other is slim. However, in longitudinal and/or time series studies, autocorrelation will almost certainly be present. In this case, seasonal effects are usually removed prior to entry in the model if some regular, cyclical effect is present. However, autocorrelation may also indicate the absence of an important explanatory variable in the model. If you are concerned about autocorrelation, you can compute the Durbin-Watson coefficient that tests the null hypothesis that there is no correlation among the residuals that are serially related. Alternatively, other regression algorithms, such as *Generalized Least Squares* (GLS), may be employed.

Note that cross-sectional designs have their own problems—the assumption of homoskedasticity is routinely violated—in which case, GLS may also be used. The model may also need further refinement and/or addition of a further IV, since *heteroskedasticity* would suggest some systematic source of variation existing in the residuals that has not been accounted for by the IVs. Stepwise regression algorithms may also be useful here, and homoskedasticity may be an important criterion in ultimate model selection.

Exercises

Multiple linear regression can be used to investigate a number of different types of research questions, as shown in the examples below.

Question

As a human resource specialist, you are interested in the motivational factors that are associated with productivity (DV) in IT teams, based on the KLOC metric (thousands of lines of code written per week). There are four motivational factors that can influence productivity, based on either intrinsic or extrinsic measures of motivating experiences per working week, which have either been self-reported or observed. The hypothesis is that self-reported measures exert a much stronger influence, since they provide a more objective view, rather than the subjective self-report measures. Thus, four IVs are derived and used:

- Intrinsic self-report (IS)
- Intrinsic observed (IO)
- Extrinsic self-report (ES)
- Extrinsic observed (EO)

Sample data is shown in Table 14-13.

Table 14-13. Four IVs (IS, IO, ES, EO) may have an impact on KLOC

Productivity (KLOC)	Intrinsic self-report (IS)	Intrinsic observed (IO)	Extrinsic self-report (ES)	Extrinsic observed (EO)
2.5	45	27	34	30
1.2	3	100	44	14
8.3	54	85	33	65
5.4	35	56	45	89
3.6	56	34	45	67
5.6	44	58	34	51
4.3	55	41	22	32
2.3	34	18	12	23
0.4	43	100	1	4
3.4	44	28	1	32

Answer

Since you have not used the data set before, you decide to create a correlation matrix to show the relationships between the variables, as shown in Example 14-6.

Example 14-6. Relationship between four IVs (IS, IO, ES, EO) and a DV (KLOC)

```
       |   KLOC      IS       IO       ES       EO
-------+-----------------------------------------------
  KLOC |  1.0000
    IS |  0.4776   1.0000
    IO | -0.0442  -0.3909   1.0000
    ES |  0.3616  -0.1884   0.1192   1.0000
    EO |  0.7796   0.3581  -0.1751   0.6115   1.0000
```

You can see that the IV with the strongest relationship with the DV is extrinsic observed (EO), $R = 0.78$, $R^2 = 0.61$. On the other hand, intrinsic observed (IO) has almost no relationship with the DV, $R = -0.04$, $R^2 = 0.0016$. In terms of potential multicollinearity, only ES and EO have a strong relationship, $R = 0.6115$, $R^2 = 0.37$.

Given the relationships revealed by the correlation matrix, you decide to run a simple linear regression analysis, with the results shown in Example 14-7.

Example 14-7. Regression analysis showing the relationship between one IV (EO) and the DV (KLOC)

```
    Source |       SS       df       MS              Number of obs =      10
-----------+------------------------------           F( 1,    8) =     12.40
     Model | 29.5743294    1  29.5743294             Prob > F      =   0.0078
  Residual | 19.0856724    8  2.38570905             R-squared     =   0.6078
-----------+------------------------------           Adj R-squared =   0.5587
     Total | 48.6600018    9  5.40666686             Root MSE      =   1.5446


-----------------------------------------------------------------------------
    KLOC |     Coef.   Std. Err.      t    P>|t|     [95% Conf. Interval]
---------+-------------------------------------------------------------------
      EO |  0.0681908  0.0193676    3.52   0.008     0.0235289    0.1128526
    _cons|  0.9246364  0.9273239    1.00   0.348    -1.213776     3.063049
```

Thus, with the model KLOC = $\beta + \alpha$EO, you calculate $\beta = 0.92$, $\alpha = 0.19$, with $F(1, 8) = 12.40$, with EO making a significant contribution to the model, $t = 3.52$, $p = 0.008$. You then decide to run the full model with all IVs, as shown in Example 14-8.

Example 14-8. Regression analysis showing the relationship between four IVs (IS, IO, ES, EO) and the DV (KLOC)

```
    Source |       SS       df       MS              Number of obs =      10
-----------+------------------------------           F( 4,    5) =      2.77
     Model | 33.5311801    4  8.38279502             Prob > F      =   0.1468
  Residual | 15.1288217    5  3.02576434             R-squared     =   0.6891
-----------+------------------------------           Adj R-squared =   0.4404
     Total | 48.6600018    9  5.40666686             Root MSE      =   1.7395


-----------------------------------------------------------------------------
    KLOC |     Coef.   Std. Err.      t    P>|t|     [95% Conf. Interval]
---------+-------------------------------------------------------------------
```

Example 14-8. Regression analysis showing the relationship between four IVs (IS, IO, ES, EO) and the DV (KLOC) (continued)

IS	0.0397661	0.049625	0.80	0.459	-0.0877989	0.1673312
IO	0.0155191	0.0206913	0.75	0.487	-0.0376695	0.0687078
ES	-0.0115671	0.0513429	-0.23	0.831	-0.1435482	0.120414
EO	0.0676238	0.0349941	1.93	0.111	-0.0223315	0.1575791
_cons	-1.230059	2.74642	-0.45	0.673	-8.289956	5.829837

Thus, with the model $KLOC = \beta + \alpha_1 EO + \alpha_2 ES + \alpha_3 IO + \alpha_4 IS$, you calculate $\beta = -1.23$, $\alpha_1 = 0.68$, $\alpha_2 = 0.68$, $\alpha_3 = 0.68$, and $\alpha_4 = 0.68$, with $F(4, 5) = 2.77$, $p = 0.15$, which is not significant. You can see from the results that the coefficient of determination has risen from 0.6078 to 0.6891, which is unsurprising, since any even the smallest amount of variation accounted for by the additional IVs will be additive to the bivariate case for KLOC and EO (i.e., R^2 will never decrease with the addition of more IVs). Adding more and more IVs—assuming they are equally likely to contribute something—will necessarily keep increasing R^2.

However, if you adopt the more conservative adjusted R^2, you can see that the proportion of variance accounted for actually decreases in the multivariate case. This is a typical indicator that adding the additional IVs have not enhanced the explanatory power of the model and thus, the additional factors should be removed.

The lack of significance of any individual factor in the multivariate model, while indicating multicollinearity, also suggests that the model is a poor one. The bivariate model, in which EO alone accounts for more than 60% of variation in KLOC, is probably the best option, and suggests that for the population concerned, extrinsic rewards that can be externally verified (such as money) are likely to have a very strong impact on productivity. The causal effect could be established further by running a within subjects design experiment, where individual team members were rewarded more or less on two different tasks. Further refinement of the "internal" measures, in terms of validity, may also lead to an enhanced contribution to the model.

Question

You are a management consultant working in the retail sector, conducting a time-in-motion study to determine which of two IVs (barcode scanner size and operator accuracy) has the greatest effect on DV (throughput), measured in items per second. The question is difficult to answer, because the units of measurement in each case are different: the scanner size is measured in cubic centimeters, while accuracy is measured as the mean time to successfully scan an item. Your client wants to increase throughput, since customers have complained that queues in the store are long. However, larger scanners are more expensive than smaller ones, and training courses for staff will not necessarily increase accuracy. The manager wants to know whether to spend money on more training (or hiring better staff) or purchasing larger scanners. The data is shown in Table 14-14.

Table 14-14. Data for the time-in-motion analysis with DV (throughput) and two IVs (scanner size and operator accuracy)

Throughput	Size	Accuracy
0.5	4	95
0.9	4	98
0.4	2	85
0.5	2	90
1.2	6	95
1.1	4	98
0.8	4	89
0.9	4	91
1.1	2	99
1.3	6	89

Answer

You decide to create a correlation matrix to show the relationships between the variables, as shown in Example 14-9.

Example 14-9. Correlation matrix for the time-in-motion analysis with DV (throughput) and two IVs (scanner size and operator accuracy)

```
           | throughput    size accuracy
-----------+---------------------------
throughput |   1.0000
      size |   0.6520   1.0000
  accuracy |   0.4415   0.0920   1.0000
```

You can see that the IV with the strongest relationship with the DV is scanner size, $r = 0.65$, $r^2 = 0.42$, while accuracy only has $r = 0.44$, $r^2 = 0.19$. In this example, the coefficients for each IV can be standardized by using the standard deviation (i.e., beta coefficients). To assist you to interpret the results, descriptive statistics for the DV and IVs are shown in Example 14-10.

Example 14-10. Descriptive statistics for the time-in-motion analysis with DV (throughput) and two IVs (scanner size and operator accuracy)

```
   Variable |    Obs     Mean    Std. Dev.      Min       Max
-----------+---------------------------------------------------
throughput |     10     0.87    0.3164034      0.4       1.3
      size |     10      3.8    1.47573        2         6
  accuracy |     10     92.9    4.748099       85        99
```

The regression analysis is shown in Example 14-11, along with the beta coefficients. The overall model is not statistically significant, with $F(2, 7) = 4.68$, $p = 0.051$. However, $R^2 = 0.57$, with adjusted $R^2 = 0.45$. The only IV to make a significant univariate contribution to the model was scanner size, $t = 2.48$, $p = 0.042$.

After examining the beta coefficients, you can see that a one standard deviation change in size would result in a throughput change of 0.385 standard deviations, and a one standard deviation change in scanner size would result in a throughput change of 0.617 standard deviations. Thus, a change in scanner size has a greater relative impact on throughput than operator accuracy, which is borne out by the significant model contribution made by scanner size. Thus, you present the argument to management that increasing the scanner size will have a greater impact on increasing throughput than operator accuracy.

To verify the argument, you perform a univariate regression analysis with only throughput and scanner size (Example 14-12). The results indicate that the model is statistically significant, $F(1, 8) = 5.92$, $p = 0.041$. Scanner size made a significant univariate contribution to the model, with $t = 2.43$, $p = 0.041$.

Example 14-11. Regression analysis for the time-in-motion analysis with DV (throughput) and two IVs (scanner size and operator accuracy)

Source	SS	df	MS		
Model	0.515343685	2	0.257671843		
Residual	0.385656317	7	0.05509376		
Total	0.901000002	9	0.100111111		

Number of obs = 10
F(2, 7) = 4.68
Prob > F = 0.0513
R-squared = 0.5720
Adj R-squared = 0.4497
Root MSE = 0.23472

throughput	Coef.	Std. Err.	t	P>|t|	Beta
accuracy	0.0256441	0.0165484	1.55	0.165	0.3848282
size	0.1322073	0.0532437	2.48	0.042	0.6166251
_cons	-2.014729	1.533835	-1.31	0.230	.

Example 14-12. Regression analysis for the time-in-motion analysis with DV (throughput) and the most correlated IV (scanner size)

Source	SS	df	MS		
Model	0.383040801	1	0.383040801		
Residual	0.517959202	8	0.0647449		
Total	0.901000002	9	0.100111111		

Number of obs = 10
F(1, 8) = 5.92
Prob > F = 0.0411
R-squared = 0.4251
Adj R-squared = 0.3533
Root MSE = 0.25445

throughput	Coef.	Std. Err.	t	P>|t|	[95% Conf. Interval]	
size	0.1397959	0.0574744	2.43	0.041	0.0072596	0.2723322
_cons	0.3387755	0.2327537	1.46	0.184	-0.1979556	0.8755066

15

Other Types of Regression

In Chapter 14, you learned about multiple linear regression. Although multiple linear regression is appropriate for many scenarios, other types of regression, including logistic and nonlinear, are best used under different circumstances. This chapter reviews scenarios in which these types of regression are the correct analytic choice, and covers how to perform these analyses, the meaning of the results, and issues to be wary of, such as overfitting.

Logistic Regression

In Chapter 14, multiple linear regression was presented as regressing a real-valued DV on two or more IVs, measured on interval or ratio scales, or categorical IVs, coded using binary variables. Logistic regression is commonly used when the DV is also categorical, typically nominal. Logistic regression is commonly used in epidemiological studies to understand the relationship between a number of risk factors (categorical or real-valued) and a categorical DV. For example, while it may be possible to use a real-valued DV from which hypertension can be deduced, a clinician is typically interested in making a diagnosis (hypertensive/not hypertensive) based on several different IVs. The reason that you need logistic regression is that the assumption of common variance in the DV is not met when the two possible values are 0 and 1. Also, any linear regression model may predict DV values less than 0 or greater than 1, which would have no meaning in terms of the nominal codings for categories. The odds ratios for each IV in the model can also be estimated by using logistic regression.

Imagine that you are a clinical epidemiologist working at a city hospital, exploring factors that predict the incidence of prostate cancer, without requiring an invasive biopsy; one possible factor is the width of the prostate gland opening.

The instrument used to measure prostate sizes estimates widths in four different ranges:

- 0–1 mm
- 1–2 mm
- 2–3 mm
- 3–4 mm

All of the patients investigated for prostate cancer in the previous 12 months have had their gland openings measured, with the results shown in Table 15-1.

Table 15-1. Likelihood of getting prostate cancer based on gland opening size

Gland size	Number of cancer cases	Number of patients	Proportion of cases to patients
0.5mm	25	30	0.83
1.5mm	40	50	0.8
2.5mm	20	40	0.5
3.5mm	10	60	0.17

Clearly, the likelihood of getting cancer increases with a decrease in the size of the prostate gland opening. Note that the form of the function is not linear but *sigmoidal*—a small increase in size leads to a significant reduction in the likelihood of getting cancer.

Rather than utilizing the proportion of cases, the log of the odds of the proportion is used as the DV, which is known as a *logit*. Where π is the proportion of cases to patients:

$$\text{Logit}(\pi) = \log_e\left(\frac{\pi}{1-\pi}\right)$$

For binary categorical variables:

$$\text{Logit}(\pi) = \log_e[p(\text{true})] - \log_e[p(\text{false})]$$

Thus, the regression model can be written as follows:

$$\log_e\left(\frac{\pi}{1-\pi}\right) = \beta + \alpha x$$

where α is the slope and β is the intercept. Maximum likelihood estimation is used to obtain estimates for α and β.

The hypothesis to be tested in this example is that there is a log-linear increase in the odds ratio as the independent variable (gland size) decreases. The null hypothesis is therefore that there is no change in the odds ratio as the gland size increases.

The algorithm for maximum likelihood estimation is quite complicated, and will not be covered here. However, noting that the test statistic for logistic regression has a chi-square distribution, the *Wald test* can be used, which is the quotient of the maximum likelihood parameter estimate and its standard error. Since the null hypothesis predicts that the parameter is zero, the quotient distribution should be standard normal, and will improve in accuracy with larger sample sizes.

When performing the regression analysis, data is coded using a standard procedure. For DVs containing ranges, the midpoint of the ranges is chosen as the value entered into the model. You now take a subset of the results for one day, shown in Table 15-2, and see how accurately the model can be fitted.

Table 15-2. Clinical data subset showing cases of prostate cancer based on gland opening size

Gland size	Cancer
0.5	1
0.5	1
0.5	1
0.5	1
1.5	1
1.5	1
1.5	1
1.5	0
2.5	1
2.5	1
2.5	0
2.5	0
3.5	0
3.5	0
3.5	0
3.5	1

Using an enter (block) method, similar to what was used for the multiple linear regression examples discussed in Chapter 14, the algorithm iterates until it reaches a solution where parameter estimates do not change by more than a certain amount.

The classification table is shown in Table 15-3, and the model significance results are shown in Table 15-4. Overall, the predicted values were only 50% correct when the patient did not have cancer, but 90% correct when the patient did have cancer. In terms of goodness of fit, gland size made a significant contribution to the model, with the result of the Wald test equal to 3.97, $p = 0.046$.

Table 15-3. Classification table for logistic regression

					Classification table
			Predicted		
			Cancer		
	Observed		0	1	Percentage correct
Step 1	Cancer	0	3	3	50.0
		1	1	9	90.0
		Overall percentage			75.0

Table 15-4. Significance results for logistic regression model

							Variables in the equation	
Step 1	Size	−1.403	0.704	3.970	1	0.046	0.246	
	Constant	3.600	1.785	4.067	1	0.044	36.584	

Logarithmic Transformations

As you now know, violating the assumptions of tests derived from the general linear model can render the results meaningless, especially where confidence is concerned. One traditional way of ensuring that data follows a normal distribution is to take the logarithms of the variable values and use these to construct a regression model.

Let's consider the problem of measuring IQs from a biased sample of college students, who all presumably have higher than normal IQ (otherwise, how would they end up in college?). Thus, you can assume from the outset that there will be a bias in the results, and a negative skew in the distribution. Does it then make sense for researchers to use college students as their mainstay of experimentation concerning human intelligence?

Possibly—as long as the values in the extreme range, which are clustered at one end of the untransformed scale, are then shifted to center in any transformation. For example, in an experiment that produces a cluster of results with the value 10, it may make sense to use a log transformation to the base 10, which will have the effect of shifting these values to 1. This also opens up the possibility of coding up categorical variables based on the transformed data.

Using logarithmic transformations is not cheating; when you obtain means and confidence intervals using log-transformed data, you will still need to calculate antilogs to compute the actual means and CIs. However, using logarithmic transformations provides a way to meet at least one of the requirements of meeting the underlying assumptions of regression fairly easily.

Depending on the direction of any skew in your data, it may also be possible to use a square root or square transformation of the data.

Polynomial Regression

So far, you have largely learned about model fitting when the relationship between a DV and one or more IVs is linear, i.e., the value of a DV can be predicted by a weighted linear sum of the IVs, plus an intercept value. In the two-dimensional plane, such relationships can be viewed as straight lines that have a nonzero slope. However, as you will no doubt realize, many phenomena in the physical sciences that you may wish to create models for, and test goodness-of-fit, may be nonlinear. Any relationship that is not entirely linear is, by definition, nonlinear, so any discussion of nonlinear modeling must be very broad indeed. In this section, you will learn about two of the most commonly used regression models, which are based on either *quadratic* or *cubic* polynomials.

A quadratic model has both a linear and squared term for the IV, while the cubic model has a linear, squared, and cubic term for the IV. Each curve has a number of extreme points equal to the *highest order term* in the polynomial, so a quadratic model will have a single maximum, while a cubic model has both a relative *maximum* and a *minimum*. Visually, a quadratic model looks similar to the left-hand side of a cubic model, as shown in Figures 15-1 and 15-2; thus, an important question to answer for a specific range of the IV is whether the model is actually quadratic or cubic.

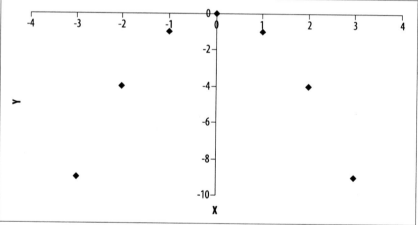

Figure 15-1. Negative quadratic model

As you can see in Figure 15-3, adding a linear component into the model tends to flatten out the nonlinear contribution (for the cubic), or shorten or lengthen the shape of the curve, as shown in Figure 15-4 for the quadratic.

Let's look at an example from sports psychology. The Yerkes-Dodson Law, first formulated in 1908, predicts a quadratic relationship between arousal (the IV) and performance (the DV). For many athletes, then, achieving the optimal level of

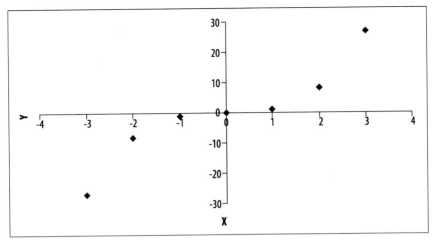

Figure 15-2. Cubic model

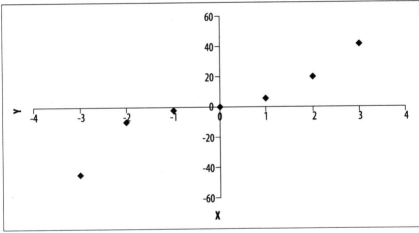

Figure 15-3. Cubic model with linear component

physiological arousal—corresponding to the single maxima of the DV—becomes the goal for producing the best performance. If athletes are not aroused enough, their performance will be poor; conversely, if athletes are over-aroused, their performance will also be poor. If the athlete is over-aroused, the coach can realistically only wait until arousal has decreased before performance increases.

However, if the relationship between arousal and performance was actually cubic, increasing arousal even further might result in improvements in performance, which would be a contrary prediction to the quadratic model. Thus, regression can be used to determine the goodness-of-fit to both the quadratic and cubic models, and the one with the best goodness-of-fit can be taken as the most accurate.

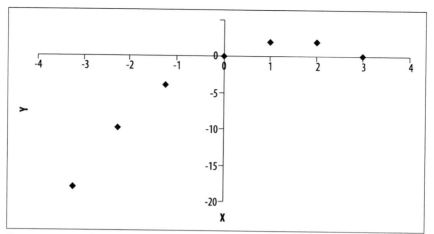

Figure 15-4. Quadratic model with linear component

Watters, Martin, and Schreter (1997)[*] designed an experiment to determine whether there was a quadratic relationship between caffeine (a drug that produces arousal) and cognitive performance on a battery of tests. The experimental setup required a dose of caffeine to be administered at regular intervals in a single session (6×100 mg); this would introduce practice effects, and would lead to an increase in performance session-to-session, independent of arousal. Any residual variation accounted by a quadratic term would then indicate the underlying relationship between arousal and performance.

You may be wondering why participants in the study were not simply invited back several times to complete the test, with the caffeine dosage randomly assigned on each occasion. The reasoning was ethical; the researchers wanted to observe any adverse reactions at low dosages, which would be impossible on the first trial in a truly randomized design, and also the researchers wanted to minimize the number of return visits. To obtain a higher degree of experimental control, a repeated measures design was used, in which each participant attended a placebo and treatment session (single blind). If the experimenter noted an adverse reaction, the experiment would be halted. The order of attendance for either the placebo or treatment condition was randomized.

As designed, the experiment had both a within-subjects and between-subjects comparison, with the former showing the dose-response relationship, and the latter confirming that the dose-response relationship observed was not the product of chance (or practice). Only the within-subjects analysis is shown below. The analysis proceeds by adding a term into the model progressively, starting with caffeine, followed by the square and cube of caffeine. Table 15-5 shows some sample data that may be obtained in this type of experiment.

[*] Watters, P.A., Martin, F. & Schreter, Z. (1997). "Caffeine and cortical arousal: The nonlinear Yerkes-Dodson Law." *Human Psychopharmacology: Clinical and Experimental*, 12, 249–258.

Table 15-5. Relationship between caffeine and cognitive performance

0mg	100mg	200mg	300mg	400mg	500mg	600mg
10.0	15.0	17.0	18.0	15.0	13.0	11.0
8.0	10.0	14.0	16.0	12.0	10.0	9.0
15.0	16.0	18.0	24.0	20.0	17.0	15.0
14.0	17.0	21.0	22.0	21.0	17.0	13.0
15.0	16.0	18.0	20.0	18.0	16.0	12.0
10.0	15.0	17.0	18.0	15.0	13.0	11.0
8.0	10.0	14.0	16.0	12.0	10.0	9.0
15.0	16.0	18.0	24.0	20.0	17.0	15.0
14.0	17.0	21.0	22.0	21.0	17.0	13.0
15.0	16.0	18.0	20.0	18.0	16.0	12.0

For the linear model $y = ax + b$, where y is performance and x is caffeine, there was virtually no relationship, $R^2 = 0.001$, showing no significant linear relationship between the two variables $F(1, 68) = 0.097$, $p = 0.757$, as shown in Table 15-6.

Table 15-6. Linear relationship between caffeine and cognitive performance

	Sum of squares	df	Mean square	F	Sig.
Regression	1.429	1	1.429	0.097	0.757
Residual	1004.057	68	14.766		
Total	1005.486	69			

For the quadratic plus linear model $y = a_{1x} + a_{2x}^2 + b$, where y is performance and x is caffeine, there was a strong relationship, $R^2 = 0.462$, showing a significant linear and quadratic relationship between the two variables, $F(2, 67) = 28.81$, $p < 0.001$ as shown in Tables 15-7 and 15-8. Note the significant contribution made by both the linear and quadratic term, with a strongly linear practice effect accompanied by a negative quadratic term, indicating the Yerkes-Dodson Law. The relative contribution that both terms make to the model, demonstrated by the beta coefficients, is comparable ($\beta_{linear} = 2.314$ versus $\beta_{quadratic} = -2.448$).

Table 15-7. Linear and quadratic relationship between caffeine and cognitive performance

	Sum of squares	df	Mean square	F	Sig.
Regression	464.971	2	232.486	28.818	0.000
Residual	540.514	67	8.067		
Total	1005.486	69			

Table 15-8. Linear and quadratic relationship between caffeine and cognitive performance, continued

	Unstandardized coefficients		Standardized coefficients	t	Sig.
	B	Std. error	Beta		
Caffeine	0.044	0.006	2.314	7.166	0.000
Caffeine ** 2	−7.429E-5	0.000	−2.448	−7.580	0.000
(Constant)	12.014	0.784		15.324	0.000

For the cubic plus quadratic plus linear model $y = a_{1x} + a_{2x}^2 + a_{3x}^3 + b$, where y is performance and x is caffeine, there was no additional variation accounted by the addition of the cubic term, thus confirming the significant linear and quadratic relationship between the two variables $F(3, 66) = 18.93$, $p = 0.000$, as shown in Tables 15-9 and 15-10. Note the significant contribution made by both the linear and quadratic term, but not the cubic term, demonstrated by the beta coefficients ($\beta_{linear} = 2.314$ versus $\beta_{quadratic} = -2.448$ and $\beta_{cubic} = 0.110$). The results are illustrated in Figure 15-5.

Table 15-9. Linear, quadratic, and cubic relationship between caffeine and cognitive performance

	Sum of squares	df	Mean square	F	Sig.
Regression	465.038	3	155.013	18.930	0.000
Residual	540.448	66	8.189		
Total	1005.486	69			

Table 15-10. Linear, quadratic, and cubic relationship between caffeine and cognitive performance, continued

	Unstandardized coefficients		Standardized coefficients	t	Sig.
	B	Std. error	Beta		
Caffeine	0.045	0.014	2.373	3.265	0.002
Caffeine ** 2	−7.929E-5	0.000	−2.613	−1.409	0.164
Caffeine ** 3	5.556E-9	0.000	0.110	0.090	0.928
(Constant)	11.981	0.872		13.740	0.000

Overfitting

One of the amazing features of modern statistical packages is that you can automatically specify and perform any number of tedious statistical tests at the click of a button, which can be useful if your *a priori* hypotheses have failed to meet expectations. This approach to statistics is commonly known as a *fishing expedition*, and when used with nonlinear regression, is known as *arbitrary curve-fitting*. Here, you can simply instruct a computer package to calculate all of the possible

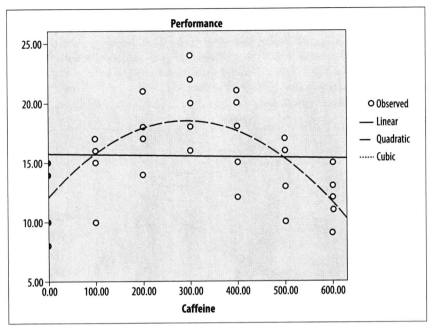

Figure 15-5. Linear, quadratic, and cubic terms in the model

nonlinear relationships between two variables, and simply select the one that gives you the best match, in terms of F values and significance results.[*]

Imagine that you are a nutritionist interested in the relationship between smoking and blood pressure, with the results obtained from a small study shown in Table 15-11. You know that there is a relationship between the two, but as an expert witness in a court case, you are under pressure to prove the strongest possible link between the two variables.

Table 15-11. Relationship between diastolic blood pressure and daily cigarette smoking

DiastolBP	DailyCigs
80.0	0.0
75.0	0.0
90.0	1.0
80.0	0.0
75.0	0.0
95.0	10.0
90.0	20.0
100.0	25.0
110.0	30.0
140.0	35.0

[*] A deeper question is whether hypothesis-driven statistics is superior to agnostic and hypothesis-free data mining, but that is beyond the scope of this book.

As you can see from the results shown in Figure 15-6, there are varying values of R^2 being generated from multiple models that were included in the analysis, including an amazing 97% of variability in diastolic BP being accounted for by a cubic model! Even more amazingly, no one had ever suspected a cubic relationship between the two variables; nonetheless, you would be able to make a very convincing argument on the basis of your analysis.

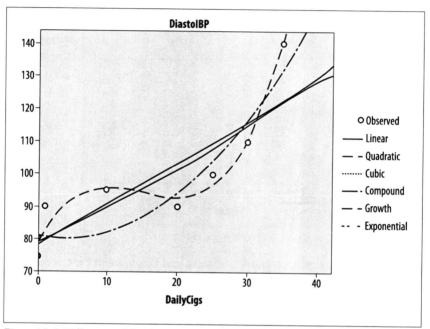

Figure 15-6. Different models relating cigarette smoking to diastolic blood pressure

Do the R^2 values computed by such an approach have any real meaning? Yes and no: the real risk with small sample sizes and fishing expeditions is *overfitting*. This means your data fits a model too well, given the specific sample that you have available. The only way to protect against overfitting is to replicate the result across a number of different samples. If only limited samples are available—such as in destructive testing environments—generalization techniques such as *cross-validation*, *bootstrapping*, and the *jack-knife* may be employed. Table 15-12 shows more data.

Table 15-12. Relationship between diastolic blood pressure and daily cigarette smoking

Equation	Model summary					Parameter estimates			
	R square	F	df1	df2	Sig.	Constant	b1	b2	b3
Linear	0.781	28.518	1	8	0.001	78.423	1.246		
Quadratic	0.869	23.118	2	7	0.001	80.984	−0.386	0.053	

Table 15-12. Relationship between diastolic blood pressure and daily cigarette smoking (continued)

Equation	Model summary					Parameter estimates			
Cubic	0.970	64.155	3	6	0.000	79.069	3.975	−0.299	0.007
Compound	0.813	34.853	1	8	0.000	79.007	1.013		
Growth	0.813	34.853	1	8	0.000	4.370	0.012		
Exponential	0.813	34.853	1	8	0.000	79.007	.0120		

The full results of the analysis are shown in Tables 15-13 through 15-24, which show models of the relationship between diastolic blood pressure and daily cigarette smoking.

Table 15-13. Linear model (1)

	Sum of squares	df	Mean square	F	Sig.
Regression	2,774.247	1	2,774.247	28.518	0.001
Residual	778.253	8	97.282		
Total	3,552.500	9			

Table 15-14. Linear model (2)

	Unstandardized coefficients		Standardized coefficients	t	Sig.
	B	Std. error	Beta		
DailyCigs	1.246	0.233	0.884	5.340	0.001
(Constant)	78.423	4.207		18.641	0.000

Table 15-15. Quadratic model (1)

	Sum of squares	df	Mean square	F	Sig.
Regression	3,085.388	2	1542.694	23.118	0.001
Residual	467.112	7	66.730		
Total	3,552.500	9			

Table 15-16. Quadratic model (2)

	Unstandardized coefficients		Standardized coefficients	t	Sig.
	B	Std. error	Beta		
DailyCigs	−0.386	0.780	−.274	−0.495	0.636
DailyCigs ** 2	0.053	0.024	1.195	2.159	0.068
(Constant)	80.984	3.681		22.003	0.000

Table 15-17. Cubic model (1)

	Sum of squares	df	Mean square	F	Sig.
Regression	3445.101	3	1148.367	64.155	0.000
Residual	107.399	6	17.900		
Total	3552.500	9			

Table 15-18. Cubic model (2)

	Unstandardized coefficients		Standardized coefficients	t	Sig.
	B	Std. error	Beta		
DailyCigs	3.975	1.053	2.819	3.774	0.009
DailyCigs ** 2	−0.299	0.080	−6.761	−3.761	0.009
DailyCigs ** 3	0.007	0.002	5.028	4.483	0.004
(Constant)	79.069	1.954		40.475	0.000

Table 15-19. Compound model (1)

	Sum of squares	df	Mean square	F	Sig.
Regression	0.276	1	0.276	34.853	0.000
Residual	0.063	8	0.008		
Total	0.340	9			

Table 15-20. Compound model (2)

	Unstandardized coefficients		Standardized coefficients	t	Sig.
	B	Std. error	Beta		
DailyCigs	1.013	0.002	2.464	474.805	0.000
(Constant)	79.007	3.000		26.333	0.000

Table 15-21. Growth model (1)

	Sum of squares	df	Mean square	F	Sig.
Regression	0.276	1	0.276	34.853	0.000
Residual	0.063	8	0.008		
Total	0.340	9			

Table 15-22. Growth model (2)

	Unstandardized coefficients		Standardized coefficients	t	Sig.
	B	Std. error	Beta		
DailyCigs	0.012	0.002	0.902	5.904	0.000
(Constant)	4.370	0.038		115.065	0.000

Table 15-23. Exponential model (1)

	Sum of squares	df	Mean square	F	Sig.
Regression	0.276	1	0.276	34.853	0.000
Residual	0.063	8	0.008		
Total	0.340	9			

Table 15-24. Exponential model (2)

	Unstandardized coefficients		Standardized coefficients	t	Sig.
	B	Std. error	Beta		
DailyCigs	0.012	0.002	0.902	5.904	0.000
(Constant)	79.007	3.000		26.333	0.000

16

Other Statistical Techniques

This chapter introduces more advanced statistical techniques by providing some specific examples; the techniques themselves will not be presented because the intent is to help the reader identify when one of these techniques is appropriate for a given research question. Methodologies covered include factor analysis, cluster analysis, discriminant function analysis, and multidimensional scaling.

Factor Analysis

Factor Analysis (FA) uses standardized variables to reduce data sets using *Principal Components Analysis* (PCA), the most widely used data reduction technique. It is based on an *orthogonal decomposition* of an input matrix to yield an output matrix that consists of a set of orthogonal components (or factors) that maximize the amount of variation in the variables from the input matrix. In turn, the process almost always produces a smaller, compact number of output components. In linear algebra terms, PCA works from the covariance matrix to produce a set of eigenvectors and eigenvalues. The components in the output matrix are linear combinations of the input variables, where the first component maximizes the variance captured, and with each subsequent factor capturing as much of the residual variance as possible, while taking on an uncorrelated direction in space. A more general version of PCA is *Hotelling's Canonical Correlation Analysis* (CCA), which—assuming multivariate normality—can be used to test whether two sets of variables are independent.

PCA is primarily used for three major purposes:

- In hypothesis testing using techniques based on the general linear model, PCA produces variables that are orthogonal, meaning that one of the major assumptions of the general linear model can be easily met.
- To compress a large number of variables into a smaller, more manageable data set.

- To identify latent variables in large data sets that are represented by highly correlated input variables.

While the first two purposes are usually achieved by PCA, the third is typically approached using Factor Analysis (FA), which is also based on orthogonal decomposition, but may involve more complex techniques such as variance maximizing rotation (*varimax*). You will learn about some of these techniques in this chapter. Note that in FA, the retained principal components are known as *common factors*, and correlations with the input variables are called *factor loadings*.

Let's look at an example from the field of psychometrics, which is a major user of FA. Historically, FA has been used to test various theories of mental performance and intelligence, including the hypothesis that a single "general" factor underlies intelligence, or that multiple, orthogonal factors comprise intelligence. In turn, the general findings derived from large-sale studies of intelligence and cognitive function in the population have allowed a very reliable understanding of individual differences to be determined from a number of different test instruments. While sampling does play a role in the development of test batteries, many countries have mandatory ability testing programs that minimize sampling error, although measurement error may still influence results for individual cases. The process of understanding individual differences—and compensating for them—was heavily influenced by the thinking of Carl Friedrich Gauss, the inventor of the "Gaussian" distribution, followed by the later work of Bessel, who developed a "personal equation" to make corrections in observations made by different astronomers.

Early attempts to understand intelligence and measurable variables started with scientists like James Cattell, who tried to quantify intelligence in terms of a set of mental tests, such as reaction times, rate of movement, grip strength, and so on. Later work showed that results from these tests were uncorrelated with actual academic performance. However, work by Spearman on the general intelligence factor g, extracted from the results of a battery of psychological tests, led to the widespread adoption of FA and PCA-like methods in psychometrics. Later work by Thurstone and others suggested that there must be at least two different and independent cognitive factors underlying intelligence: a linguistic factor L and a quantitative factor Q. Even today, the Scholastic Aptitude Test (SAT) and the Graduate Record Examination (GRE) produce 2–3 factor scores, which are distinctly correlated with different types of questions; one SAT factor is correlated highly with English and humanities topics, while another is more strongly related to mathematical and logical performance.

Let's look at a typical psychometric example, where a set of cognitive and mental performance scores is taken as the input matrix, and an output matrix is then produced, which is of lower dimension. The dimensionality could be driven by a number of different considerations; a specific psychological theory might predict two factors (e.g., L and Q), so only the two factors accounting for the highest proportion of variance would be selected. On the other hand, if the work is more exploratory, then a standard criterion could be adopted for *factor retention*. The most commonly used criterion is the *Guttman-Kaiser criterion*, which only retains eigenvalues > 1 (in the case of FA), i.e., where the variation accounted for by the factor is greater than the average for the variable if the variation was equally distributed across the input data set. Other more sophisticated criteria include the *Velicer partial correlation procedure*, *Bartlett's test*, or the *broken stick model*. A more graphical approach is to use the *scree plot* of eigenvalues to determine any "leveling" of the slope.

The results from the administration of a standard battery of tests for 10 study participants are shown in Table 16-1. A psychologist is interested in determining whether there is a general intelligence factor underlying performance across all of these different components of intelligence (e.g., reading, verbal ability, musical ability) or whether there are distinct factors on which individual variables are highly loaded. For example, is there an L factor that is strongly associated with reading and verbal ability, and a separate Q factor that is associated with arithmetical and geometrical ability?

Table 16-1. Psychometric test results

Reading	Music	Arithmetic	Verbal	Sports	Spelling	Geometry
8	9	6	8	5	9	10
5	6	5	5	6	5	5
2	3	2	6	8	6	4
8	9	10	9	8	10	6
10	7	1	10	5	10	2
9	8	4	9	1	7	2
3	9	10	2	6	4	9
8	10	3	8	5	7	2
10	9	3	10	6	10	3
7	10	1	9	6	10	2

The first way to begin exploring the data is to create a correlation matrix, as shown in Table 16-2. This displays all of the significant relationships between variables (and just as importantly, any lack of relationship).

Table 16-2. Correlations among psychometric test variables

		Reading	Music	Arithmetic	Verbal	Sports	Spelling	Geometry
Reading	r	1.000	0.535	−0.253	0.860**	−0.469	0.762*	−0.386
	p		0.111	0.481	0.001	0.172	0.010	0.270
Music	r	0.535	1.000	0.249	0.262	−0.263	0.380	0.069
	p	0.111		0.488	0.464	0.463	0.278	0.850
Arithmetic	r	−0.253	0.249	1.000	−0.501	0.206	−0.307	0.758*
	p	0.481	0.488		0.140	0.568	0.389	0.011
Verbal	r	0.860**	0.262	−0.501	1.000	−0.236	0.895**	−0.569
	p	0.001	0.464	0.140		0.511	0.000	0.086
Sports	r	−0.469	−0.263	0.206	−0.236	1.000	0.054	0.266
	p	0.172	0.463	0.568	0.511		0.881	0.458

Table 16-2. Correlations among psychometric test variables (continued)

		Reading	Music	Arithmetic	Verbal	Sports	Spelling	Geometry
Spelling	r	0.762*	0.380	−0.307	0.895**	0.054	1.000	−0.291
	p	0.010	0.278	0.389	0.000	0.881		0.415
Geometry	r	−0.386	0.069	0.758*	−0.569	0.266	−0.291	1.000
	p	0.270	0.850	0.011	0.086	0.458	0.415	

The correlations appear to support the idea of separate Q and L factors.

For L:

- Verbal performance and reading scores appear to be highly correlated ($r = 0.86**$).
- Reading and spelling scores are highly correlated ($r = 0.762*$).
- Verbal performance and spelling scores are also correlated ($r = 0.895**$).

For Q:

- Geometry and arithmetic scores are highly correlated ($r = 0.758*$).

None of the other variables (e.g., sporting or musical performance) were significantly correlated with any other variables, so you could expect that two interpretable factors will result from the FA.

The first step after computing PCA is to examine what proportion of variance is accounted for by the factor structure. This is done by examining the *communalities*, as shown in Table 16-3. Here, you can see that some variables, like music, have relatively low communality, while others, like spelling, have very high communality.

Table 16-3. Communalities

	Initial	Extraction
Reading	1.000	0.929
Music	1.000	0.779
Arithmetic	1.000	0.868
Verbal	1.000	0.955
Sports	1.000	0.943
Spelling	1.000	0.967
Geometry	1.000	0.814

Tables 16-4 through 16-6 show the initial eigenvalues, extraction sums of squared loadings, and rotation sums of squared loadings resulting from the FA. This is the most significant part of the results for interpretation. In Section A, you can see that three factors were extracted, accounting for 89.378% of the cumulative variance; thus, you can immediately see the power of PCA, since it has reduced seven variables to three factors, while still accounting for almost all of the variation within the data! Section B shows the three extracted factors before rotation, while

Section C shows the extracted factors after rotation was performed using varimax with the *Kaiser Normalization*. The varimax rotation rotates the axes of the factors in such a way that orthogonality is preserved, while maximizing the sum of variances of the loadings. Note that this does not affect the total amount of variance accounted for by the three factors, but the relative proportion of variance between factors does change.

Table 16-4. Initial eigenvalues

Component	Initial eigenvalues		
	Total	% of variance	Cumulative %
1	3.488	49.829	49.829
2	1.651	23.591	73.420
3	1.117	15.958	89.378
4	0.425	6.069	95.446
5	0.234	3.343	98.789
6	0.067	0.952	99.742
7	0.018	0.258	100.000

Table 16-5. Extraction sums of squared loadings

Extraction sums of squared loadings		
Total	% of variance	Cumulative %
3.488	49.829	49.829
1.651	23.591	73.420
1.117	15.958	89.378

Table 16-6. Rotation sums of squared loadings

Rotation sums of squared loadings		
Total	% of Variance	Cumulative %
2.846	40.653	40.653
2.066	29.517	70.170
1.345	19.208	89.378

New users of FA always feel that there must be some trickery with rotation, especially as it is used as an aid to interpreting factor loadings, and the existence of latent structure. But it really does serve a very useful purpose in trying to tease out which variables are most closely associated with each factor.

Tables 16-7 and 16-8 show the unrotated and rotated component matrices before and after rotation. For component 1, which corresponds to the latent L factor, you can see that rotation has the effect of increasing the relative loadings of the most relevant variables, such as spelling, so that spelling, reading, and verbal skills now have the highest scores. Conversely, component 2, which corresponds to the

Q factor, now has higher loadings for arithmetic and geometry, while unrelated variables such as music are now relatively decreased. Component 3 has a high loading only for sport, and while representing a distinct factor, can be disregarded in this analysis, since it doesn't reflect any latent structure.

Table 16-7. Unrotated component matrix

| | Component | | |
	1	2	3
Reading	0.902	0.328	−0.085
Music	0.386	0.775	−0.174
Arithmetic	−0.582	0.727	0.028
Verbal	0.955	0.009	0.209
Sports	−0.403	−0.059	0.882
Spelling	0.819	0.235	0.491
Geometry	−0.664	0.597	0.130

Table 16-8. Rotated component matrix

| | Component | | |
	1	2	3
Reading	0.859	−0.144	−0.412
Music	0.593	0.490	−0.433
Arithmetic	−0.158	0.917	0.050
Verbal	0.869	−0.438	−0.088
Sports	−0.046	0.176	0.954
Spelling	0.955	−0.164	0.169
Geometry	−0.246	0.846	0.195

Returning to the question of the selection of eigenvalues, Figure 16-1 shows the scree plot resulting from the analysis. Note the distinctive shallowing of slope that occurs after the second and the fourth eigenvalues; these points could be used to exclude factors, and indeed, the two components of interest (L and Q) are identified as the first two components.

Figure 16-2 shows the effect of the rotation; you can see that the variables associated with the L factor (spelling, verbal, and reading) are closely clustered in 3D space, as are the variables associated with the Q factor (arithmetic and geometry). Note that the other two variables (sports and music) are then roughly equidistant from the centroids of the two component-oriented clusters. The impact of the rotation is easier to observe in 3D space than by looking at the loading tables.

The output matrix from the FA procedure is shown in Table 16-9. This shows the scores for the three components computed for each of the study participants; if this were the GRE or SAT, these are the scores that would be reported back to the test takers. Note that the precision of the results depends on your computer package.

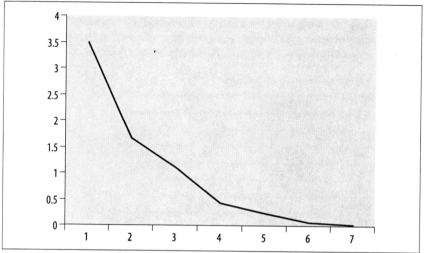

Figure 16-1. Scree plot

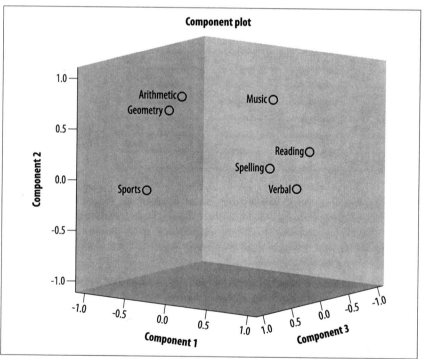

Figure 16-2. Component plot in rotated space

Table 16-9. Component scores for each participant

Component 1, L	Component 2, Q	Component 3, Sport
0.518	1.132	−0.095
−1.170	−0.128	0.084
−1.396	−1.207	1.619
1.094	1.198	1.128
0.706	−1.049	0.0139
−.225	−0.588	−2.097
−1.416	1.711	−0.309
0.109	−0.259	−0.721
1.064	−0.273	0.179
0.715	−0.536	0.198

As with all of the other techniques you have learned about in this book, PCA and FA have some basic prerequisites that need to be met in order for the results to be valid and/or reliable. As the data set grows larger, the results become more reliable. In the case of psychometrics, reliability is usually established when a test has been administered to many hundreds of thousands of individuals, across different national and linguistic groups. The other main requirement is that the number of cases must always be larger than the number of variables in the input matrix. Normally, tests for statistical significance are not performed with PCA, so outliers and other potential sources of bias are much less likely to cause problems than (say) with ANOVA. This is why PCA is often regarded as a data-cleaning tool, used before any other type of analysis or statistical test is applied.

For PCA, the assumptions of linear correlation also hold, i.e., that variables must be linearly related. The best results are obtained when all of the variables are neither zero nor perfectly correlated, as this introduces problems with sphericity and the underlying calculation of the orthogonal decomposition.

Cluster Analysis

Cluster analysis is a set of techniques that allows groupings of cases to be made on the basis of one or more variables. Some cluster analysis techniques allocate cases to groups by partition, while other techniques provide for hierarchical trees that show the taxonomic relationship between groups and their ancestors. A related technique, *Discriminant Function Analysis* (DFA), can be used to develop rules to assign cases to groups, based on an understanding of the parametric structure of the groups, and is better at predicting group membership than cluster analysis alone. Thus, the two techniques are often used in conjunction with each other; cluster analysis may be used where the number of groups is initially unknown, and once this number has been established, DFA may be used for the prediction of individual group membership for each case.

Cluster analysis is very useful for two scenarios. Firstly, you may already know how many groups you expect to find in the data, so you pass this number of groups to the algorithm, and let it take care of the allocation (*k-means*). Alternatively, you may not know how many groups exist, in which case, you can ask the algorithm to estimate how many groups there actually are.

Cluster analysis is a highly empirical tool; its success depends largely on the quality of data supplied. Cluster analysis works by taking an input vector of Y, with n cases and p variables, and allocating each of n cases to one of k groups. Each of the variables measures some aspect of an object under study; continuing with the psychometric example, each variable may represent a score on a particular type of ability test (reading, spelling, etc.). The algorithm works by randomly creating k clusters, identifying the *centroids* (or cluster centers), and then assigning each case to the closest centroid. Cases are moved between clusters to minimize within-cluster variability and maximize between-cluster variability. The process continues until it converges according to some predefined criterion, e.g., the cluster membership doesn't change after one iteration. Note that because there is some randomness introduced by the initial assignment of centroids, you don't always get the same answer.

The computational goal is to ensure that all members of groups 1...k are similar to other members of that group and dissimilar to members of other groups. Similarity—or dissimilarity—is determined by the use of a specific distance measure. A number of different measures have been developed, including:

Euclidean distance
  The geometric distance between two points in a multidimensional space.

Manhattan distance
  A "city block" distance that reduces the influence of outliers.

Mahalanobis distance
  Within-cluster distances tend to be increased while between-cluster distances are decreased.

Let's revisit the psychometric example. Having shown that there are three factors, including L and Q, the psychologist is now interested in determining whether there might be some basis for classifying students into different educational groups based on this latent structure, since the identified factors for L, Q, and *Sports* were orthogonal. The issue is specialization: if students are only "good" at *Sports*, or "Linguistic" or "Quantitative" work, then those students should be streamed appropriately into specialist classes.[*] The main problem is that some students may be "good" at more than one of these skills, and the idealized view provided by the rotated loading matrix shown previously in Figure 16-2 may not apply to all cases.

Cluster analysis is proposed to be used to determine if there are three distinct groups in this data set, corresponding to distinct members of the proposed L, Q,

---

[*] Educators' preferences for comprehensive versus specialist education seem to run along a "business cycle," in which the prevailing trend changes every 10 years or so; thus, the issue of providing evidence for one approach or the other is nontrivial.

and *Sports* classes. In this case, we pass $k = 3$ to the algorithm and ask it to identify three groups, and then assign each student to a class.

The initial cluster centers are shown in Table 16-10, and after several iterations the algorithm converges to a solution, with the final cluster membership, final cluster centers, and pairwise distances between the final clusters shown in Tables 16-11 through 16-13. The initial cluster centers are related to correlations shown in Table 16-10, and the corresponding principal components that were extracted: cluster 1 is strongly associated with reading, verbal, and spelling; cluster 2 with arithmetic and geometry; and cluster 3 with sports. While there are some changes during the iterative process, these groupings tend not to change. The resulting group allocations are simply a function of the distance from each centroid. The pairwise distances between each centroid are also reasonably consistent with each other, i.e., the between-group distances appear to have been successfully maximized, and there does not appear to have been difficulty in separating them. Also, only one student was allocated to the second cluster, which would not have been expected from the PCA. Adding more cases into the analysis would almost certainly improve the reliability of the result.

Table 16-10. Initial cluster centers

	Cluster		
	1	2	3
Reading	10.00	3.00	2.00
Music	9.00	9.00	3.00
Arithmetic	3.00	10.00	2.00
	1	2	3
Verbal	10.00	2.00	6.00
Sports	6.00	6.00	8.00
Spelling	10.00	4.00	6.00
Geometry	3.00	9.00	4.00

Table 16-11. Cluster solution: cluster membership

Case number	Cluster	Distance
1	1	6.565
2	3	2.915
3	3	2.915
4	1	7.078
5	1	4.468
6	1	5.053
7	2	0.000
8	1	3.332
9	1	2.556
10	1	4.238

Table 16-12. Cluster solution: final cluster centers

	Cluster		
	1	2	3
Reading	8.57	3.00	3.50
Music	8.86	9.00	4.50
Arithmetic	4.00	10.00	3.50
Verbal	9.00	2.00	5.50
Sports	5.14	6.00	7.00
Spelling	9.00	4.00	5.50
Geometry	3.86	9.00	4.50

Table 16-13. Cluster solution: pairwise distances between final cluster centers

Cluster	1	2	3
1		12.971	8.562
2	12.971		9.925
3	8.562	9.925	

Table 16-14 shows the ANOVA results for the significance of each variable in terms of *discriminability*. The results are not intended to be a strict test of statistical significance in terms of hypothesis testing, but are useful in examining which variables provided discriminability. Spelling, verbal, and reading scores were all significant (unsurprisingly), but the scores for the second and third clusters (arithmetic and geometry, and sports) were not significant. The first of these makes sense, since scoring highly on spelling, verbal, and reading does discriminate between the first and second groups, but the lack of discriminability for the third cluster is a surprise (although recall from PCA that this only accounted for 15% of the variance).

Table 16-14. ANOVA

	Cluster		Error		F	Sig.
	Mean square	df	Mean square	df		
Reading	28.893	2	1.745	7	16.558	0.002
Music	15.321	2	1.622	7	9.443	0.010
Arithmetic	17.000	2	9.214	7	1.845	0.227
Verbal	26.950	2	0.643	7	41.922	0.000
Sports	2.771	2	4.122	7	0.672	0.541
Spelling	17.550	2	1.786	7	9.828	0.009
Geometry	11.571	2	8.194	7	1.412	0.305

Although hierarchical clustering will not be covered in detail, the *dendrogram* that arises from the analysis is very useful in understanding the relatedness of specific cases, and is widely used in taxonomic analysis of various kinds. Figure 16-3 shows the dendrogram computed using the average *linkage* between groups. Here, you can see that the closest root relations are between cases 5 and 9, and 8 and 10, and these two groups of cases are next most closely related to each other, and then to case 6. Overall, the case with the greatest relation to all others is 7. Examining the linkages in this way can be very useful in trying to characterize and understand the relations between individual cases and the clusters to which they are allocated.

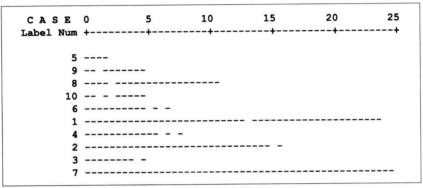

Figure 16-3. Dendrogram computed using the average linkage between groups

Discriminant Function Analysis

Discriminant Function Analysis (DFA) is used to construct rules that allow classification of cases into two or more groups using a linear combination of variables. The goal is to maximize the distance between two or more groups, which in turn maximizes discriminability through a set of one or more functions of a specific rank, i.e., the number of functions required to maximize the separation between groups. These functions are typically linear combinations of the input variables, and are called *Linear Discriminant Functions* (LDFs). DFA is related to classification analysis, where the goal is to maximize the accuracy of allocation of cases to groups. Classification functions may be either linear or nonlinear, or use a generalized function approximator, such as an *artificial neural network*.

Thus, cluster analysis and classification analysis are in some ways trying to solve the same problem but by different means, expressed as finding the maxima of different functions (e.g., maximizing distance or classification accuracy).

Returning to the psychometric example, and given the group allocations provided by cluster analysis, DFA can be used to determine a set of discriminant functions that provides maximum separation between the groups. It is then possible to test the null hypothesis of the equality of group means for each variable. In the two-group case, this can be evaluated using a *t*-test, or in the case of more than two groups, an F-test can be performed. The results shown in Table 16-15 indicate that there are significant differences for reading, $F(2, 7) = 16.558$, $p = 0.002$; music, $F(2, 7) = 9.443$, $p = 0.010$; verbal, $F(2, 7) = 41.922$, $p = 0.000$; and spelling, $F(2, 7) = 9.828$, $p = 0.009$. Thus, in terms of discriminability, you could retain reading, music, verbal, and spelling and still maximize the distance between groups.

Table 16-15. Tests of equality of group means

	Wilks's Lambda	F	df1	df2	Sig.
Reading	0.174	16.558	2	7	0.002
Music	0.270	9.443	2	7	0.010
Arithmetic	0.655	1.845	2	7	0.227
Verbal	0.077	41.922	2	7	0.000
Sports	0.839	0.672	2	7	0.541
Spelling	0.263	9.828	2	7	0.009
Geometry	0.713	1.412	2	7	0.305

Table 16-16 shows the two *canonical discriminant functions* required to classify the cases into groups. Interestingly, the first function captures 96% of the variance, while the second function only captures 4%.

Table 16-16. Canonical discriminant functions

Function	Eigenvalue	% of variance	Cumulative %	Canonical correlation
1	79.224	96.0	96.0	0.994
2	3.287	4.0	100.0	0.876

Table 16-17 shows the computed values for Wilks's Lambda, which can be used to evaluate the significance of the discriminant functions in a multivariate sense. Unfortunately, none are significant. This probably reflects the fact that function 1 accounts for such a high proportion of variance. Having a larger number of samples would provide more robust results and hopefully lead to significance.

Table 16-17. Wilks's Lambda

Test of function(s)	Wilks's Lambda	Chi-square	df	Sig.
1 through 2	0.003	23.362	14	0.055
2	0.233	5.822	6	0.443

The standardized canonical discriminant function coefficients are shown in Table 16-18.

Table 16-18. Standardized canonical discriminant function coefficients

	Function	
	1	**2**
Reading	−0.706	−0.141
Music	1.838	−0.368
Arithmetic	−0.364	−0.707
Verbal	3.686	1.409
Sports	−0.150	1.309
Spelling	−1.884	−2.030
Geometry	1.916	0.945

The structure matrix is shown in Table 16-19. Here, you can see significant loadings of reading and spelling on function 1, and significant loadings of music, verbal, arithmetic, geometry, and sports on function 2. These are slightly different from what you might have expected—say, from PCA or cluster analysis—but it's worth keeping in mind that the algorithm in each case has a different computational goal (e.g., maximizing distance between clusters or maximizing accuracy of classification).

Table 16-19. Structure matrix

	Function	
	1	**2**
Reading	0.243*	−0.140
Spelling	0.188*	0.034
Music	0.115	−0.708*
Verbal	0.379	0.433*
Arithmetic	−0.046	−0.331*
Geometry	−0.055	−0.225*
Sports	−0.043	0.121*

Finally, Table 16-20 shows the relationship between the two discriminant functions and the group centroids.

Table 16-20. Functions at group centroids

Cluster number of case	Function	
1	4.804	−0.169
2	−14.483	−3.465
3	−9.573	2.324

Multidimensional Scaling

Multidimensional scaling (MDS) is similar to PCA in the sense that it is concerned with data reduction; unlike cluster analysis, MDS is concerned with determining the underlying dimensionality of a data set, based on a measure of dissimilarity using a *proximity matrix*. The goal is to reduce a data set into a lower dimension k based on the dissimilarity between all objects in the data set, and to identify a set of principal coordinates. The ordination process uses a geometric representation to identify the principal components. Many of the same distance measures used in cluster analysis are also applicable to MDS.

A number of different algorithms are available to perform MDS, including *Prox-scal* and *Alscal*, with the former using distances in the dissimilarity measures while the latter uses squares of distances. As with the other multivariate techniques reviewed in this chapter, the psychometric example is also suitable for analysis using MDS. The final coordinates arrived at using Proxscal are shown in Table 16-21, and the resulting common space projections are shown in Figure 16-4. In the two dimensions shown, you can see that reading, verbal, and spelling are closest to each other, as are arithmetic and geometry, identifying once again the L and Q factors. However, also notice that music is approximately equidistant from reading as reading is from spelling. Once again, sports appears relatively isolated from the other variables.

Table 16-21. Functions at group centroids

	Dimension	
	1	2
Reading	−0.558	0.154
Music	−0.253	0.454
Arithmetic	0.798	0.312
Verbal	−0.630	−0.136
Sports	0.279	−0.428
Spelling	−0.461	−0.218
Geometry	0.825	−0.138

One of the nice features of MDS is that a number of metrics are available for determining deviations from *monotonicity* to satisfy a constraint relating to the form of the function that maps the distances between cases in a k-dimensional space. To determine whether the relationship between distances and similarities is nonmonotonic, the *Standardized Residual Sum of Square* (STRESS) measure can be used, which is always $0 <$ STRESS < 1. STRESS is normally minimized during MDS, and thus, a target for an excellent fit is usually $0 <$ STRESS < 0.1. In this example, STRESS $= 0.09$, so the fit of the model is excellent.

The results using Alscal are slightly different, as shown in Figure 16-5. Here, arithmetic and geometry (Q) are clustered well away from reading, spelling, and verbal (L), with sports once again being significant. Table 16-22 shows the iterative nature of Alscal, where the algorithm iterates until a satisfactory level of STRESS (in this case, Young's S-STRESS formula based on squares of distances rather than actual distances) is achieved.

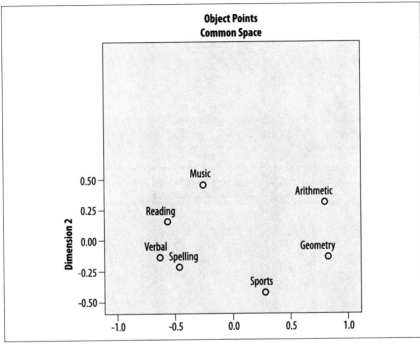

Figure 16-4. MDS common space mappings (Proxcal)

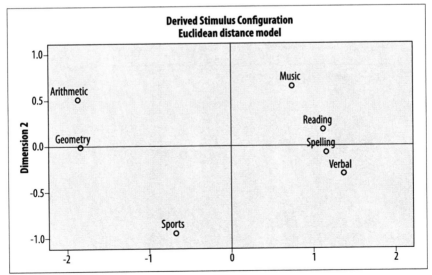

Figure 16-5. MDS common space mappings (Alscal)

Table 16-22. Alscal iterations to minimize S-STRESS

Iteration	S-STRESS	Improvement
1	0.02266	
2	0.01736	0.00530
3	0.01456	0.00280
4	0.01279	0.00178
5	0.01157	0.00121
6	0.01066	0.00091

The significance of STRESS and the monotonicity constraint is made clearer in Figure 16-6, where you can see that the function is monotonically nondecreasing in a very linear fashion.

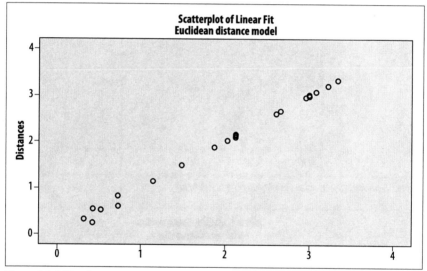

Figure 16-6. Nondecreasing monotonicity

17

Business and Quality Improvement Statistics

Many of the statistics used in business and quality improvement applications are those within the common repertoire of basic statistics, including the chi-square test (covered in Chapter 10), *t*-tests (Chapter 8), and techniques based on the General Linear Model (Chapters 12–15). However, there are also a number of techniques developed for the specific needs of business and quality improvement applications, and those will be the subjects of this chapter.

Index Numbers

Index numbers are commonly used in business to measure the change in quantity or price over time for some good or combination of goods and services, and are often the data points used in time series analyses. One example is the Consumer Price Index (CPI), which represents the average price of a quantity of consumer goods and services believed to be typical household purchases in the United States. The U.S. CPI is calculated monthly by the Bureau of Labor Statistics of the U.S. Department of Labor and is used as a measure of inflation and to calculate cost of living adjustments for pensions and wages. Although many criticisms have been made of the CPI, it has proven highly useful as a summary measure of the average cost of living and allows comparison across historical periods and geographic areas. Many other countries also calculate a CPI or similar index, including Canada, China, Israel, New Zealand, Australia, and many European countries.

Calculation of indexes can be very simple (when the index reflects the change in the price or quantity of a single commodity) or very complex (when the index reflects a weighted average of a number of goods and services, as is true for the CPI). A *simple index number* displays the change in time of the price or quantity of a single commodity, for instance the number of television sets sold or the price of an ounce of gold. To calculate a simple index, you must choose a *base period* to be used for comparison: the index will then represent the change in price or quantity relative to that base period. To calculate a simple index, three steps are required:

1. Obtain the prices or quantities for the commodity for the time period of interest.

2. Select a base period and obtain the price or quantity for that year.

3. Calculate the index number for each time period, using the formula:

$$I_t = \frac{Y_t}{Y_0} \times 100$$

where I_t = the index at time t, Y_t = the price or quantity at time t, and Y_0 = the price or quantity in the base period.

For instance, suppose we wanted track to the health of the automobile manufacturing industry in the United States over the last 20 years. As part of this research we could create an index expressing the number of automobiles manufactured each year in terms of the first year of the time period. If we had data for the years 1986–2005, 1986 would be the base year and the quantity of cars manufactured that year would be Y_0. Consider Table 17-1, which shows a reduced and entirely hypothetical data set to demonstrate calculation of a simple index.

Table 17-1. Simple index calculation

Year	Number of automobiles manufactured
1986 (base year)	5,000
2005	4,000

$$I_{2005} = \frac{4000}{5000} \times 100 = 80$$

An index of 100 represents the same quantity or price as the base period. An index less than 100 indicates a decline in quantity or price, and an index greater than 100 indicates an increase in quantity or price compared to the base period. One of the great advantages of index numbers is that they put quantities measured on different scales and with different ranges of scores into a common metric. For instance, using indexes we can easily compare the relative increase or decrease over time in production of automobiles, motorbikes, and bicycles.

A *composite index* combines information about the price or quantity of several types of goods or services. For instance, we might calculate the quantity of beer sold by the three largest breweries in Scotland by adding together the quantity sold by each manufacturer. If we performed this calculation for a number of years and selected one year to use as the base period, we could calculate an index number for each year as we did the simple index in the example above. This type of index is known as a *simple composite index* because it is calculated by combining information from several sources without using any type of weighting.

When some type of weighting is used to create the totals used to calculate the index number, this is known as a *weighted composite index*. Price indexes are often weighted by the quantity of goods sold, for instance. There are several different ways to apply a weighting scheme, because the quantities of items purchased will change from one time period to the next, and the choice of weights can have an important influence on the results of the index calculations. Once a scheme of weighting is selected, however, calculating the index numbers themselves is straightforward. The total price is calculated for each time period, and the index numbers for each time period are calculated using a procedure analogous to that used for the simple index.

A *Laspeyres index* uses the base period quantities as weights, so inflation or deflation is measured for a fixed "basket" of goods or services. The CPI is an example of a Laspeyres index: the quantities used for weighting are based on samples of purchases by over 30,000 families in the years 1982–1984. The steps in calculating a Laspeyres index are:

1. Collect price information $(P_{1t}, P_{2t}, ...P_{kt})$ for each time period for each item (1 through k) to be included in the index.

2. Collect purchase quantity information $(Q_{1t_0}, Q_{2t_0}, ...Q_{kt_0})$ for the base period for each item to be included in the index.

3. Select a base period (t_0).

4. Calculate the weighted totals for each time period, using the formula:

$$\sum_{i=1}^{k} Q_{it_0} P_{it}$$

5. Calculate the Laspeyres index, I_t, by dividing the weighted total for each time period by the weighted total for the base period and multiplying by 100, i.e.:

$$I_t = \frac{\sum_{i=1}^{k} Q_{it_0} P_{it}}{\sum_{i=1}^{k} Q_{it_0} P_{it_0}} \times 100$$

Table 17-2 shows a simple example of the calculation of the Laspeyres index for a "market basket" containing only two types of goods.

Table 17-2. Laspeyres index example

Product	Base quantity (2000)	2000 price	2005 price
Bread	10	1.00	1.50
Milk	20	2.00	4.00

The weighted total for 2000 is:

$$(10 \times 1.00) + (20 \times 2.00) = 50.00$$

The 2005 weighted total is:

$$(10 \times 1.50) + (20 \times 4.00) = 95.00$$

The Laspeyres index for this basket of goods in 2005, using 2000 as the base year, is therefore:

$$I_{2005} = \frac{95}{50} \times 100 = 190$$

A *Paasche Index* calculates weighted totals using the quantities of items purchased in each time period. This has the advantage of adjusting for changes in consumer habits: for instance, if the price of a good rises, people tend to buy less of it and purchase less expensive substitutes. An example of substitution would be if the price of beef rose faster than the price of chicken, and people responded by buying more chicken and less beef. This change in consumer habits would not be reflected in the Laspeyres index but would be in the Paasche index.

The steps to calculate a Paasche index are similar to those for a Laspeyres index. The main difference is that information about the quantities purchased in each time period must be collected and used to calculate the weighted totals.

1. Collect price information (P_{1t}, P_{2t}, ...P_{kt}) for each time period for each item (1 through k) to be included in the index.

2. Collect purchase quantity information (Q_{1t}, Q_{2t}, ...Q_{kt}) for each time period for each item to be included in the index.

3. Select a base period (t_0).

4. Calculate the weighted totals for each time period, using the formula:

$$\sum_{i=1}^{k} Q_{it} P_{it}$$

5. Calculate the Paasche index, I_t, by dividing the weighted total for each time period by the weighted total for the base period and multiplying by 100, i.e.:

$$I_t = \frac{\sum_{i=1}^{k} Q_{it} P_{it}}{\sum_{i=1}^{k} Q_{it} P_{it_0}} \times 100$$

Table 17-3 shows a simple example of calculating a Paasche index.

Table 17-3. Calculating a Paasche index

Product	2000 quantity	2000 price	2005 quantity	2005 price
Bread	10	1.00	15	1.50
Milk	20	2.00	15	4.00

The 2000 weighted total is:

$$(10 \times 1.00) + (20 \times 2.00) = 50.00$$

The 2005 weighted index is:

$$(15 \times 1.50) + (15 \times 4.00) = 82.50$$

The Paasche index for this basket of goods in 2005, using 2000 as the base year, is therefore:

$$I_{2005} = \frac{82.5}{50} \times 100 = 165$$

Note that although the prices were the same in each example, the different methods of weighting resulted in substantial differences in the two index numbers (190 versus 165). The Paasche index has the advantage of comparing prices for a basket of goods at purchase levels appropriate to each time period. It has the disadvantage of requiring that this information (quantities of each type of good purchased) be collected for each time period, which may be prohibitively expensive. Another disadvantage of the Paasche index is that because both prices and quantities may change from one period to another, it is difficult to compare Paasche index numbers for any two periods when one of the periods is not the base period.

Time Series

Time series are used frequently in business statistics to chart the changes in some quantity over time. Strictly speaking, a time series is just a sequence of measurements of some quantity taken at different times, often at equally spaced intervals. The previous example of the number of automobiles manufactured in the years 1986–2005 would qualify, as would the measurements discussed later in this chapter in the section on control charts. Time series may be used for either descriptive or inferential purposes; the latter includes *forecasting*, i.e., predicting values for time periods that have not yet occurred. The reader should bear in mind, however, that time series analysis is a complex topic with many specialized techniques, and that this section can only introduce some of the terminology and a few simple examples. Anyone planning to work in this area should consult a textbook devoted to the subject, such as Robert S. Shumway, *Time Series and Its Applications: With R Examples* (Springer). Note also that some authors (e.g., Tabachnick and Fidell) specify that at least 50 data points are required to use time series techniques.

One characteristic of time series data is that data points in sequence are assumed to not be independent, as would be required for standard General Linear Model and many other analytical techniques, but to be *autocorrelated*. This means that the value for a given time point is expected to be related to the points before and after it, and perhaps to points more distant in the series as well.

Criticisms of the U.S. Consumer Price Index (CPI)

The CPI is the principal measure of price changes in the United States and has been produced in some form by the Bureau of Labor Statistics since 1919. It is used for many purposes, including as a measurement of inflation and in calculating cost of living adjustments for negotiated wage packages and social security and civil service retirement benefits. Not surprisingly, an index used for so many purposes also comes under criticism from many quarters.

Among the principal criticisms, all of which tend to lead to the CPI overstating inflation, are:

Quality change and new product bias

The CPI does not account for the improved quality of some items, such as electronics. A DVD player that sells for $150 in 2005 may be of a substantially higher quality and therefore "worth more" to the consumer than one that cost $100 in 2000, but this increase in quality is not reflected in the CPI. Similarly, because a fixed market basket of items is used, new items are not included in the index in a timely fashion. The result is that early declines in price (typical among new electronics products, for instance) are not captured in the index.

Substitution bias

The use of a fixed basket of goods (weights are updated about once every 10 years) does not allow for changes in consumer purchasing patterns in response to changes in price. For instance, if the price of meat rises faster than that of other protein foods such as poultry or eggs, consumers may respond by purchasing more poultry and eggs and less meat, but this shift will not be reflected in the CPI.

Outlet substitution bias

Because price information is gathered from traditional sales outlets such as department stores, newer outlets such as big-box discounters or Internet sales are not fully represented in the CPI surveys.

Time series data is assumed to be *stationary*, meaning that properties such as mean, variance, and autocorrelation structure are constant over the entire range of the data. Sometimes data has to be preprocessed by *differencing* in order to achieve stationarity: this means subtracting each data point from some previous point. The distance between the two points is called the *lag*. Techniques to test for the types of differencing required, and to perform them automatically, are included in software packages dedicated to time series analysis. Other transformations, such as taking the square root or logarithm of the data to stabilize the variance, may also be applied in before the times series analysis begins.

Additive models are often used to describe the components of a time series, i.e.:

$$Y_t = T_t + C_t + S_t + R_t$$

The components of the trend Y_t in this model are:

T_t

Secular or long-term trend, i.e., the overall trend over the time studied.

C_t

The cyclical effect, i.e., fluctuations about the secular trend due to business or economic conditions, such as periods of general economic recession or expansion.

S_t

The seasonal effect, i.e., fluctuations due to time of year, for instance the summer versus the winter months.

R_t

The residual or error effect, i.e., what remains after the secular, cyclical and seasonal effect have been accounted for; it may include both random effects and effects due to rare events such as hurricanes or epidemics.

Much of time series analysis is devoted to resolving the variance observed over time into these components. The concept is similar to partitioning the variance in ANOVA models, although the mathematics involved is different.

Exact measurements plotted over time, also known as *raw time series*, will almost always show a great deal of minor variation that may obscure major trends that could help explain the pattern and make accurate future forecasts. Various types of *smoothing* have been devised to deal with this problem. They can be divided into two types: *moving average* or *rolling average* techniques, which involve taking some kind of average over a series of consecutive points and substituting this average for the raw values, and *exponential techniques*, in which an exponential series is used to weight the data points.

To calculate a *simple moving average* (SMA), take the unweighted mean of a specified number of data points (*n*) prior to the time point in question. The size of *n* is sometimes described as a *window* because the idea is that a window including *n* data points (a window of width *n*) is used to calculate the moving average. As you progress forward in time through the data, the window moves so you can "see" different data points each time, and the average is calculated using the points included in the window for each time point. For instance, a five-point SMA would be the average of a given value and the previous four data points.

The SMA for each new data point drops only one value and adds only one new value, reducing the fluctuation from point to point. This attribute gave rise to the term *rolling average*, because the last value "rolls off" the series as the new value "rolls on." This is similar to the methodology used to compute player standings on professional tennis tours, although in that case a total rather than an average is computed. Each player's total points in a given week is the sum of their points from the previous 52 weeks, and each week the total is recalculated as the oldest week's points are dropped and the newest week's points added in.

The greater the size of the window used to calculate an SMA, the greater the smoothing since each new data point has less influence relative to the total. At some point the data may become so "smoothed" that important information about the pattern is lost. In addition, the larger the window, the more data points that have to be discarded (because you need more points to calculate each average). This may be seen in the example in Figure 17-1 and Table 17-4.

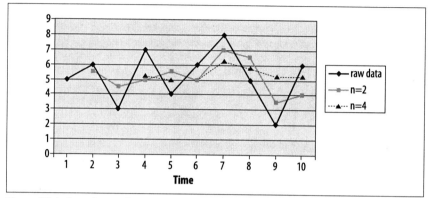

Figure 17-1. Raw data and moving averages with n = 2 and n = 4

Table 17-4. Simple moving average with different sized windows

Time	1	2	3	4	5	6	7	8	9	10
Raw data	5	6	3	7	4	6	8	5	2	6
$n = 2$		5.5	4.5	5	5.5	5	7	6.5	3.5	4
$n = 4$				5.25	5	5	6.25	5.75	5.25	5.25

As would be expected, the largest fluctuations are seen in the raw data, while by the time the window is increased to 4, there is very little fluctuation in values between time points.

When a window of 2 is used, only one data point has to be dropped from the moving average (the first, because it has no prior point to use in calculating the average. When a window of 4 is used, the first three points have to be dropped because none of them has three prior points to use in calculating the average.

The *central moving average* (CMA) is similar to the moving average but uses a window of size n with both past and future data used to calculate the average for each point. For a CMA of size 3, for instance, the value at time 2 would be 4.67 or (5 + 6 + 3)/3. Note that the future points are measured data, not forecasts: they are "future" only in that they are measured at a later time than the central data point for a given CMA. Table 17-5 shows an example.

Table 17-5. Central moving average (n = 3) for previous data

Time	1	2	3	4	5	6	7	8	9
Raw data	5	6	3	7	4	6	8	5	2
CMA ($n = 3$)		4.67	5.33	4.67	5.67	6.00	6.33	5.00	

The *weighted moving average* (WMA) uses values from a window of size n but assigns greater weight to the data points closer to the point in question. If not otherwise specified, arithmetic rather than exponential weights are used. A typical system assigns the weight n to the day whose weight is being calculated, where n is the number of days included in the weight. Every other day included in the WMA is weighted one less for each day it is removed from the day being weighted. In a five-day WMA, the day being weighted would be given a weight of five, the previous day a weight of four, and so on down to four days previous, which would have a weight of one. This weighted sum is divided by the sum of the weight factors, which will be $[n(n-1)]/2$. The WMA makes intuitive sense in any situation where consecutive points can be assumed to be the most closely related, with the relationship lessening as the length of time between data points increases.

The *exponential moving average* (EMA) also applies more weight to closer measurements, but the weights allocated to data points further from the point in question decrease exponentially rather than arithmetically. To calculate an EMA, an exponential smoothing constant α between 0 and 1 is selected. This constant is related to the number of time points included, n, by the following equation:

$$\alpha = \frac{2}{n+1}$$

so that $\alpha = 0.2$ is equivalent to $n = 9$ because ($2/10 = 0.2$). α is then applied in the following formula, which is continued until the terms become so small as to become negligible:

$$\text{EMA} = \frac{p_1 + (1-\alpha)p_2 + (1-\alpha)^2 p_3 + (1-\alpha)^3 p_4 + \dots}{1 + (1-\alpha) + (1-\alpha)^2 + \dots}$$

where p_1 is the measure at the given time point for which the EMA is being calculated, p_2 is one time point removed, p_3 is two time points removed, and so on. The denominator approaches $1/\alpha$ as the number of points included increases, and 86% of the total weight in the calculation will be included in the first n time points. n is not the number of data points included in calculating the EMA, as it is in the simple and weighted moving averages: the stopping point will be determined by the value chosen for α, and by the researcher's decision as to what constitutes a negligible value.

Decision Analysis

We all make decisions every day, but how do we go about making the best decision, particularly in a situation where a lot (for instance, a large amount of money) is at stake? *Decision analysis* is a body of professional practices, methodologies, and theories used to systematize the decision-making process in the service of improving the process of decision-making. There are many schools of thought within decision theory, and each may be useful in a particular context: this section concentrates on several of the most common decision analysis methods, which will help to introduce the student to the types of processes involved, as well as providing concrete assistance in particular decision-making contexts.

Improvement
Statistics

The decision-making process will be described in terms of financial costs and payoffs, but can be used with other metrics as well (for instance, personal satisfaction or improved quality of life) as long as they can be quantified.

In decision analysis, the process of making a decision is usually conceived of as a series of steps that is not unlike the process involved in hypothesis testing. They are also not that different, except for the selection and application of a mathematical model in steps 5–6, from the ordinary type of decision-making process we engage in every day. Besides the potential to lead to better decisions, going through these steps (and justifying and documenting them) should make the reasons for a particular decision easier to explain and justify to someone who wasn't involved in the process. The basic steps are:

1. Define the situation or context, including *states of nature* (any situation in the real world that may influence the outcomes). States of nature must be stated as mutually exclusive and exhaustive alternatives, for instance, strong/medium/weak market, or low rainfall/adequate rainfall.

2. Identify the choices at hand, i.e., the alternative decisions that could be made; these are known as *actions*.

3. Identify the possible *outcomes* or consequences.

4. Assign costs and profits associated with all possible combinations of choices and outcomes.

5. Select an appropriate mathematical model.

6. Apply the model using the information from steps 2–4.

7. Make a decision based on the best expected outcome as predicted by the model.

Choice of a decision theory methodology depends in part on how much is known about a situation. There are three types of contexts in which one may apply decision theory:

- Decision-making under certainty
- Decision-making under uncertainty
- Decision-making under risk

Decision-making under certainty means that the future state of nature is known, so the decision-making process requires only stating the alternatives and payoffs in order to be able to pick the choices that will invariably lead to the best outcome. This situation will not be further discussed because no mathematical modeling is required and because there is no uncertainty about what is the best choice.

Decision-making under uncertainty is a more common situation: we don't know the probabilities of each state of nature and must make our decision based only on the gains or losses from different actions under each state. For instance, if we are choosing from several cities in which to open a restaurant, the success of the restaurant depends in part on the economic climate in each city when the restaurant opens, but we may not have good estimates of the future economic climates in the future in these cities. Similarly, when choosing what crop or variety to plant, our success at harvest time depends partly on the amount of rainfall during the growing season, but we may feel we don't have sufficient information to estimate this in advance.

In *decision-making under risk*, we know the probabilities of each outcome (or have reasonable estimates of them) and can combine this information with that about expected payoffs to determine which decision is optimal.

Minimax, Maximax, and Maximin

The information needed to make a decision under uncertainty may be summarized in a payoff table, where each row represents a possible action taken, and each column a state of nature. The numbers within the cells of the table represent the outcome expected under different combinations of actions and states of nature. For instance, suppose we are considering whether to invest in staging an event in a large outdoor venue or a smaller indoor venue, with a third alternative to not invest in the event at all. Suppose also that the event is to be held in a climate where rainstorms are common during the season of the year when the event will take place, and we don't feel we can assign reasonable probabilities to the chance of rain on a particular day. The payoff table might look like Table 17-6.

Table 17-6. Payoff table for investing in an event

		Weather	
		Rain	No rain
	Outdoor venue	−$50,000	$500,000
Action	Indoor venue	$200,000	$200,000
	Do not invest	$0	$0

The outdoor venue is larger so if it doesn't rain that night we stand to make a large profit (gain of $500,000). If it rains, the event will be canceled and we will lose our investment as well as not making any revenue (loss of $50,000). On the other hand, the indoor venue should return about the same profit ($200,000) whether it rains or not: less than the outdoor venue if the weather is good, more than the outdoor venue if it rains. Finally, we might decide that investing in staged events is too risky and choose to apply our money elsewhere.

We can also create an *opportunity loss* table, which expresses the amount of money we lost the opportunity to make by choosing a particular course of action. For our hypothetical event-investment-in-rainy-country scheme, the opportunity loss table would look like Table 17-7.

Table 17-7. Opportunity loss table for investing in an eventt

		Weather	
		Rain	No rain
	Outdoor venue	$250,000	0
Action	Indoor venue	$0	$300,000
	Do not invest	$200,000	$500,000

Note that there are no negative numbers in an opportunity loss table. The best action for a given state of nature has a loss of $0, while the others represent the amount of money lost by not choosing the best action for that state of nature.

Three procedures have been developed for decision-making under uncertainty: *minimax*, *maximax*, and *maximin*. The *minimax* procedure involves choosing the action that will minimize opportunity loss. To make a minimax decision, we use the opportunity loss table to identify the maximum opportunity loss for each action, then choose the action with the lowest opportunity loss. In this example:

Maximum opportunity loss (outdoor venue) = $250,000
Maximum opportunity loss (indoor venue) = $300,000
Maximum opportunity loss (do not invest) = $500,000

Using the minimax procedure, we would decide to finance the event at the outdoor venue, because it has the smallest maximum opportunity loss of the three choices.

The *maximin* strategy involves choosing the action that has the largest minimal outcome. This has been described as the strategy for pessimists, because it chooses the alternative with the highest minimal gain or smallest loss, i.e., the best outcome under unfavorable conditions. In this example:

Minimum gain (outdoor venue) = −$50,000
Minimum gain (indoor venue) = $200,000
Minimum gain (do not invest) = $0

Using the maximin strategy, we would choose the indoor venue, because the worst we could do is make $200,000 regardless of weather conditions.

The *maximax* strategy involves choosing the action that has the highest maximum outcome. For this reason it might be called the strategy for optimists, because it chooses the strategies that provide the best outcome under the most favorable state of nature. In this example:

Maximum gain (outdoor venue) = $500,000
Maximum gain (indoor venue) = $200,000
Maximum gain (do not invest) = $0

Using the maximax strategy, we would choose the outdoor venue, because it offers the highest maximum outcome.

Decision-Making Under Risk

If the probabilities of different states of nature are known or can be reasonably estimated, we are in a decision-making under risk situation. Let's say that in the previous example, we also had information about the probability of rain on the night when the event is scheduled. If the probability of rain is 0.6, that means the probability of no rain is 0.4 because they are mutually exhaustive states of nature. We add this information in Table 17-8.

Table 17-8. Expected payoff from various actions, given probabilities of different states of nature

		Rain	No rain	Expected payoff
	Probability	0.6	0.4	
Actions	Outdoor venue	–$50,000	$500,000	170,000
	Indoor venue	$200,000	$200,000	200,000
	Do not invest	$0	$0	0

The expected payoff is calculated by multiplying the payoff under each combination of actions and states of nature by the probability of the state of nature. For instance, for the "outdoor venue" option:

$$E(payoff) = (.6)(-50,000) + (.4)(500,000) = -30,000 + 200,000 = 170,000$$

We choose the option with the greatest expected payoff. In this case we would choose to stage our event indoors. This method requires that we have reasonable estimates of the probability of the states of nature: if they were reversed in the above example, the highest expected payoff would come from the outdoor venue.

Decision Trees

If the probability of various outcomes, given particular actions, is known, then a decision tree can be constructed that displays the actions and payoffs under different states of nature and can be used to clarify the outcomes of different combinations. The decision tree containing the same information as the decision table in Table 17-8 is shown in Figure 17-2.

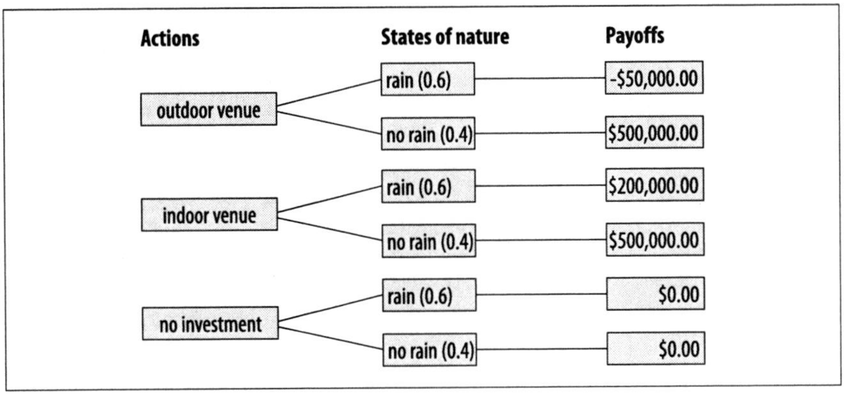

Figure 17-2. Decision tree for event venue example

The purpose of a decision tree is to display decision-making information, including available actions, states of nature, and expected payoffs in a clear and graphical manner. It does not include any rules for making decisions but can aid decision-making by presenting the relevant information in one graphical summary.

Quality Improvement

The roots of Quality Improvement (QI) date back to the 1920s, when Walter Shewhart began developing a statistical approach to studying variation in manufacturing processes. Interest in QI got a major boost in the 1950s with the work of W. Edwards Deming, who developed a statistical approach to QI based on Shewhart's work. Ironically, Deming's approach was initially rejected in his native country (the United States) but enthusiastically embraced in Japan, who applied QI techniques to manufacturing so successfully that they were able to challenge and in some cases surpass the American supremacy in manufacturing. In response, American companies began adopting QI approaches in the 1980s; Motorola and General Electric are among the best-known early adopters.

There are multiple approaches to QI, including a popular program known as Six Sigma (6σ), which is part of a general approach known as Total Quality Management (TQM). This section concentrates on the basics of QI, which are common to many such programs, and avoids getting into the specifics of jargon and acronyms of any particular program. It also concentrates on the statistical methodology used in QI, although the reader should bear in mind that most QI programs are multifaceted and include psychological and organizational strategies as well as statistical measurement and analytic techniques.

Although QI began in the manufacturing sector, it is now applied in other areas, including health care and education. "Quality" may be the buzzword of the new century, so consideration of the basic aspects of quality measurement and improvement may prove useful to people working in widely disparate fields, and, anywhere quality can be defined and measured, the field of QI may provide useful tools.

The first step in measuring anything is defining it. *Quality* in the QI context is generally defined in terms of the customer: a high-quality product satisfies the needs and preferences of the customer. In the case of manufacturing, this might mean machine parts with specified dimensions and durability. In the case of healthcare, it might mean a doctor's visit that answers the patient's concerns and does not involve excessive waiting or other aversive experiences. The customer's needs and preferences must be translated into *product variables* that can be measured. Picking up on the healthcare example, "no excessive waiting time" might be operationalized into "waiting time of no more than 10 minutes." This would allow each visit to be evaluated as to whether the standard was met. Similarly, specific dimensions can be established for machine parts, and specific parts evaluated as to whether they fall into the acceptable range as specified by the customer.

The language of QI is drawn from manufacturing, and commonly refers to *products* that are created by *processes*, which are part of a larger *system*. For instance, a company may manufacture bolts (the *product*) through a series of *processes* (such as cutting, stamping, and polishing), which are part of a larger *system* that takes *inputs* (such as metal) and transforms them into *outputs* (the bolts). An inherent fact about any process is that it is *variable*: not every bolt will have exactly the same dimensions. QI to a large extent is concerned with defining acceptable limits of variation, tracking variation within processes, and identifying causes and finding solutions when products are not within the acceptable range of variation.

Run Charts and Control Charts

Control charts, developed by Walter Shewhart of Bell Laboratories in the 1920s, are a basic graphical technique used to monitor process variation. The control chart is a refinement of the basic run chart, which is simply a time series chart displaying some characteristic of the product in question on the y-axis, and time or order of production on the x-axis. Often the data points graphed represent a statistic such as the mean that is calculated from small samples of product, rather than individual values.

Graphing sample means allows us to invoke the Central Limit Theorem and assume an underlying normal distribution for the data points (without regard to the distribution of the individual values in the population). This is essential when using the decision rules detailed below for determining when a process is going out of statistical control. If individual data points are represented in the control chart, these rules cannot be used unless the underlying process is normal, but graphing the points may still be useful as a graphical representation of the variation present in the process.

We expect to find variation in the output from any process, but do not expect the distribution of the output to change, either in location (mean or median) or variation (standard deviation or range). If the distribution of output from a process is consistent over time, we say the process is *in statistical control* or simply *in control*. If it changes, the process is said to be *out of statistical control* or simply *out of control*. The process of monitoring and eliminating sources of variation for some process in order to bring it into or keep it in statistical control is called *statistical process control*.

There are two basic sources of the total variation of any process: *common causes* and *special* or *assignable causes*. *Common causes of variation* are those that are attributable to the design of a process and affect all output of the process. For a manufacturing process, common causes might include lighting in a factory, the quality of raw materials, and worker training. If the amount of variation due to common causes is too great, the process must be redesigned. Perhaps the lighting can be improved, workers can be given more training or the tasks broken down into smaller segments that are easier to do accurately, or a more consistent source found for the raw materials used in the manufacturing process. This type of correction is generally the responsibility of management and does not figure in the type of analysis discussed in this section.

For the purposes of this section, a process that has only common causes of variation is a process that is in control. Instead, we focus on *special causes of variation*, which are actions or events that are not part of the process design. Special causes are usually temporary and affect only small parts of the process. For instance, a worker may become fatigued and fail to execute his job accurately, or a machine may get out of adjustment and start producing products outside the range of acceptable values. Control charts are used to identify when processes are going out of statistical control and may also aid in identifying special causes of variation.

Control charts usually include a *centerline* drawn at the process mean or median. The centerline acts as a reference point to evaluate the data points: for instance, use of a centerline makes it easier to evaluate whether data points are close to or distant from a central value. The value of this centerline is usually specified in advance by the analyst, and represents the expected value when the process is *in control* (running correctly, producing acceptable output) rather than the mean of the sample points. One other convention in control charts is the addition of lines connecting each consecutive point, which makes it easier to see the pattern across the sequence of measurements. Both features are displayed in the hypothetical run chart in Figure 17-3.

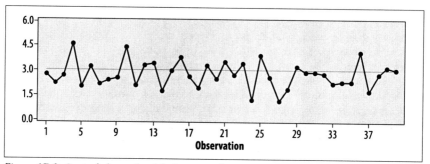

Figure 17-3. Control chart of weight in ounces for 40 screws (individual values), with a process mean of 3.0

This run chart displays the weight of 40 consecutively produced screws from a hypothetical manufacturing process. The y-axis displays the weight in ounces of each screw, while the x-axis displays the order of observation, and the green centerline displays the process mean of 3.0. We can observe therefore that the first three screws were slightly below the mean, the fourth was above, and so on. We can also see that the pattern is basically random and centered around the process mean, and that the longest run (consecutive values in the same direction) is 5 (values 29–33).

There's no particular pattern in the data presented in Figure 17-3 (not surprising, since it was created using a random number generator!), which is one of the indications that a process is in control. The charts in Figures 17-4 through 17-9 display some of the patterns that can be spotted by a run chart and might signal the need for further investigation.

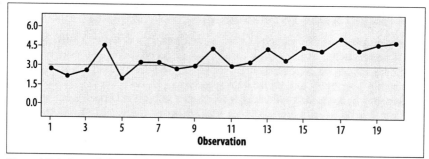

Figure 17-4. Control chart with an upward trend

Note that at this stage, because we are looking at individual data points, we are looking for general patterns rather than performing statistical tests. More formal rules are discussed shortly that may be used to determine when a data pattern cannot be attributed to random variation but should be investigated as evidence that a process is going out of control.

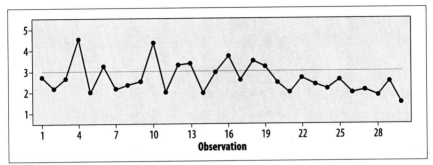

Figure 17-5. Control chart with downward trend

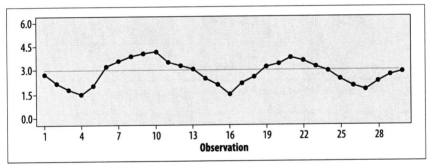

Figure 17-6. Control chart with cyclical pattern

When a control chart is based on sample means, thanks to the Central Limit Theorem we can use the normal distribution to identify values or patterns that would be highly improbable for a process in statistical control. A number of rules have been determined that indicate a process is going out of control, based on the expected distribution of values of the data points if they were based on samples drawn from a normal distribution with mean and variance specified from the process when it is in control.

Use of the standard deviation to define acceptable ranges of values for the outputs from a process is the source of the name for the Six Sigma program, because sigma (σ) is the symbol for standard deviation. The idea behind the Six Sigma program is to reduce variability sufficiently that output in the range of $\pm 3\ \sigma$ will still be acceptable to the customer.

As discussed in Chapter 8, with normally distributed data, the probability of data points within particular ranges is known. The percentage of data from a normal

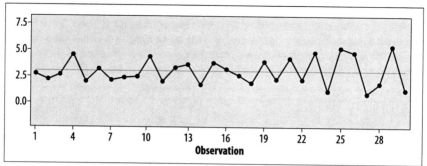

Figure 17-7. Control chart with increasing variability

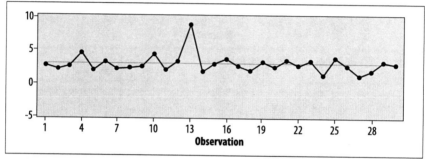

Figure 17-8. Control chart with shock or outlier (single extreme value)

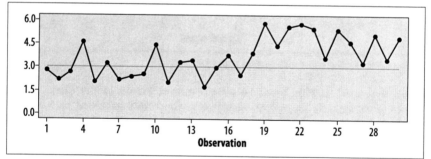

Figure 17-9. Control chart with change of level (upward shift of mean)

distribution contained in different ranges, defined by standard deviations from the mean, is displayed in Figure 17-10.

The probability of a data point within one standard deviation of the mean is about 68%. The probability of a data point in the range between one and two standard deviations above or below the mean is about 27%. The probability of a point between two and three standard deviations above or below the mean is 4% and the probability of a point beyond three standard deviations above or below the mean is about 0.2%. To look at it another way, in repeated samples from a normally distributed population, we would expect about 68% of the sample means to fall within one standard deviation of the mean, about 95% within two standard deviations,

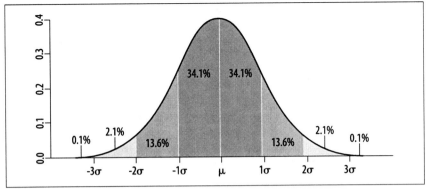

Figure 17-10. Probability of data points in particular ranges in a normal distribution

about 99% to be within 3 standard deviations, and about 0.2% beyond three standard deviations from the mean.

A control chart with the addition of control limits translates this information so the distribution of points is on the *y*-axis while the *x*-axis displays the time or order of samples charted. The different ranges are often labeled as shown in Figure 17-11.

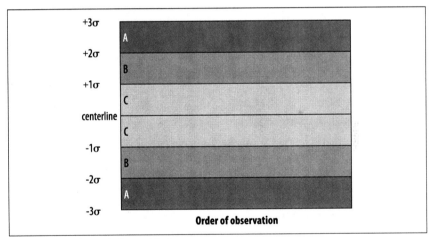

Figure 17-11. Control chart with sigma ranges

In this chart:

1. Zone A, or the three-sigma zone, is the area between two and three σ of the centerline.

2. Zone B, or the two-sigma zone, is the area between one and two σ of the centerline.

3. Zone C, or the one-sigma zone, is the area within one σ of the centerline.

These zones are used in conjunction with a set of *pattern analysis rules* to determine when a process has gone out of control.

Because both the mean value and variability of the samples are important to determining if a process is in control, control charts are usually produced in pairs, one representing mean values of the samples and one representing variability. For continuous data, an *x-bar chart* (so called because $\bar{x}$, pronounced *x*-bar, is the statistical symbol for a sample mean) is used to track the mean value. Variability is represented with either an *s-chart* displaying the standard deviation of the samples, or an *r-chart* representing the range of the samples.

The pattern analysis rules below are used to interpret data from the *x*-bar chart but could be applied to any of the various types of control charts. This list is an amalgam of several sets of rules including the "Western Electric rules" developed at the Western Electric Company (now part of AT&T) and first published in 1956, and the "Nelson rules" developed by Lloyd S. Nelson and first published in 1984.

The circumstances under which a process is judged out of control under pattern analysis rules are:

1. If any point falls outside Zone A.
2. If nine consecutive points fall in Zone C or beyond (further from the centerline) on the same side of the centerline.
3. If six consecutive points fall in the same direction, i.e., all increasing or all decreasing.
4. If 14 consecutive points alternate up and down.
5. If two out of three consecutive points fall in Zone A or beyond, on the same side of the centerline.
6. If four out of five consecutive points fall in Zone B or beyond, on the same side of the centerline.
7. If 15 points in a row fall in Zone C.
8. If 11 consecutive points fall in Zone B or beyond.

If data is binary rather than continuous (for instance, if items are simply classified as defective or acceptable), *p-charts* or *np-charts* based on the binomial distribution can be created in place of the *x*-bar chart. Note that binomial data is often referred to as *attribute data* within the field of quality control. If the interest is the *number of defects* rather than the *number of defective units* (i.e., if a unit can have more than one defect and the total count of defects is the variable of interest), then *c-charts* and *u-charts* can be created in place of the *x*-bar chart. Because these charts are usually created using computer software (as are *x*-bar charts) they will not be discussed in detail here: the key point is that the principles of interpretation are the same as for *x*-bar charts. The following set of rules should help clarify which type of chart to use for each type of data:

1. Data points represent sample means from continuous data (*x*-bar chart).
2. Data points represent the number of defective items per sample, and all samples are the same size (*np*-chart).
3. Data points represent the proportion of defective items per sample, and samples are of different sizes (*p*-chart).

4. Data points represent the average number of defects per unit, and all samples are the same size (*c*-chart).

5. Data points represent the average number of defects per unit, and samples are of different sizes (*u*-chart).

W. Edwards Deming and Japan

Japan was not always the manufacturing powerhouse we know today. In the first half of the twentieth century, Japan was noted primarily for the manufacture of inexpensive products, and the industrial infrastructure of the country was severely damaged during the Second World War. However, after the war the victorious Allied command assigned a group of engineers to help Japan rebuild their economy.

One aspect of this rebuilding was teaching Japanese manufacturers about the statistical quality control methods developed by Walter Shewhart at Bell Laboratories in the 1920s. In 1950, W. Edwards Deming (1900–1993), a statistician who had studied with Shewhart, was invited to present a series of lectures on statistical quality improvement under the auspices of the Japanese Union of Scientists and Engineers. During his visit, Deming also met with the top executives of many major Japanese companies.

Deming so impressed the Japanese industrial leaders they established two annual awards in his name for achievements in the field of quality: the Deming Prize for Individuals (awarded to individuals who have made important contributions in the study, methodology, or dissemination of TQM) and the Deming Application Prize (awarded for outstanding performance improvement through application of TQM principles). Further information about these prizes is available from the Deming Institute web site at *http://www.deming.org/demingprize/*.

Exercises

Here's a quick review of the topics covered in this chapter.

Question

Calculate the simple index for 2000, using each of the other years as a base year. What do the results tell you about the selection of the base period?

Year	Price
1970	1,000
1980	1,500
1990	2,000
2000	1,500

Answer

$I_{2000} = 150$ when 1970 is the base year, 100 when 1980 is the base year, and 75 when 1990 is the base year. This demonstrates the importance of the base year in index calculations, and in not allowing politics or other considerations to affect this choice.

Calculations when 1970 is the base year:

$$I_{2000} = (1{,}500/1{,}000) \times 100 = 150$$

When 1980 is the base year:

$$I_{2000} = (1{,}500/1{,}500) \times 100 = 100$$

When 1990 is the base year:

$$I_{2000} = (1{,}500/2{,}000) \times 100 = 75$$

Question

Calculate the Laspeyres index and Paasche index for 2000 for the following data, using 1990 as the base year. Why do they differ?

Product	1990 quantity	1990 price	2000 quantity	2000 price
Beef	100 pounds	$3.00/pound	50 pounds	$5.00/pound
Chicken	100 pounds	$3.00/pound	150 pounds	$3.50/pound

Answer

The Laspeyres index is 141.67, while the Paasche index is 87.50. The difference is due to the weighting: the Laspeyres index uses the weighting from the base year, while the Paasche index uses the weights for the index year. In this case, the same amount of meat was purchased in 1990 and 2000, but less beef and more chicken was purchased in 2000 relative to 1990. An inflation index based on the Laspeyres index would miss this change in consumer habits.

Here are the calculations for the Laspeyres index:

$$\frac{100(5.00) + 100(3.50)}{100(3.00) + 100(3.00)} \times 100 = \frac{850}{600} \times 100 = 141.67$$

And for the Paasche index:

$$\frac{50(5.00) + 150(3.50)}{100(3.00) + 100(3.00)} \times 100 = \frac{775}{600} \times 100 = 129.17$$

Question

Calculate the SMA and CMA for $n = 3$ and $n = 5$ for the sixth time point for the following table.

Time	1	2	3	4	5	6	7	8	9
Raw data	3	5	2	7	6	4	8	7	9

Answer

$SMA(n = 3) = (7 + 6 + 4)/3 = 5.7$
$SMA(n = 5) = (5 + 2 + 7 + 6 + 4)/5 = 4.8$
$CMA(n = 3) = (6 + 4 + 8)/3 = 6.0$
$CMA(n = 5) = (7 + 6 + 4 + 8 + 7)/5 = 6.4$

Notice that since there is a general trend upward in this data, the CMA estimates are higher, particularly with the larger window.

Question

Suppose you were considering whether to open a stationer's shop in a small or a large city. There is greater potential profit to be made in the large city but also greater potential loss (due to the greater expenses of setting up business there). The success of the shop will largely depend on the local business climate when you open: if other local businesses are expanding, you have a good chance to land some large orders, while if they are struggling, you may barely meet your expenses.

Here's a table of payoffs under two states of nature: calculate the minimax, maximax, and maximin decisions for this situation.

		Weather	
		Good business climate	Poor business climate
Action	Large city	200,000	10,000
	Small city	100,000	20,000

Answer

For the minimax solution, construct an opportunity loss table as follows.

		Weather	
		Good business climate	Poor business climate
Action	Large city	0	10,000
	Small city	100,000	0

The minimax solution is to choose the action that minimizes opportunity loss; in this case, we would choose to place our store in the large city.

The maximax solution is to select the action that has the highest maximum outcome, so in this case we would place our store in the large city.

The maximin solution is to select the action that has the largest minimal outcome, so in this case we would place our store in the small city.

Improvement
Statistics

Question

What pattern analysis rules are violated in the control chart in Figure 17-12?

Note that for the in-control process, the mean = 3 and the standard deviation = 0.5, so the centerline is at 3.0, the 3-sigma limits at 1.5 and 4.5, the 2-sigma limits are at 4.0 and 2.0, and the 1-sigma limits are at 3.5 and 2.5.

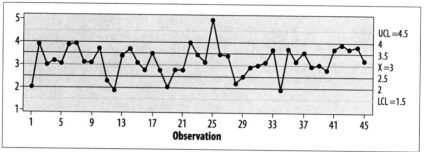

Figure 17-12. Control chart with pattern violations

Answer

The violations are listed below and identified in Figure 17-13.

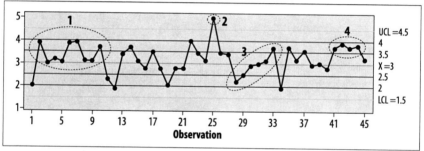

Figure 17-13. Control chart with pattern violations flagged

1. Nine points in a row on the same side of the centerline (rule 2).
2. One point outside three-sigma range, i.e., outside Zone A (rule 1).
3. Six points in a row in the same direction (rule 3).
4. Four out of five consecutive points beyond the one-sigma range (Zone B or beyond) on the same side of the centerline (rule 6).

18

Medical and Epidemiological Statistics

Many of the statistics used in medicine and epidemiology are common to other fields: examples include the *t*-test (covered in Chapter 8), correlation (covered in Chapter 9), and the various types of regression and ANOVA (covered in Chapters 12–15). But some other statistics have been developed specifically to meet the needs of medical and epidemiological research (such as the odds ratio), and others, while common to many fields, are used so frequently in medicine and epidemiology that they are covered in this chapter (for instance, standardized rates).

Measures of Disease Frequency

Before getting into specific measures of disease frequency, it is worthwhile to discuss the meanings of several terms in common usage that are often confused. We can always report disease frequency in terms of the number of cases: there were 256 cases of tuberculosis (TB) in city A and 471 in city B last year, for instance. Raw numbers are useful for people who allocate current resources and plan future monetary and space allocation, because they need to know how many cases of TB and how many hip fractures to expect in the coming year so they can allocate resources accordingly. However, for research and planning at the national and international level, disease occurrence is more usefully described in terms of relative rather than absolute occurrence, because we often want to look at trends over time or across different geographical areas with different population sizes. For instance, the hypothetical raw numbers above suggest that city B has a worse problem with TB than city A, but if city B has five times the population, the characterization would be reversed. Similarly, the number of cases of a disease may increase because the population is also increasing, so to make comparisons we often need to translate counts into other metrics.

Ratio, Proportion, and Rate

Three related types of metrics are the ratio, the proportion, and the rate. A *ratio* expresses the magnitude of one quantity in relation to the magnitude of another quantity, without making further assumptions about the two numbers, and without requiring that the two numbers share a common unit. Ratios may be expressed as A:B or A per B, and are often converted to standard metrics for easy comparison, such as 1:B or A per 10,000. For instance, we might be interested in the ratio of men to women living with AIDS in the United States. According to the Centers for Disease Control and Prevention (CDC) in 2005 there were 769,635 men and 186,383 women living with AIDS in the U.S. The ratio of men to women living with AIDS is therefore 769,635:186,383, which could also be expressed as 4.13:1. The second formulation makes it clearer that there were just over four times as many men as women living with AIDS in the U.S. in 2005.[*]

Two types of ratios often used in epidemiology and public health are the risk ratio and the odds ratio, which will be further discussed below. Ratios do not require the quantities compared to be measured in the same units: for instance, a common measure used to compare healthcare availability in different countries is the ratio of hospital beds to the population size. This is often expressed as the number of hospital beds per 10,000 people. According to the World Health Organization, in 2005 the United Kingdom had 39.0 hospital beds per 10,000 people, while Sudan had 7.0 and Peru 11.0, suggesting that hospital care was more readily available in the United Kingdom than in the other two countries.[†] This type of ratio is sometimes referred to as a *rate*, although it does not meet the strict definition of rate (discussed below) because the denominator does not include a measure of time.

A *proportion* is a particular type of ratio in which all cases included in the numerator are also included in the denominator. To return to the previous example, if we want to know what proportion of people living with AIDS in the U.S. were male, we would divide the number of males by the total number of cases (the number of cases in males plus the number of cases in females):

$$\frac{769,635}{769,635 + 186,383} = 0.805$$

Proportions are often expressed as percents, which means literally *per cent* or *per 100* (*cent* is Latin for 100). To translate proportions to percents, multiply by 100:

$$0.805 \times 100 = 80.5\%$$

The proportion of males among all people living with AIDS in the U.S. could also be expressed as 80.5 percent or 80.5%.

A *rate*, strictly speaking, is a proportion in which the denominator includes a measure of time. For instance, we commonly measure a person's heart rate in terms of beats per minute, and disease or injury occurrence in terms of the number of

---

[*] Source: *http://www.statehealthfacts.org/profileind.jsp?ind=505&cat=11&rgn=1)*

[†] Bed counts from *http://www.who.int/whosis/database/core/core_select.cfm.*

cases per week, month, or year. Morbidity and mortality (disease and death) statistics are often reported in terms of the rate per 1,000 or 100,000 per time unit: this is done because it is easier to interpret numbers like 3.57 versus 12.9 annually per 100,000 population than 0.00000357 versus 0.0000129 annually per person.

Converting rates to standard quantities facilitates comparison across populations of different sizes. For instance, the CDC reports that the annual death rate in the U.S. in 2004 was 816.5 per 100,000 population, as compared to 1,076.4 per 100,000 in 1940 and 954.7 per 100,000 in 1960. There were more deaths in 2004 than in any of the comparison years (2,397,615 in 2004 versus 1,417,269 in 1940 and 1,711,982 in 1960) but since the population of the U.S. was also increasing the annual death rate per 100,000 decreased.[*]

This may be seen by a simple example using hypothetical data (Table 18-1).

Table 18-1. Calculating annual death rate per 100,000 population

Year	Deaths	Population	Deaths per 100,000
1940	75	50,000	150.0
1950	95	60,000	158.3
1960	110	75,000	146.7
1970	125	90,000	138.9

We can see that although deaths increased each year, the population increased even faster, so the annual death rate per 100,000 decreased in each year studied. To calculate the death rate per 100,000, use this formula:

$$\frac{deaths}{population} \times 100,000$$

So for 1940, the rate is calculated as:

$$\frac{75}{50,000} \times 100,000 = 150.0$$

One issue in computing rates over a long period of time, such as a year, is to decide what number to use in the denominator. Often the population at the midpoint of the year is used.

There are several other issues involved in reporting disease incidence. One is whether the number of individuals with a condition, or the number of conditions itself, is being reported. For instance, if you were studying oral health, you might be interested in tooth decay. But a single person could have more than one cavity: are you interested in the number of people who had at least one cavity, or the total number of cavities?

A similar issue arises if you are studying a transient condition. For instance, if your topic is homelessness, are you interested in how many people had been homeless at least once over a time period, or would you count each separate

[*] Source: http://www.cdc.gov/nchs/data/nvsr/nvsr55/nvsr55_19.pdf.

instance of homelessness, with the understanding that some people might have been homeless more than once in the time period in question? These are problems of *unit of analysis*, meaning that you need to decide what entity you are studying (e.g., is it a *person* who may develop one or more cavities, or a number of individual *teeth*, each of which may develop a cavity) and collect and analyze data with that definition in mind.

Prevalence and Incidence

In epidemiology and medicine, when we speak of the number of cases of a disease, a basic distinction is made as to whether we are counting all the existing cases of a disease, or if we are only counting new cases. This may seem like hair-splitting to the average person, but is significant because we often want to separate new cases of a disease from existing cases. This lets us determine if a sanitation campaign is effective in preventing new infections, for instance. We separate existing from new cases by measuring two types of disease frequency: prevalence and incidence.

Prevalence describes the number of cases that exist in a population at a particular point in time. Prevalence describes the disease burden on a population without differentiating between new versus existing cases: a diabetic diagnosed the day the survey was conducted is counted equally as having the condition as is a diabetic who has been living with the condition for 20 years. Prevalence is particularly useful to people involved in resource allocation and planning, because they need to know the disease burden in the population as well as what it will be like in the future. Prevalence has also assumed increasing importance as the focus in epidemiology in the industrialized world has shifted from infectious to chronic diseases and conditions. This is because chronic diseases and conditions are often not curable but not rapidly fatal either, so a person can live for years with the disease or condition if appropriate medical care is provided.

Prevalence is defined as the proportion of individuals in a population who have the disease at a particular moment in time, and is calculated as:

$$P = \frac{\text{number of cases}}{\text{total population}}$$

at a given point in time.

If a survey of a city with a population of 150,000 people found that 671 were diabetics, the prevalence of diabetes at the time of the survey in that city would be 671 per 150,000 or 447.3 per 100,000. Because prevalence tells you the disease status of a population at a particular point in time, it is sometimes called *point prevalence*. Note that the "point" can be either a calendar time, such as a day, or a time in the life cycle or other course of events, such as the onset of menopause or the first day following surgery. Prevalence is sometimes referred to as *prevalence rate*, particularly when longer time intervals such as a year are used, although this is not strictly correct because there is no unit of time in the denominator.

Incidence is more complicated to calculate, because it requires three elements to be defined. Incidence describes the number of *new cases* of a disease or condition that develop in a *population at risk* during a particular *time interval*. Population at risk means people who have the potential to develop the condition: men are not at risk for pregnancy, for instance, and so would not be included in the population at risk. Similarly, once a person is infected with HIV (the virus that causes AIDS), that person cannot become infected again (or become uninfected, as far as we know), so the population at risk for HIV infection is restricted to those individuals who are not already HIV positive. Both incidence and prevalence are also used to describe health behaviors as well as diseases and conditions; for instance, we can refer to the prevalence of smoking in Mexico or the incidence of smoking onset in 2005 among teenagers at a particular school.

There are two types of incidence, cumulative incidence and incidence density. *Cumulative incidence* (CI) is the proportion of people who contract a disease during a specific time interval and is calculated as:

$$CI = \frac{\text{number of new cases}}{\text{population at risk}} \text{ for specified period}$$

CI is used to estimate the probability that an individual at risk will develop a disease or condition within a specified period, so it is important that the period be identified: the CI of a woman developing breast cancer in a one-year period following initial use of oral contraceptives will be different from the CI for a 10-year period.

The formula to calculate CI assumes the entire population at risk can be studied for the entire specified period: this means that unless otherwise qualified, incidence is a proportion. If the population at risk changes over the period included in the incidence calculations, then the *incidence density* (ID), also known as the *incidence rate* (IR), should be calculated instead. This would be necessary if people entered a study after it began, or dropped out before it was completed. Calculation of the IR requires expressing the denominator in *person-time units*, which represent the amount of time each person was observed. The time of observation is often referred to as the time each person *contributed* to the study.

The calculation of person-time units is demonstrated in Table 18-2. It represents hypothetical data on the annual rate of post-surgical infections at two hospitals. Because the hospitals serve different numbers of patients and patients are in the hospital for different lengths of time, we need to calculate the IR using person-time units in the denominator. Our statistic of comparison will be the number of complications per 100 patient-days. Each patient-day can be considered an opportunity for an infection to occur, so using patient-days in the denominator corrects for the different exposure to risk at the two hospitals.

Table 18-2. Post-surgical infection rates per 100 patient-days at two hospitals

Hospital	Patient ID	Days followed	Infection?
1	1	30	N
1	2	25	Y
1	3	15	N
Total for Hospital 1	70	1	
2	1	45	Y

Table 18-2. Post-surgical infection rates per 100 patient-days at two hospitals (continued)

Hospital	Patient ID	Days followed	Infection?
2	2	30	N
2	3	50	N
2	4	75	Y
Total for Hospital 2	200	2	

The rate of infections per 100 patient days is calculated as:

$$\frac{\text{number of infections}}{\text{person-days studied}} \times 100$$

So for this example the rates are:

$$\frac{1}{70} \times 100 = 1.43 \text{ per } 100$$

for hospital A and:

$$\frac{2}{200} \times 100 = 1.00 \text{ per } 100$$

for hospital B.

Even though hospital B had more post-surgical infections in the period studied, these occurred during proportionally more patient-days, so hospital B had a lower rate of post-surgical infections than hospital A.

The relationship between incidence and prevalence for a particular disease depends largely on the duration of the disease. If a disease has short duration (such as the common cold), prevalence will be low relative to incidence. In contrast, if a disease has a long duration (typical of many chronic diseases such as diabetes), the prevalence will be high relative to incidence. Changes in prevalence across time periods may be due either to changes in incidence or to changes in duration. For instance, incidence of a fatal disease may decrease but prevalence may increase if new treatments are developed that allow people to live for longer periods with the disease without curing it (an increase in the duration of the average case of the disease). Or the incidence of a disease may increase but the prevalence decrease if the duration of the disease is shortened through the development of new treatments that promote faster recovery.

Prevalence may be expressed mathematically as the product of incidence times average duration:

$$P = I \times \overline{D}$$

If two of the variables are known, the third can be calculated. For instance, if the incidence of a disease is 75 per 100,000 and the average annual prevalence is 45 per 100,000, the average duration can be calculated as:

$$\overline{D} = \frac{P}{I} = \frac{45/100,000}{75/100,000/\text{year}} = \frac{45}{75/\text{year}} = 0.6 \text{ years}$$

This assumes steady-state conditions for the time period under study, i.e., no major changes in disease incidence or duration. The formula can also be used to calculate how prevalence would change if either incidence or duration changes. For instance, if incidence of a particular disease remains steady at 125 per 100,000, but duration drops from 0.6 years to 0.1 years, prevalence will decrease from 75 per 100,000 per year to 12.5 per 100,000 per year. Similarly, if duration increases, prevalence will increase. If incidence of some disease remains steady at 200 per 100,000 per year, but duration increases from 0.5 years to 2 years, prevalence will increase from 100 per 100,000 per year to 400 per 100,000 per year.

Crude, Category-Specific, and Standardized Rates

If not otherwise qualified, the term rate usually means the *crude rate*. The crude rate is the rate for the entire population under study, with no particular weighting or adjustment. For instance, according to the CDC, the overall death rate for cancer in the U.S. in 2003 was 195.5 per 100,000. However, these mortality rates were not constant across ethnic group, age group, or gender, nor were they constant across different types of cancer. To examine these differences, we need to look at the *category-specific* rates, in which both the numerator and denominator represent one population group or one type of disease. For instance, in the U.S. in 2003, the cancer mortality rate for men was 201.4/100,000, while for women it was 182.0/100,000, and the crude mortality rate for lung cancer was 76.9/100,000, while for skin melanomas it was 2.7/100,000.

For white Americans in 2003 the crude cancer death rate was 203.8/100,000, while for African-Americans it was 164.3/100,000, a finding that may seem paradoxical until we consider that increased life expectancy is often associated with increased cancer mortality. Someone who dies as an infant is unlikely to have died of cancer, for instance, while someone who lives into their 80s has a much higher probability of a cancer-related death. This is true of general mortality as well: under most circumstances a person who is 90 years old has a much higher probability of dying in the next year than a person who is 12 years old. For this reason, death rates used to make comparisons across different populations or time periods are usually standardized by age, and may also be standardized by categories such as ethnicity or gender.

The importance of age-adjustment can be seen by comparing the crude and age-adjusted cancer mortality figures for the U.S. in 2003 in Table 18-3.

Table 18-3. Crude and age-adjusted cancer mortality rates (per 100,000) for the U.S in 2003

	Crude	Age-adjusted
Overall	191.5	190.1
White	203.8	188.3
African American	164.3	234.5
Asian/Pacific Islander	79.4	114.3
American Indian/ Alaska Native	69.3	121.0
Hispanic	60.3	127.4

This makes it clear that although the crude death rate from cancer is highest among white Americans, this is due in part to a longer life expectancy. A longer life expectancy means that there are more white Americans in the older age categories, where mortality from cancer is higher. When age-adjustment is considered, African-Americans have the highest death rate from cancer.[*]

There are two types of standardization, *direct* and *indirect*. Both are used to compare morbidity and mortality in different populations while removing the influence of other population characteristics, such as age or gender distribution. In *direct standardization*, a population is chosen to serve as the standard and adjusted rates for the populations to be compared are calculated using weights from the standard population. For instance, consider the hypothetical example of the occurrence of arthritis by employment status in Table 18-4.

Table 18-4. Arthritis diagnosis by employment status

Employment status	Population	Arthritis	Rate per 1000
Employed	10,000	387	38.7
Unemployed	5,000	892	178.4

The rate (really the proportion) of arthritis is over twice as high among persons not employed as among employed persons, according to this data. Could this be due to people being forced out of the labor market due to severe arthritis? Possibly, but a more logical explanation is that people over the age of 65 are more likely to not be employed, and also more likely to have a diagnosis of arthritis. To test the hypothesis that age distribution is the reason for the observed differences in rate of arthritis diagnosis by employment status, we need to compute age-adjusted rates of arthritis using a standard population. First, we need to calculate age-specific rates for employed and unemployed individuals, as in Table 18-5.

Table 18-5. Age-specific rates of arthritis diagnosis

Age	Employed Population	Unemployed Diagnoses	Rate/1000	Population	Diagnoses	Rate/1000
18–44	5,000	127	25.4	1,000	32	32.0
45–64	4,500	260	57.7	1,500	100	66.7
65+	500	105	210.0	2,500	760	304.0
Total	10,000	387	38.7	5,000	892	178.4

Looking at the age distribution and age-specific rates for the employed versus unemployed populations, we see that the age-specific rates are somewhat higher in the unemployed group. We also see that a much higher proportion of the unemployed group (50%, versus 5% for the employed group) is in the 65+ age category, where the rates of arthritis diagnosis are highest.

---

[*] Source: *http://apps.nccd.cdc.gov/uscs/Table.aspx?Group=TableAll&Year=2003&Display=n.*

We used very broad age categories (basically young working adult, older working adult, and retirement age) in this table for ease of calculation. Often smaller categories are used, such as 10-year age ranges. We can use these age-specific rates to calculate expected numbers of diagnoses in each age category for the two employment groups, using the age distribution from a hypothetical standard population. Usually a standard source would be used in these calculations, for instance U.S. population in 2000 as determined by the U.S. Census Bureau. The calculations are shown in Table 18-6.

Table 18-6. Expected numbers of diagnoses by age category and employment category

Age group	Standard population Population	Employed Rate/1,000	Unemployed Expected diagnoses	Rate/1000	Expected diagnoses
18–44	100,000	25.4	2,540.0	32.0	3,200.0
45-64	70,000	57.7	4,039.0	66.7	4,669.0
65+	30,000	210.0	6,300.0	304.0	9,120.0
Total	200,000		12,879		16,989

The expected diagnoses are calculated by applying the age-specific rates for each population to the number of people in that age category in the standard population. This may be considered a type of weighting, and is equivalent to saying how many arthritis diagnoses we would expect to see in each population if the age distribution was the same as in the standard population. For instance, for the 18–44 age group in the employed population, the calculation is:

$$E = \frac{25.4}{1000} \times 100,000 = 2540$$

For the 65+ age category in the unemployed population, it would be:

$$E = \frac{304}{1000} \times 30,000 = 9120$$

We can immediately see that, if the two populations had the same age distribution, employed people would have fewer arthritis diagnoses (12,879) than people who were unemployed (16,989). We can further refine this finding by calculating the age-adjusted arthritis diagnosis rates for each population by dividing the number of expected diagnoses by the total size of the reference population. For employed people this would be:

$$\frac{12,879}{200,000} = 64.4 \text{ per } 1000$$

For unemployed people it would be 84.9 per 1,000. So the rate of arthritis is slightly higher in unemployed than employed persons, but the difference is much less than the crude rate would suggest. Note that the age-adjusted rates calculated through direct standardization do not represent the actual rates in any population: they represent what rate would be expected in one or more particular populations, *if* they had the age distribution of some reference population.

Indirect standardization takes the reverse approach: it takes the category-specific rates from some standard population, and applies them to the actual category distribution in two or more populations. Applying indirect standardization to our arthritis example, we will calculate the expected number of arthritis diagnoses if both populations had the same age-specific rate of diagnosis but kept their own specific population age distribution. The rates (which are hypothetical) are shown in Table 18-7.

Table 18-7. Indirect method of standardization

Age group	Standard Rate/1,000	Employed Population	Unemployed Expected diagnoses	Population	Expected diagnoses
18–44	30.0	5,000	150	1,000	30
45-64	60.0	4,500	270	1,500	90
65+	200.0	500	100	2,500	500
Total		10,000	520	5,000	620

We can use these numbers to calculate the *standardized morbidity ratio* (morbidity means disease) by dividing the observed number of diagnoses (from Table 18-5) by the expected number of diagnoses (from Table 18-7). The standardized morbidity ratio for employed people is:

$$\frac{\text{observed diagnoses}}{\text{expected diagnoses}} = \frac{387}{520} = 0.744 \text{ or } 74.4\%$$

For the unemployed group, it would be 69.5%.

If we were dealing with deaths rather than diagnoses of arthritis, we could use the same technique to calculate the standardized mortality ratio (SMR), a statistic commonly used to compare mortality (death) across populations.

The Risk Ratio

Many medical and epidemiological studies are concerned with the relationship between two dichotomous variables. A common example is the exposure to some risk factor (such as asbestos or tobacco smoke) and the development of some disease or condition (such as asbestosis or lung cancer). The exposure can be an inherent quality, such as gender or ethnicity, and need not be negative; for instance, engaging in regular physical activity is an exposure that has a positive influence on health.

The relationship between two dichotomous variables is often presented in a *crosstabulation* or *contingency table*, also called a 2×2 or "two by two" table because of its dimensions (two rows and two columns). The standard way to set up such a table is illustrated in Table 18-8.

Table 18-8. The classic 2x2 table

		Disease			Total
		D+	D–		
Exposure	E+	a	b		$a+b$
	E–	c	d		$c+d$
Total		$a+c$	$b+d$		$a+b+c+d$

E+ means the person had the exposure, E– that they did not. D+ means they have the disease, D– that they do not. Individuals in a study are classified by their exposure and disease status, and the cells labeled a, b, c, and d contain the frequencies for each combination of exposure and disease. For instance, cell a holds the frequency for people who have the exposure and have the disease, while cell d holds the frequency for people who have neither exposure nor disease.

The frequencies in the four cells a, b, c, and d are sometimes referred to as *joint frequencies* because the people in those cells are classified on both exposure and disease. On the margins of the table are the row and column totals, often referred to as *marginal* frequencies. For instance, $a + c$ is the total number of people in the study with the disease regardless of exposure status, while $a + b$ is the total number of people with the exposure regardless of disease status. The total number of people in the study is $a + b + c + d$.

The *risk ratio*, also called the *relative risk*, estimates the likelihood of developing the disease for people with the exposure, relative to people without the exposure. It is the ratio of the proportion of the exposed that develop the disease to the proportion of the unexposed that develop it. The risk ratio is calculated as:

$$RR = \frac{a/(a+b)}{c/(c+d)}$$

The risk ratio can also be thought of as the ratio of disease incidence in the exposed (I_e) versus unexposed (I_0) populations:

$$RR = \frac{\text{incidence in exposed group}}{\text{incidence in unexposed group}} = \frac{I_e}{I_0}$$

For studies in which the denominator is person-time units, the calculation is analogous but uses the ratio of the incidence densities (incidence rates) from the two populations:

$$RR = \frac{ID_e}{ID_0}$$

Let's look at data from a hypothetical study to see if there is a relationship between consumption of a high-fat diet (the exposure) and Type II diabetes (the disease). The data is presented in Table 18-9.

Table 18-9. Relationship of a high-fat diet to Type II diabetes

	D+	D–	
E+	350	1200	1550
E–	200	1900	2100
	550	3100	3650

The risk of Type II diabetes, given consumption of a high-fat diet, is:

$$\frac{a}{a+b} = \frac{350}{1550} = 0.226$$

The risk of Type II diabetes for someone on a normal or low-fat diet, i.e., not consuming a high-fat diet, is:

$$\frac{c}{c+d} = \frac{200}{2100} = 0.095$$

The relative risk of developing diabetes, given consumption of a high-fat diet versus nonconsumption of a high-fat diet, is the ratio of these two risks (hence the term *risk ratio*), or:

$$RR = \frac{RR_1}{RR_2} = \frac{a/(a+b)}{c/(c+d)} = \frac{0.226}{0.095} = 2.38$$

A relative risk greater than 1 indicates that the exposure increases the risk of the disease. If there is no relation between exposure and risk, the relative risk will be 1, while if the exposure is protective (associated with lower risk of disease), the risk ratio will be less than 1. In this case we would say that people consuming a high-fat diet have 2.38 times the risk of Type II diabetes, compared to people consuming a low-fat or normal diet.

Like many other statistics, risk ratios are usually reported along with their confidence interval (CI). These calculations must take into account the fact that the risk ratio is right-skewed because it has a bound of 0, but no upper bound. To deal with this skew, we take the natural logarithm (*ln*) of the risk ratio, which transforms it to an approximately normal distribution. The procedure for calculating the CI for an RR requires taking the natural log of the RR, finding the confidence interval for this *ln*(RR) and then taking the natural antilogarithm of the confidence interval limits to return to the original units. Note that in statistical notation, e^x is often written as exp(x) for the sake of convenience.

There are several different ways to calculate the confidence interval for a risk ratio, the most common being to use statistical software. However, the calculation can also be done by hand. A simple computational formula, which is applicable when the rare disease assumption holds and the odds ratio is a good estimate of the risk ratio (see below), is:

$$CI = \frac{ad}{bc} \exp[\pm z \sqrt{\text{Variance}(\ln RR)}]$$

where z is the value of the standard normal distribution associated with the desired confidence level, usually 1.96, which results in a 95% confidence interval. When the RR is estimated using the odds ratio (discussed below) from a case-control study, the CI may be calculated using values from the 2×2 table using this formula:

$$CI = \frac{ad}{bc}\exp\left(\pm z\sqrt{\frac{1}{a} + \frac{1}{b} + \frac{1}{c} + \frac{1}{d}}\right)$$

Using values from Table 18-9, this translates to:

$$CI = \frac{350 \times 1900}{200 \times 1200}\exp\left(\pm 1.96\sqrt{\frac{1}{350} + \frac{1}{1200} + \frac{1}{200} + \frac{1}{1900}}\right)$$

$$= 2.77\exp(\pm 1.96\sqrt{0.00286 + 0.0008 + 0.005 + 0.00053})$$

$$= 2.77\exp[\pm 1.96(0.0959)]$$

$$= 2.77\exp(\pm 0.188) \text{ or } 2.77e^{\pm 0.188}$$

So the upper bound is $(2.77)e^{188} = (2.77)(1.207) = 3.34$ and the lower bound is $(2.77)e^{188} = (2.77)(.829) = 2.30$, giving us a confidence interval of (2.30,3.34). Because this CI does not include the null value of 1.0, we conclude that the relationship between consumption of a high-fat diet and diagnosis with Type II diabetes is significant.

The time period over which data is collected is significant in interpreting relative risk. The risk of developing many chronic diseases increases with duration of exposure, for instance, so the risk of a high-fat diet for development of Type II diabetes would be expected to be higher in a 10-year study than in a 5-year study. This is particularly true for studies of mortality, because if a study is continued long enough, the probability of mortality for all the subjects is 100%!

Because there is often some risk of disease for people without the exposure being studied, epidemiology also uses the concept of *attributable risk* (AR). Attributable risk is the absolute effect of the exposure on disease occurrence, meaning the excess risk of disease in the exposed versus the unexposed group. AR is useful as a measure of the public health cost or benefit of some exposure, because it subtracts from the exposed group the cases that would be assumed to have occurred anyway. It can also be used to estimate the impact of a proposed intervention to remove an exposure by calculating how many cases of disease would be "saved," i.e., would not occur, if the exposure were eliminated. Attributable risk is calculated by subtracting the incidence rate in the unexposed from the rate in the exposed. In our example, this would be:

$$AR = I_e - I_0 = 0.226 - 0.095 = 0.131$$

Therefore, a high-fat diet accounts for about 131 excess cases of Type II diabetes per 1,000 people. If there is no relationship between exposure and disease, there would be no excess cases in the exposed group, and the AR would equal 0.

The *attributable risk percentage* (AR%; also called the *etiologic fraction*) is the proportion of cases in the exposed population that can be attributed to the exposure and are assumed would be prevented by eliminating the exposure. It is calculated, continuing with our example, as:

$$AR\% = \frac{AR}{I_e} \times 100 = \frac{I_0 - I_e}{I_e} \times 100 = \frac{0.226 - 0.095}{0.226} \times 100 = 58.0\%$$

We would interpret this by saying that 58.0% of the cases among the exposed groups are due to the exposure. The AR% can also be calculated using the RR, as follows:

$$AR\% = \frac{RR - 1}{RR} \times 100 = \frac{2.38 - 1}{2.38} \times 100 = 58.0\%$$

The Odds Ratio

The odds ratio was developed for use in *case-control studies*. Case-controls were invented in epidemiology to facilitate research into diseases that are rare or slow to develop, so a conventional prospective study would be impractical. In the case-control study, individuals are selected on the basis of their disease status and then their exposure status is determined. Risk ratios cannot be calculated in case-control studies: the reason is that risk ratios are sensitive to the number of people without the disease, and this number is determined in case-control studies by the study design rather than the rate of disease in a population. As will be demonstrated below, the odds ratio has the beneficial quality of being insensitive to the number of controls (persons without the disease), while the risk ratio does not share this property.

The *odds ratio* is the ratio of the odds of disease for the exposed group to the odds of disease for the unexposed group. In a 2×2 table, the odds of disease given exposure are a/c, and the odds of disease given no exposure are b/d. The odds ratio is calculated using this formula:

$$OR = \frac{\text{odds for exposed}}{\text{odds for unexposed}} = \frac{a/c}{b/d} = \frac{ad}{bc}$$

Let's suppose we have a case-control study examining the effect of smoking on breast cancer. The hypothetical data is shown in Table 18-10.

Table 18-10. Relationship of smoking and breast cancer

	D+	D−	
E+	50	2,000	2,050
E−	25	1,900	1,925
	75	3,900	3,975

The odds ratio may be calculated as:

$$OR = \frac{50/25}{2000/1900} = 1.90$$

The risk ratio for this data is similar:

$$RR = \frac{50/2050}{25/1925} = 1.88$$

If a disease or condition is rare (a rule of thumb is that it must be less than 10% in all exposure groups), the odds ratio provides a reasonable estimate of the risk ratio. The reason for the "rare disease" requirements is that as a disease becomes more common, the odds ratio diverges further from the risk ratio. This is demonstrated in the data presented in Table 18-11, which represents data from a hypothetical case-control study of smoking and lung cancer.

Table 18-11. Smoking and lung cancer

	D+	D−	
E+	50	50	100
E−	20	100	120
	70	125	195

The disease is common in both exposed and unexposed subjects: 50% of the exposed subjects have lung cancer, as do 16.7% of the unexposed. The odds ratio is:

$$OR = \frac{50 \times 100}{20 \times 50} = \frac{5000}{1000} = 5.0$$

And the risk ratio is:

$$RR = \frac{50/100}{20/120} = 3.0$$

The RR is sensitive to changes in the number of controls, while the OR is not. Suppose that because controls are easier to find than cases, we increased the number of controls 10-fold (unlikely, because diminishing returns set in for control-case ratios at about 4:1, but useful to demonstrate this point). This would give us the data shown in Table 18-12.

Table 18-12. Smoking and lung cancer, 10-fold increase in controls

	D+	D−	
E+	50	500	550
E−	20	1,000	1,020
	70	1,500	1,570

The odds ratio does not change:

$$OR = \frac{50 \times 1000}{20 \times 500} = \frac{5000}{1000} = 5.0$$

But the RR does:

$$RR = \frac{50/550}{20/1020} = 4.64$$

Confidence intervals for the OR may be calculated using the method described in the RR section.

Confounding, Stratified Analysis, and the Mantel-Haenszel Common Odds Ratio

Confounding is a condition in which an observed statistical association is due at least in part to differences in the study groups other than the exposure of interest in the study. Confounding is sometimes described as the "third variable" problem: the relationship between two variables, say exposure and disease, is mixed up or confounded with the influence of a third variable related to both of them. More than one variable can be involved in confounding, but for the sake of simplicity we will demonstrate methods to deal with a single confounding variable.

Confounding is always of concern in epidemiology, particularly in observational studies where group membership is not under the control of the investigator. For instance, studies of the effects of smoking on health have to take into account the fact that smoking is a voluntary behavior (people choose to smoke or not to smoke) and people who smoke may differ in many other ways (such as alcohol consumption, diet, or level of education) from those who do not.

If possible, it is preferable to control for confounding in the study design. *Randomization* is the method of choice for intervention studies because it theoretically controls for all potential confounders at once. This is because, on average, random assignment to groups should result in approximately the same distribution of any potential confounder in each group, including confounders of which the researcher is not aware.

Two other methods that may be used in observational studies to control for known or suspected confounders are restriction and matching. Both have the disadvantage that they only implement control over the confounders used in the design. With *restriction*, the researcher studies only a subset of the population, selected based on their values on the potential confounder. For instance, medical studies are sometimes done only on men, or only on women, to remove the influence of gender on the relationship between the exposure and disease. This has the disadvantage of restricting the applicability of study results: if a relationship between alcohol consumption and psychopathology is found in a group of men, that does not immediately justify generalizing the conclusions to women, because women were not included in the study.

Matching is another technique that attempts to control for known confounders, by a different method. Matching includes all levels of the confounders but controls enrollment in the study or assignment to groups so that the confounders will be equally distributed across the groups. Matching is commonly used in case-control studies, in which controls are selected to match the cases already enrolled in the study. There are different systems for matching, but the basic concept is that categories are constructed for the confounding variables and assignment to groups is controlled so that the distribution of the confounders is the same in each group.

There are two ways to implement matching. In *direct matching*, individuals are matched on a one-to-one basis. In *frequency matching*, assignment to the groups is monitored so that equal numbers of the confounders are present in each group. If the confounders are gender and age category, in direct matching a woman of age 60–70 years (for instance) in the treatment group would be matched by a woman of age 60–70 years in the control group. In frequency matching, the project manager would monitor enrollment to see that an equal number of females and persons in the different age categories were included in the treatment and control groups.

If it is not possible to control for confounding in the research design, it must be dealt with during the analysis. There are numerous statistical methods to control for confounding after the fact, including multivariable methods that can become quite complex. However, confounding is often treated more simply in epidemiology and regression, particularly in studies focused on a single exposure and disease. This presentation demonstrates one of the most common methods to evaluate and control for confounding: computation and comparison of the crude and Mantel-Haenszel common odds ratio.

There is no implication of causality in classifying a variable as a confounder; in fact, many of the most common confounders are only correlates of another factor. For instance, studies of the influence of physical activity on health must consider age and sex as confounders, because young people and males are more likely to engage in leisure-time exercise. To qualify as a confounder, a variable must meet three requirements:

1. It must be related to the exposure.
2. It must be related to the disease, independently of its association with the exposure.
3. It must not be wholly intermediate in the causal pathway between exposure and disease.

A fourth requirement, which is practical rather than theoretical, is that in order to function as a confounder in a particular study, a variable must be unequally distributed among groups in the present study. For instance, we know that age could serve as a confounder for mortality, but if in a particular study the age distribution is the same among all groups studied, then age cannot act as a confounder in that particular study.

Let's take as an example a study of the protective effect of voluntary leisure-time physical activity (exposure) on the occurrence of heart attack (myocardial infarction or MI), which may be confounded by age. All three requirements are met:

1. Age is related to physical activity (young people exercise more than older people).
2. Age is a risk factor for MI, independent of physical activity (older people are more likely to have an MI).

3. Age is not wholly intermediate in the causal pathway between physical activity and MI (there is no way physical activity could affect a person's age, which would then affect their probability of MI).

One method to control confounding is the use of stratified analysis, in which the groups to be studied are divided into *strata* or subgroups based on values of the confounding variable. Stratification by age category is a common example. Populations of different countries have different age structures: some countries have relatively more young people, others relatively more older people. Age is related to mortality and many types of morbidity. For these reasons, comparison of morbidity and mortality between populations is often accomplished by stratifying by age category, then standardizing so the age distribution is comparable in the populations being compared.

An example should demonstrate the need to evaluate confounding. In 2007 mortality rate in the United States was 8.26 deaths per 1,000, while in Ecuador it was 4.21 per 1,000. Should this be interpreted as evidence that Ecuadorans lead more salubrious lifestyles than Americans? That's an intriguing possibility, but is not supported by examination of detailed life tables, which show that Ecuadorans have higher death rates than Americans in each specific age category. For instance, for the 45–49 age group the probability of death for Americans is 0.00341, while for Ecuadorans it is 0.00513.

The difference in mortality is due to the age structure of the two populations. Ecuador, like most developing countries, has a higher percentage of its population in younger age groups. The United States, like most industrialized countries, has a higher percentage of people in the older age categories, where the risk of mortality increases. This distinction would be missed if only crude mortality rates were considered, but becomes clear when a stratified analysis removes the influence of the confounding variable (age) from the outcome (mortality).*

There is no absolute test for confounding, but there are ways to examine the effects of potential confounders on the relationship of interest and make a reasoned decision about whether confounding is present. The general steps to follow in assessing confounding are as follows:

1. Calculate the crude measure of association, ignoring the confounding variable.

2. Stratify the study population by the confounding variable, i.e., divide the population into smaller subgroups based on values of the confounding variable.

3. Calculate an adjusted measure of association.

4. Compare the crude and adjusted measures: a difference of 10% or more is generally considered evidence of confounding.

---

* Sources: *https://www.cia.gov/library/publications/the-world-factbook/rankorder/2066rank.html*, *http://www.who.int/whosis/database/life/life_tables/life_tables_process.cfm?path=whosis,life_tables&language=english*, and *http://www.who.int/whosis/database/life/life_tables/life_tables_process.cfm?country=ecu&language=en.*

The appropriate measure of association depends on the study design; we will demonstrate stratified analysis using the crude odds ratio and the Mantel-Haenszel adjusted odds ratio. Note that in order to use the Mantel-Haenszel method, two assumptions must be met: the overall sample size must be large, and the association between exposure and outcome should be in the range of approximately 0.5 to 2.5.

The Mantel-Haenszel (MH) estimator of the common odds ratio for stratified data allows information to be combined from a series of two or more 2×2 tables, using the following formula:

$$ OR_{MH} = \frac{\displaystyle\sum_{i=1}^{k} (a_i d_i)/n_i}{\displaystyle\sum_{i=1}^{k} (b_i c_i)/n_i} $$

where there are k individual tables, i represents one of the tables (i.e., one strata of the population), n_i is the sample size for that table, and a_i, b_i, c_i, and d_i are the values of cells within that table. Suppose we are interested in the relationship between smoking and liver disease. We know that people who smoke are also more likely to consume alcohol, alcohol consumption is an independent risk factor for liver disease, and alcohol consumption is not wholly intermediate in the hypothesized causal chain between smoking and liver disease. Alcohol consumption is therefore a potential confounder in this study, which we will examine by stratifying our study population on alcohol consumption (as a dichotomy: those who drink alcohol versus those who don't) and examining the difference (if any) between the crude and adjusted odds ratios for our population.

The data looks like Table 18-13 before we consider the effect of alcohol consumption.

Table 18-13. Smoking/liver disease data before stratification

	D+	D–	
E+	50	100	150
E–	30	120	150
	800	220	300

The crude odds ratio is:

$$ OR = \frac{ad}{bc} = \frac{50 \times 120}{30 \times 100} = 2.00 $$

This is a strong positive OR and indicates that smoking is positively associated with liver disease. Smokers are twice as likely to have liver disease as nonsmokers. To examine if alcohol consumption is a confounding factor, we construct separate 2×2 tables for those who do and don't consume alcohol (Tables 18-14 and 18-15).

Table 18-14. Smoking/liver disease, for those who don't consume alcohol

	D+	D–	
E+	40	35	75
E–	30	45	75
	70	80	150

Table 18-15. Smoking/liver disease, for those who do consume alcohol

	D+	D–	
E+	60	15	75
E–	50	25	75
	110	40	150

We can compute the MH common odds ratio as follows:

$$OR_{MH} = \frac{\sum_{i=1}^{k}(a_i d_i)/n_i}{\sum_{i=1}^{k}(b_i c_i)/n_i} = \frac{(40 \times 45)/(150)}{(30 \times 35)/150} + \frac{(60 \times 25)/150}{(50 \times 15)/150} = 3.71$$

Since this is more than 10% different from the crude odds ratio of 2.00, we conclude that alcohol consumption is a confounder in the relationship between smoking and liver disease and should be included as such in our analyses.

Power Analysis

This section deals with the theory of *power and sample size*, and presents a few simple examples. Sample and power calculations are frequently simple, but they are also specific: every type of research design uses a different formula, and there's no point in listing them all when they are available in reference books. For those working in medicine and epidemiology, one particularly recommended source is the chapter on sample size calculation in the *Handbook of Epidemiology* (Springer). Many software packages, such as SAS and Minitab, include packaged routines to do power and sample size calculations, and there are various power and sample size calculators on the Web as well; a good collection of links to online calculators may be found at *http://statpages.org*.

The practice of doing inferential statistics always includes the possibility of making a wrong decision, because inferential statistics uses calculations on a sample to make conclusions about a population. As discussed in Chapter 7, there are two kinds of common errors in inferential statistics:

1. Type I error or α, when you incorrectly reject the null hypothesis.

2. Type II error or β, when you fail to reject the null hypothesis when you should have rejected it.

Another way to look at this is to say that Type I error is finding significance where none exists, while Type II error is failing to find significance when it does exist.

Power is 1-β and is the probability of accepting the null hypothesis when you should reject it. We'd all like to have high power all the time, but practical considerations, in particular the cost and availability of subjects, usually force us to compromise. A rule of thumb is that you should have at least 80% power, i.e., 80% chance of finding significant results in your sample if they exist in the population. That means that 20% of the time, you won't find significance when you should. The standard of 90% power is regularly used as well.

Four main factors affect power:

1. α level, i.e., P(Type I error) (higher α increases power)
2. Difference in outcome between the populations (greater difference increases power)
3. Variability (reduced variability increases power)
4. Sample size (larger sample size increases power)

A change in any one of these factors, while the others are held constant, will change the power level for a given design. The α level is usually chosen to be 0.05 or less (for instance 0.01); a larger value of α translates into more power. A greater difference in outcome between the populations increases power. Differences in outcome can be increased by improving the intervention so it has a stronger effect, or by choosing study groups to increase the expected difference in outcomes between them. Reduced variability also increases power. Variability can sometimes be decreased by improving measurement or through selection of study subjects (such as restricting them to a particular age range or income level).

That leaves us with *sample size*, the one factor primarily under the control of the experimenter at the planning stages of his research project. All things being equal, more subjects = greater power. However, recruiting more subjects usually costs more money and requires more effort on the part of the research team. The goal of power analysis is to find a reasonable compromise in which you have acceptable power but are not going bankrupt or collecting more data than is necessary.

The basic concepts of power are illustrated in Figure 18-1.

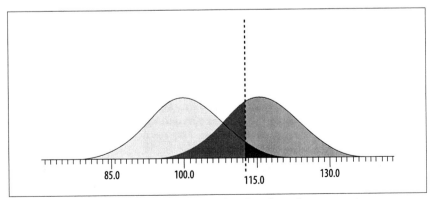

Figure 18-1. Power diagram for two normally distributed populations

Figure 18-1 illustrates aspects of a power calculation in which the null hypothesis is that the mean of the population is 100, while the alternative hypothesis is that the population mean is 115. In this figure, the leftmost (light gray) distribution is the null population, which represents the distribution if the null hypothesis is true. The right-most (dark gray) distribution is the alternative population, which represents the data distribution if the alternative hypothesis is true and the population mean is 115.

Power calculations are always carried out with respect to a particular alternative hypothesis. In this case, the alternative is not simply that the population mean is greater than 100, but that it is 115. Note that hypothesis testing involves the location of population means, although the hypotheses are tested using means calculated from samples. For simplicity's sake, in this example both populations are assumed to have equal standard deviations of 15.

The hypothesis being tested is one-tailed, so a single cutpoint or critical value, represented by the dotted line, is established. If the sample mean is above this cutpoint, the null hypothesis will be rejected. If the sample mean is below this cutpoint, the sample mean will be accepted. The location of the cutpoint, 112.5, was set with regard to the null population, which has a mean of 100 and a standard deviation of 15: it is the critical value for a significance test when $\alpha = 0.05$, because 95% of the null population lies to the left of 112.5, and 5% to the right.

The area of the null population above the cutpoint represents the P(Type I error) or the probability of rejecting the null hypothesis when it is true. In this example P(Type I error) is 0.05.

The area of the alternative population, to the left of the cutpoint, represents β or the P(Type II error) if the alternative hypothesis (population mean = 115) is true. This is the probability that if the true mean is 115, the sample value will be below the cutpoint of 112.5.

The area of the alternative population to the right of the cutpoint is the power of the test for this specific null hypothesis. This represents the probability that if the alternative hypothesis is true and the population mean is 115, the sample mean will be above the cutpoint of 112.5 and we will conclude that the population mean is significantly greater than 100.

Let's consider how each of the four factors cited above could increase power in this example, assuming that only one factor can change at once:

1. If α were increased to 0.10, the cutpoint would be lower (further to the left) and the power would increase, while the P(Type II error) would decrease. The area below the cutpoint would decrease, representing a reduction in P(Type II error).

2. If the effect size were greater, for instance if the mean of the alternative population were 120 instead of 115, the distribution for the alternative population would be shifted up the number line. The result would be a decrease in P(Type II error), and an increase in power.

3. If the standard deviation were decreased, the two populations would overlap less. This would result in a reduction in the probability of Type II error and an increase in power.

4. If sample size were increased, this would have a similar effect to decreasing standard deviation and would result in a reduction in the probability of Type II error and an increase in power.

One good way to become familiar with the influence of different factors on power is to experiment with a graphical power calculator. A good online example is available at *http://wise.cgu.edu/power/power_applet.html*.

Sample Size Calculations

As mentioned before, each type of power or sample size calculation requires that the appropriate formula be used. However, if the principles of research design, as well as power analysis, are understood, finding the correct formula is not difficult. Two simple examples of sample size calculations will be demonstrated here because they are a good illustration of the principles at work and are easily performed using only a hand calculator.

Confidence Interval for a Proportion

One common sample size calculation is determining the sample size required to calculate a proportion with acceptable precision. For instance, you may be calculating agreement among different employees assigned to do medical chart reviews, and you want an estimate of the proportion in agreement, plus or minus five percentage points. Or you may be conducting a survey of the proportion of adults immunized against influenza in a population, and want to estimate the proportion immunized plus or minus 10 percentage points. This is not a power calculation because no hypothesis is being tested, but it is a sample size calculation because you need to determine the minimum sample size required for a specified level of precision.

The formula used for a two-sided confidence interval is:

$$n = \left(\frac{Z_{1-\alpha/2}}{\omega}\right)^2 [\pi(1-\pi)]$$

π (Greek letter pi) is the hypothesized population proportion. Z is determined by the alpha level; the value of Z for the chosen level of alpha can be found using a standard normal distribution table. ω (Greek letter omega) is the half-width of the desired confidence interval. The half-width is half the confidence interval: if we use a confidence interval of 10 percentage points, the half-width is 5 percentage points.

We want to calculate a two-sided confidence interval with $\alpha = 0.05$, so $Z = 1.96$. We believe π to be 0.8, and we want a confidence interval of 10 percentage points, so $\omega = 0.05$ (5/100). Plugging these values into the equation gives us:

$$n = \left(\frac{1.96}{0.05}\right)^2 [0.8(0.2)] = 245.9$$

We round this estimate up to 246 since there generally are no fractional subjects available! So we need 246 subjects, if our estimate of π is correct, to have an estimate with a 95% confidence interval of 0.10 (0.05 above and 0.05 below the estimate).

Power for the Test of the Difference Between Two Sample Means (Independent Samples t-Test)

For an example of a simple power calculation, let's assume we want to calculate how many subjects per group we need to conduct a two-tailed independent samples t-test with acceptable power. The formula is:

$$n = \frac{2(Z_{1-\alpha/2} + Z_{1-\beta})^2}{\delta^2}$$

where δ is effect size, calculated as:

$$\frac{\mu_1 - \mu_2}{\sigma}$$

σ in this case is determined using whichever method of calculating the standard deviation for a t-test is appropriate for the data in question (see Chapter 8 for details). We need Z-values for both α and β to use this formula. We will stick with the 95% confidence interval for a two-tailed test used in the previous example, so the Z-value for $1 - \alpha/2$ will be 1.96. We will compute the sample size required for 80% power, so the Z-value for 1-β will be 0.84. Note that if we were doing a one-tailed test, Z_α would be 1.645, and if we were calculating 90% power, $Z_{1-\beta}$ would be 1.28.

The effect size is the difference between the two populations, divided by the standard deviation. If $\mu_1 = 25$, $\mu_2 = 20$, and $\sigma = 10$, the effect size is 0.5. We plug these numbers into the formula as follows:

$$n = \frac{2(1.96 + 0.84)^2}{0.5^2} = \frac{15.6}{0.25} = 62.72$$

So we need at least 63 subjects per group to have an 80% probability of finding a significant difference between two groups, when the effect size is 0.5.

Exercises

Here's a set of questions to help you review the topics covered in this chapter.

Question

A classic example of the use of contingency tables in epidemiology is investigation of outbreaks. When a number of people become ill after eating at a restaurant, the public health department will launch an investigation to try to identify the food or foods responsible. This effort is complicated by the fact that the people who got sick probably ate many different foods, and some people who ate the same foods may not have gotten sick. So one approach is to interview the customers to ascertain what they ate and whether or not they got sick.

How to Lie with Percentages

You can't work in statistics for very long before someone demonstrates their cleverness by quoting some form of the aphorism attributed to the British politician Benjamin Disraeli and popularized in the United States by Mark Twain, that there are three kinds of lies: lies, damned lies, and statistics. There's even a popular book called *How to Lie with Statistics* by Darrell Huff (Norton), which is sometimes said to be the most-read statistics book in the world. One purpose of Huff's book, and this one as well, is not to teach you how to lie with statistics, but to help you spot other people lying.

One of the easiest ways to lie (or mislead, if you prefer) with statistics is to quote percentages without reference to the raw numbers underlying them, a practice beloved of politicians but not exclusively practiced by them. For instance, if you heard that there was a 100% increase in cholera cases in the United States, you might find that cause for alarm, until you learned that the increase was from one case to two. Similarly, a 50% increase in cancer risk for some rare exposure (affecting, say, only 15 people nationally) may not have as much public health significance as a 5% increase of a common exposure (which might affect 200,000,000 people).

People often forget that percentage increases and decreases are not symmetrical. If you increase the number of college graduates by 10% one year, then decrease it by 10% the next year, you are not back to your original total. Say you have 100,000 college graduates to begin with. A 10% increase gives you 110,000. A 10% decrease of the new total gives you 99,000 (110,000×0.9), which is fewer than you started with.

The data is then arranged into a series of 2×2 tables, as in Tables 18-16 and 18-17, in which the exposure is the particular food in question and the disease is food poisoning. Calculate the risk ratios for the two foods below and justify a decision as to whether or not they were a cause of food poisoning.

Table 18-16. Contingency table for roast beef and food poisoning

	D+	D−
E+	15	85
E−	20	80

Table 18-17. Contingency table for chicken salad and food poisoning

	D+	D−
E+	80	20
E−	20	80

Answer

The OR for roast beef is:

$$RR = \frac{a/(a+b)}{c/(c+d)} = \frac{15/100}{20/100} = 0.75$$

The OR for chicken salad is:

$$RR = \frac{a/(a+b)}{c/(c+d)} = \frac{80/100}{20/100} = 4.0$$

It appears that the culprit is chicken salad, because people who ate it had four times the risk of food poisoning compared to people who didn't eat it. Roast beef seems to have a slightly protective effect, perhaps because people who ate the roast beef were less likely to eat the chicken salad. Anyway, people who ate roast beef had only 75% of the risk of food poisoning as compared to people who did not eat roast beef.

Question

Compute the odds ratio and confidence interval for the data shown in Table 18-18, from a case-control study of oral contraceptive use and breast cancer. Does the data show a significant relationship between the two?

Table 18-18. Contingency table for oral contraceptive use and breast cancer

	D+	D−
E+	30	70
E−	20	80

Answer

The odds ratio is:

$$OR = \frac{ad}{bc} = \frac{30 \times 80}{20 \times 70} = \frac{2400}{1400} = 1.71$$

To see if this is significantly different from the null value of 1.0, compute the 95% confidence interval as:

$$CI = \frac{ad}{bc}\exp\left(\pm z\sqrt{\frac{1}{a}+\frac{1}{b}+\frac{1}{c}+\frac{1}{d}}\right)$$

$$= 1.71\exp\left(\pm 1.96\sqrt{\frac{1}{30}+\frac{1}{70}+\frac{1}{20}+\frac{1}{80}}\right)$$

$$= (-0.206, 3.626)$$

The CI of (.61, 2.81) includes the null value of 1.0, so we conclude that this study does not demonstrate a significant relationship between oral contraceptive use and breast cancer.

Question

Calculate the sample size needed to estimate a proportion plus or minus 10 percentage points, when the hypothesized proportion is 0.70.

Answer

Use the sample size formula for a proportion and plug in the numbers:

$$Z_{1-\alpha/2} = 1.96$$

$$\omega = 0.10$$

$$\pi = 0.70$$

$$n = \left(\frac{Z_{1-\alpha/2}}{\omega}\right)^2 [\pi(1-\pi)] = \left(\frac{1.96}{0.1}\right)^2 [0.7(0.3)] = 80.7$$

In other words, 81 subjects.

Question

Calculate the sample size needed for an independent samples t-test with a one-tailed hypothesis, 90% power, and an effect size of 0.4.

Answer

Use the sample size formula presented above and plug in the numbers:

$$Z_\alpha = 1.645$$

$$Z_{1-\beta} = 1.28$$

$$\delta = 0.4$$

$$n = \frac{2(Z_\alpha + Z_{1-\beta})^2}{\delta^2} = \frac{2(1.645 + 1.28)^2}{0.4^2} = \frac{17.11}{0.16} = 106.9$$

In other words, 107 subjects per group.

19

Educational and Psychological Statistics

Many statistical techniques used in education and psychology are common to other fields of endeavor: these include the *t*-test (covered in Chapter 8), various regression and ANOVA models (covered in Chapters 12–15) and the chi-square test (covered in Chapter 10). The discussion of measurement in Chapter 1 will also prove useful since much of educational and psychological research involves constructs that cannot be observed directly and have no obvious units of measurement. Examples of such constructs include mechanical aptitude, self-efficacy, and resistance to change. This chapter concentrates on statistical procedures used in the field of *psychometrics*, which is concerned with the creation, validation, and use of tests and measurements applied to human intelligence, knowledge, abilities, and psychological characteristics such as personality traits.

The first question you may ask with regard to the use of statistics in education and psychology is why they are necessary at all. After all, isn't every person an individual, and isn't the point of education and psychology to perceive that person in all their individual richness, not to reduce them to a set of numbers or place them in comparison with others who may not really be comparable at all?

This is a valid concern and underscores what anyone working in the human sciences knows already: doing research on human beings is in many ways much more difficult than doing research in the hard sciences or in manufacturing, because people are infinitely more varied than lug nuts or chemical molecules. It's the diversity and individuality of people, which we certainly value as students of the human sciences, that makes research in those fields particularly difficult. It's also true that while some educational and psychological research is aimed toward making general statements about groups of people, a great deal of it is focused on understanding and helping individuals who come embedded with social circumstances, family histories, and all kinds of other complications, making direct comparisons between one person and another very difficult.

But standard statistical procedures are useful even in the most specific and individual therapeutic circumstances; for instance, when the goal of an encounter is to devise an appropriate educational plan for one student or therapeutic regimen for one patient. Making such decisions is difficult, but would be even more so without the aid of formal educational and psychological tests that yield numeric values and can be compared to scores for other individuals. No one would suggest that only formal, standardized tests and questionnaires be used in these contexts: interviews and observational testing play an important role in educational and psychological evaluations as well. But the advantages of including formal testing procedures and standardized tests in clinical and educational evaluations include:

1. Objective comparisons are facilitated by the use of a normative group: for instance, is this patient recovering from trauma experiencing more side effects than is common among others who have experienced the same injury? Are the reading skills of this pupil comparable to others of his age and grade level?

2. Answers can be gained quickly; you needn't wait for the end of the school term to discover which pupils are struggling because of poor language proficiency, for instance, and you don't need a lengthy interview or practical examination to discover that a patient is suffering from serious memory deficits.

3. Standardized tests are presented in a regulated situation and under specified conditions, and can be scored objectively, so the only issue being evaluated is the student's or patient's performance, not their appearance, sociability (unless that is germane to the context), or other irrelevant factors.

4. Most standardized tests do not require great skill to administer (unlike clinical interviews, for instance) and can be given to groups of people at once, making them particularly useful as screening procedures and as an adjunct to more personalized evaluation. They are also less open to manipulation by either the subject or the test administrator, so the test results can be interpreted with confidence that they were not affected by the choice of administrator or irrelevant qualities of the subject.

Percentiles

In many countries, school-age children are evaluated by tests that report their results in *percentiles*, also known as *percentile ranks*. So Johnny's parents may be informed that he scored at the 90th percentile in reading and the 85th percentile in math, while Susie's parents learn that she scored at the 88th percentile in reading and the 92nd percentile in math. Percentiles are a form of *norm-referenced* scoring, so called because an individual score is placed in the context of a *norm group*, meaning people similar to the test-taker. For school-age children, the norm group is often other children in the same grade within their country. Norm-referenced scoring is used in all kinds of testing situations in which an individual's rank in relation to some comparison group is more important than an absolute score.

The percentile rank of an individual score refers to the percentage in the norm group that scored lower than that individual score. So 90% of the students in Johnny's norm group scored below him in reading and 85% scored below him in math: this indicates he is doing very well in both subjects in comparison to other students of his age and grade level. Here's a brief example illustrating how to find percentile ranks for scores on an exam that was given to 100 students (on national exams the norm group would be much larger, and the scores reflect a greater range, but this example will illustrate the point).

The first step in translating from raw scores to percentiles is to create a frequency table that includes a column for cumulative percentage, as illustrated in Table 19-1. To find the percentile rank for a particular score, use the cumulative percentage from the next-highest score, i.e., from the row just above in the table. In this example, someone who scored 96 on the exam was at the 75th percentile rank (meaning 75% of the test-takers scored below 96), while someone who scored 85 was at the 25th percentile rank. There can be no 100th percentile rank because, logically speaking, 100% of the test-takers couldn't have scored below a score that is included in the table. You can have a 0th percentile, however: the person who scored 53 is in the 0th percentile because no one achieved a lower score.

Table 19-1. Scores of 100 students on an exam

Score	Frequency	Percentage	Cumulative percentage
53	1	1.0%	1.0%
55	2	2.0%	3.0%
58	1	1.0%	4.0%
61	2	2.0%	6.0%
65	3	3.0%	9.0%
67	1	1.0%	10.0%
70	2	2.0%	12.0%
71	3	3.0%	15.0%
78	2	2.0%	17.0%
80	4	4.0%	21.0%
82	2	2.0%	23.0%
84	2	2.0%	25.0%
85	5	5.0%	30.0%
86	4	4.0%	34.0%
88	3	3.0%	37.0%
90	5	5.0%	42.0%
91	7	7.0%	49.0%
92	8	8.0%	57.0%
93	7	7.0%	64.0%
94	5	5.0%	69.0%
95	6	6.0%	75.0%
96	4	4.0%	79.0%
97	3	3.0%	82.0%
98	7	7.0%	89.0%
99	6	6.0%	95.0%
100	5	5.0%	100.0%

In situations such as standardized testing at the national level, the norm group used to map the scores to percentiles is much larger, and generally calculation of percentiles for individual students is not necessary. Instead, the test manufacturer provides a chart that relates raw scores to percentile ranks.

Standardized Scores

The *standardized score*, also known as the *normal score* or the *Z-score*, (further discussed in Chapter 7), transforms a raw score into units of standard deviation above or below the mean. This translates the scores so they may be evaluated in reference to the standard normal distribution. Standardized scores are frequently used in education and psychology because they place a score in the context of other scores, and can therefore be considered a type of *norm-referenced* scoring. For frequently used scales such as the Wechsler Adult Intelligence Scale (WAIS), population means and standard deviations are known and may be used in the calculations: for the WAIS, the mean is 100 and the standard deviation is 15. To convert a raw score to a standardized score, use the following formula:

$$Z = \frac{X - \mu}{\sigma}$$

where X is the raw score, μ is the population mean, and σ is the population standard deviation. The conversion to Z-scores puts all scores on a common scale, which is the standard normal distribution that has a mean of 0 and a variance of 1, and score probabilities are distributed with the known properties of the normal distribution (for instance, about 66% of the scores will be within one standard deviation of the mean). We can convert a raw score of 115 on the WAIS to a Z-score as follows:

$$Z = \frac{115 - 100}{15} = \frac{15}{15} = 1.0$$

We can consult a table of the standard normal distribution (Z distribution) to learn that a Z-score of 1.0 means that 84.1% of individuals score at or below that individual's raw score. Standardized scores are particularly useful when comparing scores on tests with different scales. For instance, let's say we also administer a test of mathematical aptitude that has a mean of 50 and a standard deviation of 5. If a person scores 105 on the WAIS and 60 on the mechanical aptitude, we can easily compare those scores in terms of Z-scores. For the WAIS:

$$Z = \frac{105 - 100}{15} = 0.33$$

For the mechanical aptitude test:

$$Z = \frac{60 - 50}{5} = 2.0$$

So this person scored slightly above average in intelligence, but far above average in mechanical aptitude.

Some people find standardized scores confusing, particularly the fact that a person can have a Z-score that is 0 or negative (and in the standard normal distribution, half the scores are below average and therefore negative). For this reason, Z-scores are sometimes converted to *T-scores* for presentation to parents or other nonstatistical audiences. A *T*-score expresses a score in terms of a population with a mean of 50 and a standard deviation of 10, using the following formula:

$$T = Z(10) + 50$$

If a person has a Z-score of 2.0 (meaning they scored two standard deviations above the mean), this can be converted to a *T*-score as follows:

$$T = 2(10) + 50 = 70$$

Similarly, someone with a Z-score of −2.0 would have a *T*-score of 30. Because hardly anyone ever scores five standard deviations or more below the mean, *T*-scores are almost always positive, which makes them easier for many people to understand. For instance, the clinical scales of the Minnesota Multiphase Personality Inventory-II (MMPI-II), commonly used to identify and evaluate psychiatric conditions, are reported as *T*-scores.

Test Construction

Most tests in psychology and education are used for what is called *subject-centered measurement*, in which the purpose is to place individuals on a continuum with respect to particular characteristics such as language-learning ability or anxiety. Creating and validating a test is a huge amount of work (when I was in graduate school, students were barred from writing a dissertation that required creating and validating a new test, out of fear that they would never complete the process), and the burden is entirely on the test's creator to convince others working in the same field that the test scores are meaningful. Therefore, the first move for someone beginning to investigate a field is to check and see if there are any existing, validated tests that would be adequate. However, particularly if you are researching a new topic or dealing with a previously ignored population, there may be no existing test adequate to your purpose, in which case the only option is to create and validate a new test.

Anyone seriously contemplating constructing a test from scratch needs to consult a psychometrics textbook: one commonly used is *An Introduction to Classical and Modern Test Theory* by Crocker and Algina (Wadsworth). Crocker and Algina list 10 steps in the process of test construction, many of which have little to do with statistics but much to do with subject expertise and knowledge of the target population (the people to whom the test will be administered). Other authors break down the process into different numbers of steps, but all agree that it is a multi-stage, iterative process. Some of the major tasks involved are defining the purpose of the test, defining the area of knowledge or behaviors to be covered, creating a pool of items or behaviors to evaluate the target knowledge or behaviors, revising the items (multiple times), creating the final form of the test, and evaluating it for reliability and validity. The discussion in this book concentrates on the statistical aspects of test construction; for a discussion of nonstatistical issues, consult a textbook such as Crocker and Algina.

Tests may be either *norm-referenced* or *criterion-referenced*. Norm-referenced tests have already been discussed; their purpose is to place an individual in the context of some group. In contrast, the purpose of a criterion-referenced test is to compare an individual to some absolute standard; for instance, to see if they have obtained minimum competency in an academic subject. In a criterion-referenced test, everyone taking the test could receive a high score, or everyone could receive a low score, because the individuals are evaluated with reference to a predetermined standard rather than in reference to each other. Although criterion-referenced tests may yield a continuous outcome (for instance, a score on a scale of 1–100), a *cutpoint* (single score) is often established as well such that everyone who achieves that score or above passes, while everyone with a score below it fails.

Most tests are composed of numerous individual *items*, often written questions, which are combined (often simply added together) to produce a *composite* test score. For instance, a test of language ability might be constructed of 100 items, with each correct item scored as a 1 and each incorrect item as a 0. The composite score for an individual could then be determined by adding up the number of correct items. Many of the statistical procedures used in examining tests have to do with the relationship among individual items, and the relation between individual items and the composite score.

Although composite test scores are commonly used, they can be misleading measures of ability or achievement. One major difficulty is that typically all items are assigned the same weight toward the total score, while they may not all be of equal difficulty. The distinction between someone who misses some easy questions but gets more difficult questions correct versus someone who gets the easy questions correct but can't answer the difficult questions is lost when a composite score is formed by simply summing the scores of items of differing homogeneity.

The mean and variance of dichotomous items (those scored as either right or wrong) is calculated using the concept of *item difficulty*, signified as p. Item difficulty is the proportion of examinees who answer a question correctly. If N people are in the group of examinees used to establish item difficulty, p is calculated for one item as:

$$p_j = \frac{\text{number who answer item correctly}}{N}$$

With dichotomous items scored 0–1 (0 for incorrect, 1 for correct), the mean is the same as the proportion answering the item correctly:

$$p_j = \mu_j = \frac{\sum X_{ij}}{N}$$

where X_{ij} are the individual items and N is the number of examinees.

Variance for an individual dichotomous item p_j may be calculated as:

$$\sigma_j^2 = p_j(1 - p_j)$$

This is sometimes written as:

$$\sigma_j^2 = p_j q_j$$

because there is an alternative convention of writing $(1 - p_j)$ as q_j. We will not follow that convention, because it requires introducing a new symbol q_j, which is unnecessary.

The correlation coefficient between two dichotomous items p_j and p_k, also called the *phi coefficient*, may be calculated as:

$$\text{phi} = \frac{p_{jk} - p_j p_k}{\sqrt{p_j(1 - p_j)p_k(1 - p_k)}}$$

where p_{jk} is the proportion of examinees answering both p_j and p_k correctly, p_j is the item difficulty of item j (the proportion answering item j correctly), and p_k is the item difficultly for item k. An alternative method of calculation uses the frequency counts from a 2×2 table (whose setup differs slightly from the arrangement commonly used in medical and epidemiological statistics) showing the number correct and incorrect on the pair of items, using the standard cell designations shown in Table 19-2.

Table 19-2. Standard notation for a 2×2 table

		Item j	
		+	−
Item k	−	a	b
	+	c	d

+ and − signify whether each item is answered correctly, so:

- a is the number of people who got j right and k wrong.
- b is those who got both items wrong.
- c is those who got both items right.
- d is those who got j wrong and k right.

The formula to calculate phi using cell counts is:

$$\text{phi} = \frac{bc - ad}{\sqrt{(a + b)(a + c)(c + d)(b + d)}}$$

Further explanation of phi, including a worked example, is included in Chapter 9.

Computing the variance of a composite score requires knowing both the individual item variances and their covariance. Unless all pairs of variables are completely uncorrelated or are negatively correlated, the variance of a composite score will always be greater than the sum of the individual item variances. Although composite variance is usually computed using statistical software, the formula is useful to know because it underlines the relationship among the relevant quantities. The covariance for a pair of items j and k (whether the items are dichotomous or continuous) may be computed as:

$$\sigma_{jk} = \rho_{jk}\sigma_j\sigma_k$$

where σ_{jk} is the covariance of the two items, ρ_{jk} is the correlation between the two items, and σ_j and σ_k are the individual item covariances. Because there are two covariance pairs for each item pair (the covariance of j with k, and the covariance of k with j, which are identical), the covariance of a composite Y may be calculated as:

$$\sigma^2{}_Y = \sum\sigma_i^2 + 2\sum_{i<j}\rho_{ij}\sigma_i\sigma_j$$

The stipulation $i > j$ in the above formula stipulates that we only compute unique covariance terms. To get the right number of covariance terms, we then multiply each unique covariance by 2.

This formula underlines the fact that as items are added to a test, the number of covariance terms increases more quickly than the number of variance terms; for instance, if we add 5 items to a test that has 5 items to start with, the number of variance terms increases from 5 to 10, but the number of covariance items increases from 20 to 90. It helps to remember that the number of unique covariance terms for n items is calculated as $n(n-1)$; in our example, 5 items yields $5(4) = 20$ covariance terms. A 10-item test would have $10(9) = 90$ covariance terms. The number of *unique* covariance terms is $[n(n-1)]/2$, so 5 items yield 10 unique covariance terms and 10 items yield 45 unique covariance terms.

In most cases, adding items to a composite increases the variance of the composite, because the variance of the composite is increased by the variance of the individual item plus its covariance with all the existing items on the test. The proportional increase is greater when items are added to a short test than to a long test, and is greatest when items are highly correlated, because that results in larger covariances among items. All else being equal, the greatest composite variance is produced by items of medium difficulty ($p = 0.5$ produces the largest covariance scores) that are highly correlated with each other.

Classical Test Theory: The True Score Model

In an ideal world, all tests would have perfect reliability, meaning that if the same individuals were tested under the same conditions for some stable characteristic, they would receive identical scores each time. In this case, we would have no problem saying that a person's *observed score* on the test was the same as the person's *true score*, and that the observed score was an accurate reflection of that person's score on whatever the test was designed to measure. In the real world, however, many factors can influence observed scores, and repeated tests on the same material taken by the same individual often yield different scores. For this reason, we must differentiate between the true score and the observed score. We do this by introducing the concept of *measurement error*, which is that part of the observed score that causes it to deviate from the true score.

Measurement error can be either random or systematic. *Random measurement error* is the result of chance circumstances such as room temperature, variance in administrative procedure, or fluctuation in the individual's mood or alertness.

We would not expect random error to consistently affect an individual's score in one direction or the other. Random error makes measurement less precise but does not systematically bias results because it can be expected to have a positive effect on one occasion and a negative effect on another, thus "canceling itself out" over the long run. Because there are so many potential sources for random error, we have no expectation that it can be completely eliminated, but we desire to reduce it as much as possible in order to increase the precision of our measurements. *Systematic measurement error*, on the other hand, is error that consistently affects an individual's score in a particular direction but has nothing to do with the construct being tested. An example would be measurement error on a mathematics exam that is caused by poor language skills, so that the examinee cannot read the directions to take the exam properly. Systematic measurement error is a source of bias and should be eliminated from testing whenever possible.

The psychologist Charles Spearman introduced the classical concepts of true and error scores in the early twentieth century. Spearman described the observed score X (the score actually received by an individual on a testing occasion), which is composed of a true component (T) and a random error component (E):

$$X = T + E$$

Over an infinite number of testing occasions, the random error component would be assumed to cancel itself out, so the mean or *expected value* of the observed scores is the same as the true score. For individual j, this can be written as:

$$T_j = E(X_j) = \mu_{X_j}$$

where T_j is the true score for individual j, $E(X_j)$ is his expected observed score over an infinite number of testing occasions, and μ_{x_j} is the mean observed score for individual j over the same occasions. Error is therefore the difference between an individual's observed score and his true score:

$$E_j = X_j - T_j$$

Over an infinite number of testing occasions the expected value of the error for one individual is 0. Because "error" in these definitions means random error only, true and error scores are assumed to have the following properties:

- Over a population of examinees, the mean of the error scores is 0.

- Over a population of examinees, the correlation between true and error scores is 0.

- The correlation between error scores between scores by two randomly chosen examinees on two forms of the same test, or two testing occasions using the same form, is 0.

Reliability of a Composite Test

When we administer a test to an individual, one of our concerns is how well the observed score on that test represents the person's true score. In theoretical terms, what we seek is the *reliability index* for a composite test, which is the ratio of the standard deviation of the true scores to the standard deviation of the observed scores. The reliability index is calculated as:

$$\rho_{XT} = \frac{\sigma_T}{\sigma_X}$$

where σ_T is the standard deviation of the true scores for a population of examinees and σ_X is the standard deviation of their observed scores. This is sometimes expressed as the proportion of total variation on the test scores, which is explained by true variation.

In practice, the true scores are unknown, so the reliability index must be estimated by using observed scores. One way to do this is to administer two parallel tests to the same group of examinees and use the correlation between their scores on the two forms, known as the *reliability coefficient*, as an estimate of the reliability index. Parallel tests must satisfy two conditions: *equal difficulty* and *equal variance*.

The reliability coefficient is an estimate of the ratio of true score variance to observed score variance and can be interpreted similarly to the coefficient of determination (r^2) in the General Linear Model. So if a test reports a reliability coefficient of 0.88, we can interpret this as meaning that 88% of the observed score variance from administrations of this test is due to true score variance, while the remaining 0.12 or 12% is due to random error. To find the correlation between true and observed scores for this test, we take the square root of the reliability coefficient, so for this test the correlation between true and observed scores is estimated as $\sqrt{0.88}$ or 0.938.

The reliability coefficient can be estimated using one of several methods, which are discussed in more detail in Chapter 1. If we estimate the reliability coefficient by administering the same test to the same examinees on two different occasions, this is called the *test-retest method* and the correlation between test scores in this case is known as the *coefficient of stability*. We could also estimate the reliability coefficient by administering two equivalent forms of a test to the same examinees on the same occasion, also known as the *alternate form method*, in which case the correlation between scores would be called the *coefficient of equivalence*. If both different forms and different occasions of testing are used, correlation between the scores under these conditions is called the *coefficient of stability and equivalence*. Because this coefficient has two sources of error, *forms* and *occasions*, it is generally assumed to be lower than either the coefficient of stability or the coefficient of equivalence would be for a given group of examinees.

Measures of Internal Consistency

A different approach to estimating reliability is to use a measure of internal consistency that can be calculated from a single administration of a test to a single group of examinees. The reason internal consistency measurements are used to estimate reliability is that a composite test is often conceived of as being composed of test items sampled from a large domain of potential items. An internal consistency estimate is a prediction of how similar an individual's score would be if a different subset of items from that domain had been chosen.

Consider the task of creating an exam to test student competence in high school algebra. The first steps in creating this test would be to decide what topics to cover. Then a pool of items would be written that evaluate student mastery of those topics. A subset of items would then be chosen to create the final test. The purpose of this type of exam is not merely to see how well the students score on the specific items included in the test they took, but how well they mastered all the content considered to be within the domain of high school algebra. If the items used on the test are a fair selection from this content domain, the test score should be a reliable indicator of the students' mastery of the material. Item homogeneity is also a valued characteristic of this type of test because it is an indication that the items are testing the same content and do not have technical flaws such as misleading wording or incorrect scoring, which would cause student performance on an item to be unrelated to mastery of algebra.

Split-Half Methods

Split-half methods to measure internal consistency require that a test be split into two parts or forms, usually two halves of equal length, which are intended to be parallel. All items on the full-length test are completed by each examinee. The split may be achieved by several methods, including alternate assignment (even-numbered items to one form, odd-numbered to the other), content matching, or random assignment. Whatever method is used, if the original test had 100 items, the two halves will each have 50 items. The correlation coefficient between examinee scores for the two forms is the *coefficient of equivalence* for the two halves. The coefficient of equivalence is an underestimate of the reliability for the full-length test, because longer tests are usually more reliable than shorter tests. The *Spearman-Brown prophecy formula* can be used to estimate the reliability of the full-length test from the coefficient of equivalence for the two halves, using the following formula:

$$\hat{\rho}_{XX'} = \frac{2\rho_{AB}}{1 + \rho_{AB}}$$

where $\rho_{XX'}$ is the estimated reliability of the full-length test and ρ_{AB} is the observed correlation, i.e., coefficient of equivalence, between the two half-tests. For this formula to be accurate, the two half-tests must be strictly parallel. If the coefficient of equivalence for the two half-tests is 0.5, the estimated reliability of the full-length test is:

$$\hat{\rho}_{XX'} = \frac{2(0.5)}{1 + 0.5} = 0.67$$

A second method to estimate reliability of a full-length test using the split-half method is to calculate the difference between scores on the two halves for each examinee. The variance of that difference score is an estimate of error variance of reliability, so the 1 minus the ratio of error variance to total variance may also be used as an estimate of reliability. This is the formula to use for the second method:

$$\hat{\rho}_{XX'} = 1 - \frac{\sigma_D^2}{\sigma_X^2}$$

where

$$\sigma_D^2$$

is the variance of the difference scores, and

$$\sigma_X^2$$

is the variance of the observed scores.

Estimates of reliability using either method will be identical when the variance of the two half-tests is identical. The more dissimilar the two variances, the larger the estimate using the Spearman-Brown formula relative to estimates using the difference-score method. Estimation of reliability by either method depends on how the items are chosen for the two halves, because a different split will result in different correlations between the halves or a different set of difference scores. In fact, using these methods, there are as many different reliabilities for a given test as there are ways to split its items into two halves. If there are k items in the full-length test, the number of splits possible is:

$$\frac{\frac{1}{2}k!}{\left[\frac{k!}{2}\right]^2}$$

where ! signifies factorial, so $k! = (k)(k-1)(k-2)...(1)$. This may not be a problem in practice because if the test items are basically homogeneous, reliability estimated using any particular split should be similar to reliability estimated using a different split.

Coefficient Alpha

There are several methods of estimating reliability using item covariances that avoid the problem of multiple split-half reliabilities, three of which are presented below. *Cronbach's alpha* may be used for either dichotomous or continuously scored items, while the two *Kuder Richardson formulas* are for dichotomous items. The measure of internal consistency computed by any of these methods is commonly referred to as *coefficient alpha*, and is equivalent to the mean of all possible split-half coefficients computed using the difference-score method. Coefficient alpha is, strictly speaking, not an estimate of the reliability coefficient but of its lower bound (sometimes called the *coefficient of precision*). This nicety is often ignored in interpretation, however, and coefficient alpha is usually reported without further interpretation.

Note that computing coefficient alpha for a test of any considerable length is tedious and therefore generally accomplished using computer software. Still, it is useful to know the formulas and work through a simple calculation in order to understand what factors affect coefficient alpha.

Cronbach's alpha is the most common method for calculating coefficient alpha, and is the name often given for coefficient alpha in computer software packages designed for reliability analysis. It is computed using the following formula:

$$\hat{\alpha} = \frac{k}{k-1}\left(1 - \frac{\sum \hat{\sigma}_i^2}{\hat{\sigma}_X^2}\right)$$

where k is the number of items,

$$\hat{\sigma}_i^2$$

is the variance of item i, and

$$\hat{\sigma}_X^2$$

is the total test variance.

Suppose we have a 5-item test, with a total test variance of 100 and individual item variances of 10, 5, 6.5, 7.5, and 13. Cronbach's alpha would be calculated as:

$$\hat{\alpha} = \frac{5}{5-1}\left(1 - \frac{42}{100}\right) = 0.725$$

There are several Kuder Richardson formulas to calculate coefficient alpha: two useful for dichotomous items are presented here. Note that KR 21 is a simplified version of the KR 20 formula; it assumes all items are of equal difficulty. KR 20 and KR 21 yield identical results if all items are of equal difficulty; if they are not, KR 21 yields lower results than KR 20. The KR 20 formula is:

$$KR_{20} = \frac{k}{k-1}\left(1 - \frac{\sum p(1-p)}{\hat{\sigma}_X^2}\right)$$

where k is the number of items, p is the difficulty for a given item so $p(1-p)$ is the variance for one item, and:

$$\hat{\sigma}_X^2$$

is the total variance. Note that the KR 20 formula is identical to the Cronbach's alpha formula, with the exception that the item variance term has been restated to take advantage of the fact that KR 20 is used for dichotomous items, so:

$$\hat{\sigma}_i^2$$

has been replaced by $p(1-p)$.

The KR 20 formula can be simplified by assuming all items have equal difficulty, so it is not necessary to compute and sum the individual item variances. This simplification yields the KR 21 formula:

$$KR_{21} = \frac{k}{k-1}\left(1 - \frac{\hat{\mu}(k-\hat{\mu})}{k\hat{\sigma}_X^2}\right)$$

---

where k is the number of items, and:

$$\hat{\mu}$$

is the overall mean score for the test, and:

$$\hat{\sigma}^2_X$$

is the total variance for the test.

Item Analysis

Test construction often proceeds by creating a large pool of items, pilot-testing them on examinees similar to those for whom the test is intended, and selecting a subset for the final test that makes the greatest contributions to test validity and reliability. *Item analysis* is a set of procedures used to examine and describe examinees' responses to the items under consideration, including the distribution of responses to each item and the relationship between responses to each item and other criteria.

One of the first things usually computed in an item analysis is the mean and variance of each item. For dichotomous items, the mean is also the proportion of examinees that answered the item correctly, and is called the *item difficulty* or p, as discussed above. The total test score for one examinee is the sum of the item difficulties, which is the same as the sum of questions answered correctly. The average item difficulty is the total score divided by the number of items:

$$\mu_X = \sum_i p_i$$

and:

$$\mu_p = \frac{\mu_X}{k}$$

Because item difficulty is a proportion, the variance for an individual item is:

$$\sigma^2_i = p_i(1 - p_i)$$

Often items are selected to maximize variance, in order to increase the test's efficiency in discriminating among individuals of different abilities. Variance is maximized when $p = 0.5$, a fact that you can prove to yourself by calculating the variance for other values of p:

If $p = 0.50$, $\sigma^2_i = 0.5(0.5) = 0.2500$

If $p = 0.49$, $\sigma^2_i = 0.49(0.51) = 0.2499$

If $p = 0.45$, $\sigma^2_i = 0.45(0.55) = 0.2475$

If $p = 0.40$, $\sigma^2_i = 0.40(0.60) = 0.2400$

and so on. Because p and $1 - p$ are reciprocal, the variance is the same for $p = 0.4$ and $p = 0.6$, because in the first case:

$$\sigma_i^2 = 0.4(0.6)$$

and in the second:

$$\sigma_i^2 = 0.6(0.4)$$

In many common test formats, including multiple choice, examinees may raise their scores by guessing if they don't know the correct answer. This means that the p value of an item will often be higher than the proportion of examinees who actually know the material tested by the item. To put it another way, the observed scores will be higher than the true scores because the observed scores have been raised by successful guessing. The test creator's goal is usually to maximize true score variation on the test, so when an item format allows guessing (for instance, multiple choice items that carry no penalty for incorrect answers), an additional step is necessary to calculate the observed difficulty of an item to maximize item variance. This is done by adding the quantity $0.5/m$ to the item difficulty, where m is the number of choices for an item. The items are assumed to be equally likely to be selected if the examinee doesn't know the correct answer. The observed difficulty p_0 of an item that is assumed to have a true difficulty of 0.5 (half the examinees know the correct answer without guessing), for different values of m, would be as shown in Table 19-3.

Table 19-3. Item difficulties, corrected for guessing

Number of choices	p_0
2	$0.5 + 0.5/2 = 0.75$
3	$0.5 + 0.5/3 = 0.67$
4	$0.5 = 0.5/4 = 0.625$

Maximum score variance for an item with two choices is obtained if the observed item difficulty is 0.75, while maximum variance for three-choice items requires an observed difficulty of 0.67, and for four choices the observed difficulty should be 0.625. If examinees are not guessing at random, i.e., if they can eliminate one or more wrong choices before guessing, the observed item difficulty will be even higher than shown in Table 19-3. The probability of nonrandom correct guessing can't be calculated directly, but simulation studies have shown that for items whose true difficulty (without guessing) is 0.5, observed difficulty may be as high as 0.85 for a two-choice item, 0.77 for a three-choice item, and 0.74 for a four-choice item. All of which means that if you are using an item format that allows guessing, to maximize score variance you will want to choose items with an observed difficulty higher than 0.5: how high depends on the number of item choices and whether you think guessing will be random or nonrandom.

Item discrimination refers to how well an item differentiates between examinees that have high versus low amounts of the quality being tested, whether it is knowledge of geography, musical aptitude, or depression. Normally the test creator selects items that have *positive discrimination*, meaning they have a high probability of being answered correctly or positively by those who have a large amount of the quality, and incorrectly or negatively by those who have a small amount. For instance, if you are measuring mathematical aptitude, you want questions that students with high mathematical aptitude are likely to answer correctly, while those with low mathematical aptitude are unlikely to answer correctly. The reverse would be *negative discrimination*, e.g., a student with little mathematical aptitude would be more likely to answer a question correctly than someone with high aptitude. Negative discrimination is usually grounds to eliminate an item from the pool, unless it is being retained to catch people who are faking their answers (for instance, on a mental health inventory). Five indices of item discrimination are discussed in this section, followed by an index of item discrimination that can be related either to total test score or to an external criterion. If all items are of moderate difficulty (which is typical of many testing situations), all five discrimination indices will produce similar results.

The *index of discrimination* is only applicable to dichotomously scored items and compares the proportion of examinees in two groups that answered the item correctly. The two groups are often formed by examinee scores on the entire test; for instance, the upper 50% of examinees is often compared to the lower 50%, or the upper 30% to the lower 30%. The formula for the index of discrimination (D) is:

$$D = p_u - p_l$$

where p_u is the proportion in the upper group that got the item correct and p_l is the proportion in the lower group that got it correct. If 80% of the examinees in the upper group got an item correct, while only 30% of those in the lower group got it correct, the index of discrimination would be:

$$D = 0.8 - 0.3 = 0.5$$

The range of D is $(-1, +1)$. $D = 1.0$ would mean that everyone in the upper group got the item correct and no one in the lower group did. The index of discrimination is affected by how the upper and lower groups are formed; for instance, if the upper group was the top 20% and the lower group the bottom 20%, we would expect to find a larger index of discrimination than if the upper 50% and lower 50% were used.

There are no significance tests for the index of discrimination and no absolute rules about what constitutes an acceptable value. A rule of thumb suggested by Ebel (1965) is that $D > 0.4$ is satisfactory (items can be used), $D < 0.2$ is unsatisfactory (items can be discarded), and the range between suggests that the items should be revised to raise D above 0.4.

The *point-biserial correlation* can be used as a measure of how closely the response on a dichotomous item is related to the total test score or some other continuous quantity. The point-biserial correlation is calculated as:

$$\rho_{pbis} = \frac{\mu_+ - \mu_X}{\sigma_X} \sqrt{\frac{p}{1 - p}}$$

where μ_+ is the average total test score for examinees who answered the item correctly, μ_X is the average total score for the entire examinee group, σ_X is the standard deviation on the total score for the entire group, and p is the item difficulty. With a small group of items (a rule of thumb is 25 or less), performance on the individual item should be removed from the total score using this formula:

$$\rho_{i(X-i)} = \frac{\rho_{Xi}\sigma_X - \sigma_i}{\sqrt{\sigma_i^2 + \sigma_X^2 - 2\rho_{xi}\sigma_X\sigma_i}}$$

where $\rho_{i(X-i)}$ is the correlation between an individual item and the total score minus that item, ρ_{Xi} is the correlation between an individual item and the total score, σ_i is the item standard deviation, and σ_X is the total score standard deviation. A worked example of the point-biserial correlation is included in Chapter 9.

The *biserial correlation coefficient* may be used with dichotomous items, if it is assumed that performance on the item is due to a latent quality that is normally distributed. The formula to calculate the biserial correlation coefficient is:

$$\rho_{bis} = \frac{\mu_+ - \mu_X}{\sigma_X}\left(\frac{p}{Y}\right)$$

where μ_+ is the average total test score for examinees who answered the item correctly, μ_X is the average total score for the entire examinee group, σ_X is the standard deviation on the total score for the entire group, p is the item difficulty, and Y is the Y-coordinate (height of the curve) from the standard normal distribution for the item difficulty. Y values can be found from a table of the standard normal distribution: for instance, if $p = 0.5$, $Y = 0.3989$, while if $p = 0.6$, $Y = 0.3867$. Suppose for a given item, $\mu_+ = 80$, $\mu_X = 78$, $\sigma_X = 5$, and $p = 0.5$. The biserial correlation coefficient for this item would be:

$$\rho_{bis} = \frac{80 - 78}{5}\left(\frac{0.5}{0.3989}\right) = 0.5014$$

The value of the biserial correlation is systematically higher than the point-biserial correlation (discussed in Chapter 10) for the same numbers, and the difference increases sharply if $p < 0.25$ or $p > 0.75$, a fact that should be kept in mind when comparing item analyses using the two procedures. The biserial correlation coefficient is the preferred item difficulty statistic when the dichotomous item is assumed to reflect an underlying normal distribution and the goal is to select items that are very easy or very difficult, or if the test will be used with future groups of examinees with a wide range of ability.

The *phi coefficient*, discussed above, expresses the relationship between two dichotomous variables. If the variables are not true dichotomies but have been created by dichotomizing values from a continuous variable with an underlying normal distribution (such as a pass/fail score determined by establishing a single cutpoint for a continuous variable), the *tetrachoric correlation coefficient* is preferred over the phi coefficient because the range of phi is restricted when the item difficulties are not equal. Tetrachoric correlations are also used in factor analysis and structural equation modeling. The tetrachoric correlation coefficient is rarely computed by hand, but is included in some of the standard statistical software packages, including SAS and R. Further discussion of the tetrachoric correlation coefficient, including links to different computer programs (some of which are free) and macros to calculate it, can be found at *http://ourworld. compuserve.com/homepages/jsuebersax/tetra.htm.*

The *item reliability index* is the correlation between item performance and either total test score or an external, continuous criterion, weight by item variance. The logic behind using the item reliability index is that, all else being equal, items with greater variance do a better job at discriminating among individuals. This means that if two items have the same correlation with the criterion, the item with greater variance is preferred. If all items are of similar difficulty, little is gained by using the item reliability index rather than the item-total or item-criterion correlation.

If the item is being related to the total test score, the item reliability index for item i is calculated as:

$$\sigma_i \rho_{ix}$$

where σ_i is the variance for item i, and ρ_{ix} is the item-total correlation. If the item is dichotomous, the item reliability index is computed as:

$$\sqrt{p_i 1 - p_i}(\rho_{ix})$$

If the item is being related to some external criterion (for instance, overall grade point average), the item reliability index can be written as:

$$\sigma_i \rho_{iY}$$

where σ_i is the variance for item i, and ρ_{iY} is the correlation between the item and the external criterion. If the item is dichotomous, this can be written as:

$$\sqrt{p_i 1 - p_i}(\rho_{iY})$$

Item Response Theory

Although analyses based on classical test theory are still used in many fields, *Item Response Theory* (IRT) offers an important alternative approach. Anyone working in psychometrics should be aware of IRT, and it is increasingly being used in other fields from medicine to criminology. IRT will probably be used even more in the future, as IRT capabilities are implemented into commonly used statistical packages. IRT is a complex topic and can only be briefly introduced here; those who wish to pursue it should consult a textbook such as Hambleton, Swaminathan, and Rogers (1991) or a similar introductory textbook. An inventory of computer packages for IRT is available from the Rasch SIG at *http://winsteps.com/rasch.htm.*

IRT addresses several failings of classical test theory, chief among which is the fact that methods based on classical test theory cannot separate examinee characteristics from test characteristics. In classical theory, an examinee's ability is defined in terms of a particular test, and the difficulty of a particular test is defined in terms of a particular group of examinees. This is because the difficulty of a test item is defined in classical theory as the proportion of examinees getting it correct: with one group of examinees an item might be classified as "difficult" because few got it correct, while for another group of examinees it might be classified as "easy" because most got it correct. Similarly, on one test an examinee might be rated as having high ability or having mastered a body of material because he got a high score on the test, while on another test ostensibly covering the same basic material he might be rated as having low ability or mastery because he got a low score.

The fact that estimates of item difficulty and examinee ability are intertwined in classical test theory means that it is difficult to make an equivalent estimation of ability comparing examinees that take different tests, or to rate the difficulty of items administered to different groups of examinees. Classical test theory has tried various procedures to try to deal with these issues, such as including a common body of items on different forms of a test, but the central problem remains:

- Performance of a given examinee on a given item can be explained by the examinee's ability on whatever the item is testing, and ability is considered to be a latent, unobservable trait.

- An item characteristic curve (ICC) can be drawn to express the relationship between the performance of a group of examinees on a given item and their ability.

Ability is usually represented by the Greek letter theta (θ), while item difficulty is expressed as a number from 0.0 to 1.00. The ICC is drawn as a smooth curve on a graph in which the vertical axis represents the probability of answering an item and the horizontal axis represents examinee ability on a scale in which θ has a mean of 0 and a standard deviation of 1. The ICC is a monotonically increasing function, so that examinees with higher ability (higher value of θ) will always be predicted to have a higher probability of answering a given item correctly. This is shown in the theoretical ICC in Figure 19-1.

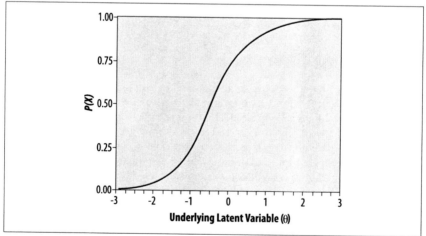

Figure 19-1. Theoretical ICC

IRT models, in relation to classical test theory models, have the following advantages:

1. IRT models are *falsifiable*, i.e., the fit of an IRT model can be evaluated and a determination made as to whether a particular model is appropriate for a particular set of data.

2. Estimates of examinee ability are not *test-dependent*; they are made in a common metric that allows comparison of examinees that took different tests.

3. Estimates of item difficulty are not *examinee-dependent*; item difficulty is expressed in a common metric that allows comparison of items administered to different groups.

4. IRT provides individual estimates of standard errors for examinees, rather than assuming (as in classical test theory) that all examinees have the same standard error of measurement.

5. IRT takes item difficulty into account when estimating examinee ability, so two people with the same number of items correct on a test could have different estimates of ability if one answered more difficult questions correctly than did the other.

One consequence of points 2 and 3 is that in IRT, estimates of examinee ability and item difficulty are *invariant*. This means that, apart from measurement error, any two examinees with the same ability have the same probability of answering a given item correctly, and any two items of comparable difficulty have the same probability of being answered correctly by any examinee.

Note that although in this discussion we are using the context of questions for which answers are scored as right or wrong (hence language such as "the probability of answering the item correctly"), IRT models can also be applied in contexts where there is no right or wrong answer. For instance, in a psychological questionnaire measuring attitudes, the meaning of item difficulty could be described as "the probability of endorsing an item" and θ as the degree or amount of the quality being measured (such as favorable attitude toward civic expansion).

There are several different models commonly used in IRT, which differ according to the item characteristics that they incorporate. Two assumptions are common to all IRT models:

Unidimensionality
> Whose strict definition is that items on a test measure only one ability, and which is defined in practice by the requirement that performance on test items be explicable with reference to one dominant factor

Local independence
> Which means that if examinee ability is held constant, there is no relationship between examinee responses to different items, i.e., responses to the items are independent

The simplest IRT logistic model includes only one characteristic of the item, item difficulty, signified by b_i. This is called the one-parameter logistic model, and is also called the *Rasch model* because it was developed by the Danish mathematician Georg Rasch. The ICC for the one-parameter logistic model is computed using the following equation:

$$P_i(\theta) = \frac{e^{\theta - b_i}}{1 + e^{\theta - b_i}}$$

where $P_i(\theta)$ is the probability that an examinee with ability θ will answer item i correctly, and b_i is the difficulty parameter for item i.

Item difficulty is defined as the point on the ability scale (*x*-axis) where the probability of an examinee getting the item correct is 0.5. For more difficult items, greater examinee ability is required before half the examinees are predicted to get it right, while for easier items, a lower level of ability is required to reach that point. In the Rasch model, the ICCs for items of differing difficulties have the same shape and differ only in location. This may be seen in Figure 19-2, which displays ICCs for several items of equal discrimination that vary in difficulty.

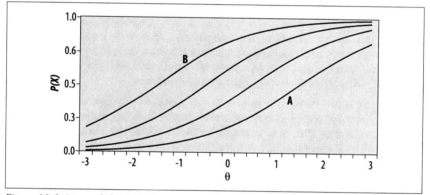

Figure 19-2. Items of identical discrimination but varying difficulty: item A is the most difficult, item B the easiest

Bearing in mind that θ is a measure of examinee ability, it can be seen that a greater amount of ability is required to have a 50% probability of answering item A correctly, compared with items further to the left. It is also clear that among the items graphed here, the least amount of θ is required to have a 50% chance of answering item B correctly. So we would say that item B is the easiest among these items, and item A the most difficult.

If this is not clear, draw a horizontal line across the graph at $y = 0.5$ and then a vertical line down to the *x*-axis where the horizontal line intersects each curve. The point where each vertical line intersects the *x*-axis is the amount of θ required to have a 50% probability of answering the item correctly: for curve B, $\theta = -1.5$, while for curve A, $\theta = 1.5$ (higher value of θ required for a 50% probability of answering an item correctly = more difficult item).

The two-parameter IRT model includes an item discrimination factor, a_i. The item discrimination factor allows items to have different slopes. Items with steeper slopes are more effective in differentiating among examinees of similar abilities than are items with flatter slopes, because the probability of success on an item changes more rapidly relative to changes in examinee ability.

Item difficulty is proportional to the slope at the point where $b_i = 0.5$, i.e., where half the examinees would be expected to get the item correct. The usual range for a_i is $(0, 2)$, because items with negative discrimination (those which an examinee with less ability has a greater probability to answer correctly) are usually discarded, and because in practice item discrimination is rarely greater than 2. The two-parameter logistic model also includes a scaling parameter, D, which is added to make the logistic function as close as possible to the cumulative normal distribution.

The ICC for a two-level logistic model is computed using the following formula:

$$P_i(\theta) = \frac{e^{Da_i(\theta - b_i)}}{1 + e^{Da_i(\theta - b_i)}}$$

Two items that differ in both difficulty and discrimination are illustrated in Figure 19-3.

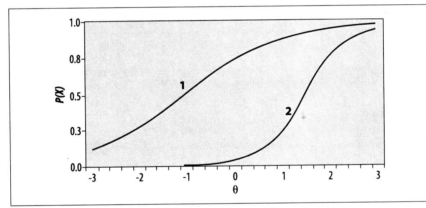

Figure 19-3. Two items that differ in both difficulty and discrimination

It may be easier to compare the two items by drawing a horizontal reference line at $b_i = 0.5$. In this example, the items clearly differ in difficulty: more ability is required to have a 50% probability of answering item 2 correctly as compared to item 1.

The three-level logistic model includes an additional parameter c_i, which is technically called the *pseudo-chance-level parameter*. This parameter provides a lower asymptote for the ICC that represents the probability of examinees with low ability answering the item correctly due to chance. c_i is often called the "guessing parameter" because one reason low-ability applicants could get a difficult question correct is by guessing the right answer. However, often c_i is lower than would be expected by random guessing because of the skill of test examiners in devising wrong answers that may seem correct to an examinee of low ability. The ICC for the three-parameter logistic model is calculated using this formula:

$$P_i(\theta) = c_i + (1 - c_i)\frac{e^{Da_i(\theta - b_i)}}{1 + e^{Da_i(\theta - b_i)}}$$

A three-parameter model is shown in Figure 19-4; it has a substantial guessing parameter, which can be seen from the fact that the curve intersects the x-axis around 0.20. This means that a person with very low θ would still have about a 20% chance of answering this item correctly.

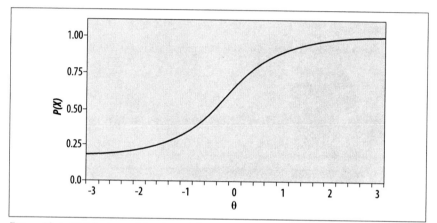

Figure 19-4. ICC for item with substantial guessing parameter

Exercises

Here is a set of questions to review the topics covered in this chapter.

Question

Given the data distribution in Table 19-1:

1. What is the percentile rank for a score of 80?
2. What score corresponds to a score at the 75th percentile?

Answer

You find the percentile by looking at the cumulative probability for the score just above the score you are interested in. To find a score corresponding to a percentile rank, reverse the process:

1. A score of 80 is in the 17th percentile.
2. A score of 96 is in the 75th percentile.

Question

Assume you are working with a published test whose mean is 100 and whose variance is 20. Convert the following individual scores to Z-scores and then to T-scores:

1. 70
2. 105

Answer

1. For 70, $Z = -1.5$ and $T = 35$.
2. For 105, $Z = 0.25$ and $T = 52.5$.

The computations for a score of 70:

$$Z = \frac{70 - 100}{20} = -1.5$$

$$T = -1.5(10) + 50 = 35$$

The computations for a score of 105:

$$Z = \frac{105 - 100}{20} = 0.25$$

$$T = 0.25(10) + 50 = 52.5$$

Question

Compute phi for the following table, both the probability and the cell count methods, and interpret the results.

			Item j	
			+	−
Item k		−	10	10
		+	25	5

Answer

phi = 0.36

The computations are as follows. Using the probabilities method, calculate each item difficulty (the proportion getting it correct) and the proportion that got both items correct, then put those values into the following formula:

$$phi = \frac{P_{jk} - P_j P_k}{\sqrt{P_j(1 - P_j)P_k(1 - P_k)}}$$

$P_{jk} = 25/50 = 0.5$

$P_j = 35/50 = 0.7$ and $1 - P_j = 0.3$

$P_k = 30/50 = 0.6$ and $1 - P_k = 0.4$

$$phi = \frac{0.5 - (0.7)(0.6)}{\sqrt{0.7(0.3)0.6(0.4)}} = \frac{0.08}{\sqrt{0.504}} = 0.36$$

Using the cell count method, the cell frequencies are:

$a = 10, b = 10, c = 25, d = 5$

Plug them into the following formula:

$$phi = \frac{bc - ad}{\sqrt{(a+b)(a+c)(c+d)(b+d)}}$$

$$phi = \frac{(10 \times 25) - (10 \times 5)}{\sqrt{(10+10)(10+25)(25+5)(10+5)}}$$

$$= \frac{250 - 50}{\sqrt{(20)(35)(30)(15)}}$$

$$= \frac{200}{\sqrt{315000}}$$

$$= 0.36$$

A

Review of Basic Mathematics

You don't need to be an ace in mathematics to learn statistics, and nowadays pocket calculators and computer programs can do much of the mathematical drudgery for you. However, a good understanding of how numbers work, including the basic laws of arithmetic and algebra, is a prerequisite to being able to reason statistically. And that's what learning statistics is about for most people: while anyone can learn to churn out calculations, a process made that much easier with the ready availability of dedicated statistical computing packages, if you don't understand the meaning of the numbers thus produced, your efforts may be useless or counterproductive. Besides, it's always more fun to understand what you are doing, and if you truly understand numbers and can explain them to others, you'll find you have a great advantage over other candidates, whether in school or at work.

If the math you learned in school has faded to a distant memory, don't worry: you have lots of company! Even if you did well in high school algebra, a brief review of the basic concepts may ease your path into statistics, and working through some elementary problems will sharpen your mind before you take on more complex calculations. Running through simple calculations is also a good way to get acquainted with a new calculator or a new software program: start by working with calculations where you already know what the right answer is, and you'll be much more confident with the technology when you use it to tackle more complex problems.

I had a calculus teacher in college who told us that most of the errors people made in class were errors in algebra: not only was he right, but many of the mistakes we made were on material we had learned in junior high school! The same is true in statistics: there's nothing complicated about the math you need, at least at the beginning level, but you need to be very comfortable with the material, and you need it fresh in your mind. So here's a friendly review of some basic mathematics, which I hope will reduce the anxiety and refresh the memories of those who don't quite remember the last time they multiplied exponents or plotted Cartesian coordinates.

If you want to see how much you remember, you can go straight to the quiz at the end of the chapter: if you do well on all the topics, you don't need to do this review section. On the other hand, if you do poorly, you might want to find an algebra review text (aimed at the high school or college freshman market) and work through it, because you may need more of a brush-up than this section can provide. And if you are really ambitious, or if you discover that you like statistics so much you want to major in it, you will eventually need to take several semesters of calculus as well as calculus-based statistics courses. Should that prove to be the case, you should probably start brushing up on your math skills using a good precalculus or college algebra textbook.

Laws of Arithmetic

It's often helpful to think of numbers as points along a number line, in which lower numbers are to the left and higher numbers are to the right. You may remember the number line from primary school (Figure A-1).

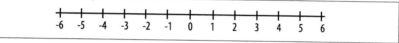

Figure A-1. Number line

The concept of the number line is useful in probability, because we often refer to a value in a distribution as being "further to the right" when what we really mean is "of a higher value." The statement that a value is "at least as extreme" or "at least as far from the mean" as another value, which you will frequently encounter in hypothesis testing, also refers to the number line: for a distribution with one most common central value (in the case of the normal distribution, that value would be the mean), as values get further from that central value (further to the left or right) they become less likely.

Numbers may be written with either a positive or negative sign; if no sign is included, positive value is assumed. The *absolute value* of a is written $|a|$ and means the distance a occupies on the number line, whether in a negative or positive direction. This means that if $a = -5$ and $b = 5$, the absolute value of a and b are identical, i.e., $|a| = |b| = 5$. Another way to look at it is that the absolute value of the number is the same as the number, after any negative sign is removed. By this rule, $|-5|$ is larger than $|4|$ even though 4 is larger (further to the right) than -5, because 5 (the absolute value of $|-5|$) is larger than 4, the absolute value of $|4|$.

To add numbers with like signs, add their absolute values and keep the sign:

$$3 + 5 + 8$$
$$-3 + -5 = -8$$

To add two numbers with different signs, subtract their absolute values and keep the sign of the number with the larger absolute value:

$$-3 + 5 = 2$$
$$3 + -5 = -2$$

To add more than two numbers with different signs, group them by signs, add the absolute values of each set, then subtract the negatives from the positives:

$$-3 + 5 + -2 + 4 = (5 + 4) - (3 + 2) = 4$$

As you can see, adding a negative number is the same as subtracting a positive number. This is formalized in the following law:

$$a - b = a + -b$$

So:

$$2 - 5 = 2 + (-5) = -3$$

To multiply numbers with like signs, multiply their absolute values. If all values are positive, the result is positive. If all are negative, count the number of negative signs. If there are an even number of negative signs, the result is positive; if an odd number, the result is negative:

$$4(2) = 8$$
$$-4(-2) = 8$$
$$-4(-2)(-3) = -24$$

To multiply numbers with unlike signs, multiply the absolute values and then count the number of negative signs: if even, the result is positive; if odd, the result is negative:

$$-4(2)(-3) = 24$$
$$-4(2)(3) = -24$$

To divide numbers with like signs, divide the absolute values and make the result positive. To divide numbers with unlike signs, divide the absolute values and make the result negative:

$$10/5 = 2$$
$$-10/-5 = 2$$
$$10/-5 = -2$$

Order of Operations

In general, we solve arithmetic expressions from left to right, but perform arithmetic operations within an expression in the following order:

1. Anything in parentheses
2. Exponents and roots
3. Multiplication and division
4. Addition and subtraction

Legions of school children have learned this by the mnemonic "Please excuse my dear Aunt Sally," i.e., parentheses, exponents and roots, multiply, divide, add, subtract. If there are multiple layers of parentheses, you solve each expression beginning with the innermost parentheses. Table A-1 shows some examples.

Table A-1. Order of operations examples

Expression	Rule	Result
$2 + 5 \times 10$	Multiplication before addition	52
$(2 + 5) \times 10$	Expressions in parentheses first	70
10×2^2	Exponents before multiplication	40
$(10 \times 2)^2 + 5$	Expressions in parentheses first, then exponents, then addition	405
$10 - 4/(2 + 2)$	Expressions in parentheses first, then division, then subtraction	9
$[5 + 3(4 + 6)]/(3 + 2)$	Innermost parentheses first	7

Properties of Real Numbers

Real numbers are the type of numbers familiar from everyday life and that are used most often in math and statistics. They can be written using decimals and therefore include rational numbers such as 4 and 7/5, and irrational numbers such as π (3.1415....) and the square root of 2 (1.4142...), but not imaginary or complex numbers (numbers that are negative when squared). Unless otherwise specified, real numbers are assumed throughout this review. Some properties of real numbers include:

- The *associative property* for addition and multiplication:

  $(a + b) + c = a + (b + c)$ so $(1 + 2) + 3 = 1 + (2 + 3) = 6$
  $a(b \times c) = (a \times b)c$ so $2 \times (3 \times 4) = (2 \times 3) \times 4 = 24$

- The *commutative property* for addition and multiplication:

  $a + b = b + a$ so $5 + 4 = 4 + 5 = 9$
  $a \times b = b \times a$ so $2 \times 3 = 3 \times 2 = 6$

- The *distributive property* of multiplication:

  $a(b + c) = ab + ac$ so $5(2 + 3) = 5(2) + 5(3) = 10 + 15 = 5(5) = 25$

- The *additive identity* of 0: any number plus 0 = the number itself:

  $a + 0 = a$ so $5 + 0 = 5$

- The *multiplicative identity* of 0: any number times 0 = 0:

  $a \times 0 = 0$ so $5(0) = 0$

- The *multiplicative identity* of 1: any number times 1 = the number itself:

  $a(1) = a$ so $5(1) = 5$

- The *inverse property of addition*: the sum of any number and its inverse is 0:

  $a + {-a} = 0$ and $-a + a = 0$ so $5 + {-5} = 0$ and $-5 + 5 = 0$

- The *rule of double negatives*: pairs of negatives cancel each other out:

  $-(-a) = a$ so $-(-5) = 5$

- The *inverse property of multiplication*:

  $a \times (1/a) = 1$ if $a \neq 0$ (because division by 0 is undefined) so $5 \times (1/5) = 1$

Exponents and Roots

An exponent tells you to multiply the base number by itself as many times as the exponent says:

- $a^n = a \times a \times a \ldots n$ times, where a is the *base* and n the *exponent*, so $2^4 = 2 \times 2 \times 2 \times 2 = 16$
- a^2 is often referred to as "*a* squared" and a^3 as "*a* cubed"; they can also be read as "*a* to the second power" or "*a* to the second" and so on, and this system is used for powers above 3 (a^7 would be read as "*a* to the seventh power").
- *Multiplying exponential numbers* with a common base: add the exponents and keep the base:
  $$a^m \times a^n = a^{m+n} \text{ so } 3^2 \times 3^3 = 3^{2+3} = 3^5 = 243 = 9 \times 27$$
- *Power rules* for exponents:
  $$(a^m)^n = a^{mn} \text{ so } (2^2)^3 = 2^6 = 64 = 4 \times 4 \times 4 \text{ or } 4^3$$
  $$(ab)^n = a^n b^n \text{ so } (5 \times 4)^2 = 5^2 \times 4^2 = 400 = 25 \times 16$$
  $$(a/b)^n = a^n/b^n \text{ so } (3/4)^2 = 3^2/4^2 = 9/16 \text{ assuming } y \neq 0$$
- *Zero exponent*: any number other than 0, with an exponent of 0, = 1:
  $$a^0 = 1 \text{ so } 245^0 = 1 \text{ and } -8^0 = 1 \text{ } (0^0 \text{ is undefined})$$
- A *negative exponent* is the same as dividing by the base raised to the power of the exponent:
  $$a^{-1} = 1/a \text{ and } a^{-2} = 1/a^2 \text{ so } 2^{-1} = 1/2 \text{ and } 2^{-2} = 1/2^2 = 1/4$$
  $$(a/b)^{-n} = (b/a)^n \text{ so } (5/3)^{-2} = (3/5)^2 = 9/25$$
- When *dividing exponential numbers* with a common base, subtract the exponents:
  $$a^m/a^n = a^{m-n} \text{ (assuming } a \neq 0) \text{ so } 3^5/3^2 = 3^{5-2} = 3^3 = 27$$

Taking the root of a number is the inverse of raising it to an exponential value: the *n*th root of x is the number such that $a^n = x$. This may be easier to understand if we consider the *square root*, which is the second root of a number. So we can say that the square root of 9 is 3, because $3^2 = 9$. This can also be written or, if not specified, the second root is assumed. Similarly, the third root of 125 is 5 because $5^3 = 125$. This can be written formally as third roots are often referred to as *cube roots*; beyond 3, the usual terminology is "fourth root," "fifth root," and so on.

Properties of Roots

$$\sqrt[n]{ab} = \sqrt[n]{a}\sqrt[n]{b} \text{ so } \sqrt{4 \times 16} = \sqrt{4}\sqrt{16} = 2 \times 4 = 8$$

$$\sqrt[n]{\frac{a}{b}} = \frac{\sqrt[n]{a}}{\sqrt[n]{b}} \text{ so } 3\sqrt[3]{\frac{27}{64}} = \frac{\sqrt[3]{27}}{\sqrt[3]{64}} = \frac{3}{4} = 0.75 \text{ where } b \neq 0$$

$$\sqrt[n]{a^m} = \sqrt[n]{a}^m = a^{\frac{m}{n}} \text{ so } \sqrt[3]{8^2} = \left(\sqrt[3]{8}\right)^2 = 8^{\frac{2}{3}}$$

$$= 4 \text{ if } n \text{ and } m \text{ are positive integers and } a \geq 0$$

Note that every positive number has two square roots, one positive and one negative. For instance, 2 and –2 are both square roots of 4, because $2^2 = 4$ and $–2^2 = 4$. The positive square root is called the *principal square root* and, unless otherwise specified, is the one usually taken.

A logarithm is the power you need to raise a given base to, in order to produce a particular number. For instance, log_{10} 100 = 2 because $10^2 = 100$. In this example, 10 is the base and 2 is the logarithm. Although any number can serve as a base, in statistics we often work with base-e exponential functions. These are also called natural logarithms or Naperian logarithms and are written ln x which means log_e x. The base e is the irrational number 2.718… and is useful to describe many processes in the natural sciences, hence the name "natural log." Scientific calculators usually have an LN key to calculate natural logs, and many computer programs have built-in functions for the same purpose. Be forewarned, however: sometimes the function to compute a natural log is abbreviated LOG rather than LN because natural logs are more common than base-10 logs in most scientific work.

The equation ln x = 1.5 is equivalent to writing $e^{1.5}$ = x. In this case, x = 4.48 (rounded) because $e^{1.5}$ = 4.48 and we can say that the natural log of 4.48 is 1.5. The following principles hold for logarithms of whatever base (the base is signified by b in these examples):

- log_b 1 = 0 because $b^0 = 1$ (any number to the 0th power = 1)
- log_b b = 1 because $b^1 = b$ (any number to the first power equals itself)
- log_b b^x = x (because by definition the log of b^x is x if the base is b)
- $b^{log_b x}$ = x (where x > 0) because log_b x is the exponent to which you raise b to get x

The following properties of logarithms are also useful in statistics:

- log_b MN = log_b M + log_b N (the logarithm of a product is the sum of the logarithms)
- log_b M/N = log_b M – log_b N (the logarithm of a quotient is the difference of the logarithms
- log_b M^p = p log_b M

You can easily prove these principles to yourself using a pocket calculator. For instance, using natural logs:

ln (2×4) = ln 8 = ln 2 + ln 4 = 0.693 + 1.386 = 2.079
ln(2/5) = ln 0.4 = ln 2 – ln 5 = 0.693 – 1.609 = –.916
ln 2^3 = = ln 8 = 3 ln 2 = 3(0.693) = 2.079

Note that logarithms for numbers between 0 and 1 are negative, and logarithms for numbers less than 0 are undefined (you'll get an error message on your calculator if you try to find ln –1).

Solving Equations

The following *properties of equality* will help you solve equations:

- If $a = b$, then $a + c = b + c$ (adding a constant to both sides of an equality does not change the equality)
- If $a = b$, then $a - c = b - c$ (subtracting a constant from both sides of an equality does not change the equality)
- If $a = b$, then $ac = bc$ (multiplying both sides of an equality by a constant does not change the equality)
- If $c \neq 0$, then $a/c = b/c$ (dividing both sides of an equality by a nonzero constant does not change the equality)

These properties come in handy, as do the properties of real numbers listed above, when solving linear equations. For instance, to solve:

$$5(x - 4) = 40$$

Multiply out the left side:

$$5x - 20 = 40$$

Then "isolate" x by adding 20 to both sides:

$$5x = 60$$

Then divide both sides by 5:

$$x = 12$$

To check the solution, we substitute 12 back into the original equation:

$5(12 - 4) = 5(8) = 40$, which is correct.

For more complex problems, we need to *combine like terms* as follows:

$$2(3x + 1) = 5(x + 2)$$
$$6x + 2 = 5x + 10 \qquad \text{Multiply out both sides}$$
$$x + 2 = 10 \qquad \text{Subtract } 5x \text{ from both sides}$$
$$x = 8 \qquad \text{Subtract 2 from both sides}$$
$$2(24 + 1) = 5(8 + 2) = 50 \qquad \text{Check: substitute 8 into the original equation}$$

Logarithms are useful for solving equations that include exponents: you take the log of both sides, then use the properties of logarithms to solve for the unknown. For instance, assuming a base of 10:

$$5^x = 3$$
$$\log_{10} 5^x = \log_{10} 3$$
$$x \log_{10} 5 = \log_{10} 3$$
$$x = \log_{10} 3 / \log_{10} 5 = 0.683$$
$$\text{Check: } 5^{.683} = 3$$

Systems of Equations

A *system of equations,* also known as a *system of simultaneous equations,* is a set of algebraic equations. The goal is to find a common solution, i.e., values for the variables that will be correct for all equations in the system. If there is a common solution (which is the case with all the systems presented here), the system is called *consistent*; if not, the system is called *inconsistent*. Systems of equations can be solved by graphing (by drawing the lines represented by the equations: the solution is the point of intersection) or by using algebra: we will present only the latter method here.

Solving systems of equations with two unknowns is a good review of algebra and logical reasoning, which will stand you in good stead even if you plan to do all your statistics using a computer package. A simple approach to solving systems of equations, which will work for the examples presented here, is to simplify each equation as much as possible, then use either the method of substitution or the method of addition (or subtraction) to solve the system. We'll demonstrate with systems of *two equations in two unknowns,* although the same principles can be used to solve larger systems, such as three equations in three unknowns. That's about the point, however, when it becomes more convenient to solve more complex problems using matrices, a topic that is beyond this basic review.

Here is a demonstration of the *method of substitution* used to solve a system of two equations in two unknowns (the unknowns are x and y):

$$2x + y = 6$$
$$3x - 2y = 16$$

Solve the first equation for y:

$$y = 6 - 2x$$

Substitute this value into the second equation:

$$3x - 2(6 - 2x) = 16$$

Solve the second equation for x:

$$3x - 12 + 4x = 16$$
$$7x = 28$$
$$x = 4$$

Substitute this value into the first equation to solve for y:

$$y = 6 - (2 \times 4) = -2$$

So the solution is $(4, -2)$ i.e., $x = 4$, $y = -2$. Check by substituting these values into the equations:

$$2(4) + (-2) = 6$$
$$3(4) - (2 \times -2) = 16$$

To use the *method of addition* (or subtraction) to solve the same system of equations, you add or subtract the like terms from the two equations so that one of the variables drops out, then solve for the other variable. An additional step is often necessary, which is to multiply one or both equations by a constant so that one of the variables (x or y) will drop out when the systems are added or subtracted. In this case, we multiply the first equation by 2:

$$2[2x + y = 6] = 4x + 2y = 12$$

We then substitute this equation (which is equivalent to the original expression, since all we have done is multiply both sides by a constant) in the system and add it to the second equation:

$$4x + 2y = 12$$
$$+3x - 2y = 16$$
$$7x = 28 \text{ so } x = 4$$

We can then use this value to solve either equation for y:

$$2(4) + y = 6 \text{ so } y = -2$$
$$3(4) - 2y = 16 \text{ so } y = -2$$

This gives us the same solution as with the substitution method: $(4, -2)$.

Graphing Equations

Points in multidimensional space are often described using *Cartesian coordinates*, also called *rectangular coordinates*, which are simply the values on each dimension in a system that locate a particular point. We will demonstrate this system using two dimensions, because that is easier to display on a printed page, but the same concepts can be applied to higher numbers of dimensions.

Identifying the location of points in two-dimensional space is done using a plane with two axes, x and y; each axis is a number line, and they intersect at 0. The y-axis is vertical, the x-axis is horizontal, and together they divide the plane into four quadrants, as shown in Figure A-2.

The point (3, 5) has an x-value of 3 and a y-value of 5, as shown in Figure A-2. Locating a point in this way is similar to finding a location on a map using rectangular coordinates: run a straight line vertically at $x = 3$, another horizontally at $y = 5$, and where the two lines intersect will be the point (3, 5). The values of Cartesian coordinates are always given in the order (x, y) and for this reason are sometimes called *ordered pairs*. The location of any point in the plane can be uniquely identified in this way, by specifying values for x and y. It can be seen that points in Quadrant I will have positive values for both x and y, those in Quadrant II will be negative on x and positive on y, those in III will be negative on both x and y, and those in IV will be positive on x and negative on y.

Linear equations are sometimes written in the form $y = mx + b$, where m is called the *slope* and b is the *y-intercept*: not surprisingly, this method of notation is called the *slope-intercept* form of a line. To plot a linear equation (one that does not include squares or higher-order terms) using Cartesian coordinates, find two or more pairs of coordinates that satisfy the equation and draw a straight line connecting them. Here's a simple example:

$$y = 2x + 4$$

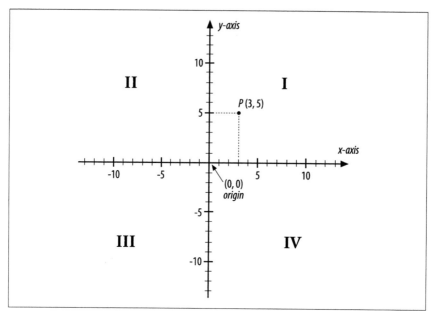

Figure A-2. Cartesian coordinate system

Here are some possible solutions (there are an infinite number!):

$x = 0, y = 4$
$x = 1, y = 6$
$x = -2, y = 0$

Graphing these solutions can be done as in Figure A-3.

The interpretation of the line's components, which are the same as will be used to interpret linear regression equations, is:

Slope
　　The amount of increase in y for a one-unit increase in x

Intercept
　　The value of y when $x = 0$, i.e., the value when the line crosses the y-axis

Even without drawing the graph, you can interpret the equation and predict new values of y given x. Take the following equation:

$y = -3x + 6$

Because the slope is negative, we know that the line will run from the upper left to lower right of the graph (the opposite of the above graph, which had a positive slope. We also know that as x increases, y decreases, and vice versa. The intercept (6) also tells us that the line will cross the y-axis at 6. We can calculate some points on the line as follows (it's often easier to find the x-intercept and y-intercept immediately). Table A-2 shows some possible values.

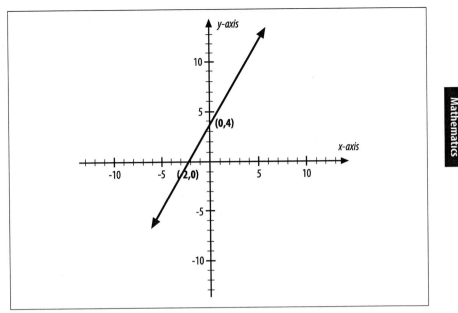

Figure A-3. Line representing the equation y = –2x + 4

Table A-2. Some values for the line y = –3x + 6

x	y
2	0
0	6
1	3

The graph of this equation is shown in Figure A-4.

Another way to write the equation of a straight line is by using what is called the *point-slope* form. This format relies on the fact that if we know the slope of the line and one point on it, we can draw the line and calculate the coordinates of any point on the line. Similarly, if we know two points on the line, we can calculate the slope). To put it another way, a straight line can be uniquely identified by two points, or by one point plus its slope. The point-slope form of a line is written as:

$$y - y_1 = m(x - x_1)$$

where m is the slope of the line and (x, y) and (x_1, y_1) are two points on the line. We can find the slope, given two points on the line, using the formula:

$$m = \frac{y - y_1}{x - x_1}$$

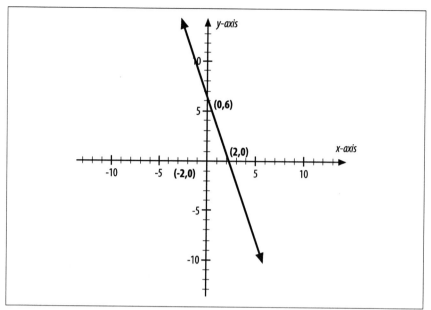

Figure A-4. Graph of the equation y = –3x + 6

You may remember this as "slope = rise over run" where rise is the change in *y*-values (the change on the vertical axis) and run is the change in *x*-values (the change on the horizontal axis) between the two points. If we have the points (0, 6) and (2, 0), the slope of the line that contains them is:

$$m = \frac{6 - 0}{0 - 2} = \frac{6}{-2} = -3$$

This corresponds to what we found in the previous example. If instead our line included the points (6, 6) and (4, 2), its slope would be:

$$m = \frac{6 - 2}{6 - 4} = \frac{4}{2} = 2$$

Continuing with this example, if we knew that a line with slope 2 ran through the point (6, 6), we could find the *y*-coordinate for 4 using the point-slope equation:

$$y - y_1 = m(x - x_1)$$
$$6 - y_1 = 2(6 - 4)$$
$$-y_1 = 4 - 6 = -2$$
$$y_1 = 2$$

Linear Inequalities

An equation is defined by the fact that it connects two expressions with an equals sign, e.g., $y = mx + b$ is the equation of a line. Often we want to connect two expressions with inequalities, stating they are not equal, as in Table A-3.

Table A-3. Commonly used inequalities

Sign, abbreviation	Meaning	Examples
≠, NE	Not equal	$a \neq b, a$ NE 5
<, LT	Less than	$a < b, a$ LT 5
>, GT	Greater than	$a > b, a$ GT 5
≤, <=, LE	Less than or equal	$a \leq b, a <= b, a$ LE 5
≥, >=, GE	Greater than or equal	$a \geq b, a >= b, a$ GE 5
≈	Approximately equal	$a \approx b, a \approx 5$

The alphabetical abbreviations such as "GE" and "LT" are often used in computer programming. Inequalities are often evaluated for their logical or truth value. For instance, if $a = 5$ and $b = 6$, then $a < 6$ and $a < b$ are both true and $a > 5$ and $a > b$ are both false. The following laws govern linear equalities:

1. If the same number is added or subtracted from both sides of an inequality, the inequality remains in the same direction.

   If $a < b$ then $a + x < b + x$ and $a - x < b - x$
   $6 < 10$, so $(6 + 4) < (10 + 4)$ and $(6 - 1) < (10 - 1)$

2. If the same positive number is used to multiply or divide both sides of an inequality, the inequality remains in the same direction.

   If $a > b$ then $ax > bx$ and $a/x > b/x$
   $5 > 3$ so $(5 \times 2) > (3 \times 2)$ and $(5/2) > (3/2)$

3. If the same negative number is used to multiply or divide both sides of an inequality, the direction of the inequality is reversed.

   If $a < b$ then $a(-x) > b(-x)$
   $2 < 4$ so $2(-3) > 4(-3)$ and $2/-3 > 4/-3$, i.e., $-6 > -12$ and $-2/3 > -4/3$

A linear inequality can be solved using the same steps used to solve linear equations. For instance:

$4(3x + 2) < 20$
$12x + 8 < 20$
$12x < 12$
$x < 1$

Fractions

A fraction is simply a way of expressing one number divided by another. The top number is called the numerator and the bottom number is the denominator:

$$\frac{\text{numerator}}{\text{denominator}}$$

The following properties of fractions should be kept in mind (all assume no division by 0):

1. $\dfrac{a}{b} = \dfrac{c}{d}$ if and only if $ad = bc$

2. $\dfrac{a}{1} = a$ and $\dfrac{a}{a} = 1$

3. $\dfrac{a}{b} = \dfrac{ac}{bc}$

4. $-\dfrac{a}{b} = \dfrac{-a}{b} = \dfrac{a}{-b}$

Note that property 3 follows from property 2: anything divided by itself = 1, so multiplying by c/c as in this case is simply multiplying by 1 and does not change the fraction. This property also allows us to *simplify* fractions by dividing out common factors. For instance:

$$\frac{8}{24} = \frac{8 \times 1}{8 \times 3} = \frac{1}{3}$$

$$\frac{4x^3y^2}{2xy^3} = \frac{2x^2}{y} = 2x^2y^{-1}$$

Remembering from our review of exponents that $y^1 = 1/y$.

To *add or subtract fractions*, they need to have a common denominator. You may remember from grade school an exercise called "finding the least common dominator" or "finding the LCD" but for our purposes any common denominator will do. Then you just add the numerators and keep the denominator:

$$\frac{a}{c} + \frac{b}{c} = \frac{a+b}{c}$$

If the fractions don't have a common denominator, you need to multiply or divide as necessary, then do the addition and simplify the result by dividing out the common factors. Therefore:

$$\frac{5}{6} + \frac{2}{4} = \frac{10}{12} + \frac{6}{12} = \frac{16}{12} = \frac{4}{3} \text{ or } 1\frac{1}{3}$$

1 1/3 is called a *mixed number* because it has both an integer part and a fractional part. 4/3 is called an *improper fraction* because its numerator is larger than its denominator. To convert an improper fraction to a mixed number, remove as many whole units (1s in this case) as possible so the final expression is the whole units with the remainder expressed as a fraction:

$$\frac{4}{3} = \frac{3}{3} + \frac{1}{3} = 1\frac{1}{3}$$

To *multiply fractions*, just multiply the numerators and denominators separately and simplify the result:

$$\frac{a}{b} \times \frac{c}{d} = \frac{ac}{bd}$$

$$\frac{9}{5} \times \frac{10}{27} = \frac{9 \times 10}{5 \times 27} = \frac{90}{135} = \frac{2}{3}$$

To *divide fractions*, simply *invert and multiply*. This is because dividing by x is the same as multiplying by $1/x$ (i.e., dividing is the same as multiplying by the *reciprocal* of the divisor). Therefore:

$$\frac{a}{b} \div \frac{c}{d} = \frac{a}{b} \times \frac{d}{c} = \frac{ad}{bc}$$

$$\frac{3}{4} \div \frac{1}{2} = \frac{3}{4} \times \frac{2}{1} = \frac{6}{4} = 1.5 \text{ or } 1\frac{1}{2}$$

Fractions can also be expressed as *decimals* or *percents*. A percent is just a fraction in which the denominator is 100 (*cent* = 100 in Latin). With calculators, it's absurdly easy to convert any fraction to a decimal, then convert it to percent by multiplying by 100, and some calculators even have a special key to return divisions automatically as percents. So:

> 1/4 = 0.25 = 25% (.25 × 100)
> 6/4 = 1.5 = 150%

To take a percent of a number, multiply by the decimal equivalent of that number. For instance, 40% of 30 = 0.4(30) = 12. To calculate an increase over some base number, multiply by 1.0 plus the increase; for instance, calculate a 20% increase by multiplying by 1.2, because multiplying by 1.0 gives you the original number, and multiplying by 0.2 gives you the 20% increase. For this reason, a 100% increase, which is the same as doubling, means multiplying by 2.0 (1.0 for the original number, 1.0 for the increase). To find a decrease from a total, multiply by 1 – the decrease; for instance, to find the number that represents a 10% decrease from 100, multiply 100 by 0.9, so 100(.9) = 90, which is a 10% decrease from 100.

Factorials, Permutations, and Combinations

The *factorial* of a number is simply that number multiplied by all the smaller integers until you get down to 1. The factorial of n is written $n!$ and means $n(n-1)(n-2)$ …(1), so:

> 5! = 5(4)(3)(2)(1) = 120

and:

> 10! = 10(9)(8)(7)(6)(5)(4)(3)(2)(1) = 3,628,800

Calculators often have a factorial key, usually indicated by *!* or *x!*, as well as permutation and combination keys, often indicated by *nPr* and *nCr*: if your calculator has these keys, experiment with them as you work through this section. Fractions that include factorials can often be simplified by canceling common factors, a useful property since factorials quickly become very large numbers, as we saw in the example of 10! above. The utility of canceling common factors should be clear from this example:

$$\frac{10!}{8!} = \frac{(10)(9)(8)(7)(6)(5)(4)(3)(2)(1)}{(8)(7)(6)(5)(4)(3)(2)(1)} = 10(9) = 90$$

Factorials are useful in problems in which you are concerned with arranging a finite number of objects in order. For instance, how many ways are there to arrange five books on a shelf? You have five choices for the first book, four for the second (because the first book has already been "used" and can't be chosen again), three for the third, two for the fourth, and one for the fifth. The answer is therefore 5! = 120.

If you are interested in the number of ways to arrange a subset of objects from a finite set of distinct objects (i.e., they are all different), you can use *permutations* to calculate the answer. In fact, the number of ways to arrange five out of five objects, as in the previous paragraph, is a permutation problem in which the subset is the same as the entire set. But more typically a permutation question deals with something like the number of ways to arrange three books from a set of five. There are several different conventions in permutation notation, so you may see any of the following denoting the number of ways to arrange r objects chosen from a set of n:

$$P(n,r) = nPr = {}^nP_r = \frac{n!}{(n-r)!}$$

The number of ways to arrange three objects selected from five is therefore:

$$5P3 = \frac{5!}{(5-3)!} = \frac{5!}{2!} = 5(4)(3) = 60$$

Note that by convention 0! is defined as 1, not 0, to avoid the problem of division by 0.

In a permutation, the order of objects is significant. If we were arranging sets of three from the first five letters of the alphabet, for instance, (a, b, c) would be a different permutation than (a, c, b). If order is not a concern, we are dealing with combinations rather than permutations. In a *combination*, we are interested in the number of distinct sets of r objects that can be selected from a set of n objects, but do not count different orders of the same objects as a different set. When choosing sets of three from the first five letters of the alphabet, (a, b, c) would be considered the same combination as (a, c, b). Like permutations, there is not one standard notation for combinations and you may see any of the following to denote the number of combinations of r objects from a set of n:

$$C(n,r) = nCr = {}^nC_r = \binom{n}{r}$$

There is a simple relationship between the number of permutations and number of combinations possible when choosing r items from n items:

$$nPr = (r!)nCr$$

Therefore:

$$\binom{n}{r} = \frac{n\mathrm{P}r}{r!} = \frac{n!}{r!(n-r)!}$$

This relationship should be clear if we consider the ways to arrange a selection of two from the first three letters of the alphabet. The six permutations are:

ab ac ba bc ca cb

While the three combinations are:

ab ac bc

because *ab* and *ba* are different permutations but the same combination, as are *ac* and *ca*, and *bc* and *cb*. Mathematically:

$$n\mathrm{P}r = \frac{3!}{(3-2)!} = 3! = 6$$

$$\binom{n}{r} = \frac{3\mathrm{P}2}{2!} = \frac{3!}{2!(3-2)!} = \frac{3!}{2!} = 3$$

Exercises

Here's a review of the concepts in this appendix.

Laws of Arithmetic and Real Numbers

You will get a better diagnosis of your mathematical understanding if you do the first seven sections without using a pocket calculator, i.e., if you use your knowledge of algebra to solve them by hand. In the case of answers with unresolved variables (such as x or y) just restate them in simplest form.

1. $3 + (-8) =$
2. $6/-3 =$
3. $(-8y)(-6z) =$
4. $2 + 5/10 =$
5. $(2 + 5)/10 =$
6. $6 + 3^2 - 5 =$
7. $(3 + 2)^2 =$
8. $[12(5) - 2(3)] / (3 \times 2) =$
9. $-(3 - 5x) =$
10. $6(4 + 2x) - x(5) =$
11. $3(4/x) =$
12. $5x(4 - 2) =$
13. $(5x + 6)(3) =$

Exponents, Roots, and Logarithms

1. $2^0 =$
2. $(1/4)^2 =$
3. $(-x)^4 =$
4. $(x^3)^2 =$
5. $2^2 (2^3) =$
6. $x^5(x^{-2}) =$
7. $(4 \times 2)^2 =$
8. $2^{-1} =$
9. $x^2/x^4 =$
10. $(2/3)^2 =$
11. $(7y^2)^1 =$
12. $(5/9)^{-1} =$
13. $x^5/x^{-2} =$
14. $(27/8)^{-1/3} =$
15. $(4/9)^{1/2} =$

16. $\sqrt{x^4} =$

17. $\sqrt[3]{27y^3} =$

18. $\sqrt{4 \times 16} =$

19. $\sqrt{\dfrac{25}{81}} =$

20. $\sqrt[4]{\dfrac{x^4}{y^6}} =$

21. $e^0 =$
22. $\ln 1 =$
23. $\log_{10} 100 =$
24. $\log_{10} (5 \times 2) =$
25. $\ln e^3 =$

Solving Equations for x

1. $3x + 7 = 20$
2. $(1/3)x = 6$

3. $3(x + 2) = 2(x + 1)$

4. $4x = 3(x - 2) + 7$

Systems of Linear Equations

1. $3x - 2y = 6$ and $x + 2y = 14$

2. $x + 3y = -1$ and $2x + y = 3$

Linear Equations and Cartesian Coordinates

1. Given a line with the equation $y = 3x + 2$, fill in the following table.

x	y
0	
	0
1	
−1	

2. In the equation $y = -x + 5$, what is the slope and what is the y-intercept?

3. Given the equation $y = 6 - 2x$, if x increases by 2, what happens to y?

4. Find the slope for the following pair of points: $(5, 3)$ and $(2, -1)$

5. Given a line with slope −1 that runs through the point $(2, 4)$, find the y-coordinate for the line when it passes through $x = -3$.

Linear Equalities

1. If $a < b$, what is the relationship of $3a$ to $3b$?

2. If $a < b$, what is the relationship of $-2a$ to $-2b$?

3. Solve down to an inequality for x: $5(2x - 1) > 8$

4. Solve down to an equality for x: $3x(2)$ GE 4

Fractions, Decimals, and Percents

1. $\dfrac{3x^2 y}{1} =$

2. $\dfrac{5xy^3 z^2}{6y^5} =$

3. $\dfrac{8}{10} + \dfrac{3}{15} =$

4. $\dfrac{8y^3}{2y} + \dfrac{9y^2}{3} =$

5. $\dfrac{5}{4} \times \dfrac{7}{3} =$

6. $\dfrac{3x}{7} \times \dfrac{2}{x} =$

7. $\dfrac{7}{5} \div \dfrac{14}{10} =$

8. $\dfrac{x}{3} \div \dfrac{2}{3x} =$

9. What is 20% of 75?

10. What is the decimal equivalent of 7/21?

11. If we sold 500 units last year, and sales increased by 10% this year, how many units did we sell this year?

12. If sales declined by 20% this year, how many units did we sell?

Factorials, Permutations, and Combinations

It's OK to use a calculator for this section.

1. 7! =

2. 6P4 =

3. 8C3 =

4. $\dfrac{x!}{(x-1)!} =$

5. How many ways are there to choose a batting lineup (9 players) from 15 players total (order does count)?

6. How many unique combinations (order does not count) of 5 items can you select from 10 unique items?

Answers

Laws of Arithmetic and Real Numbers

1. $3 + (-8) = -5$

2. $6/-3 = -2$

3. $(-8y)(-6z) = 48yz$

4. $2 + 5/10 = 2.5$ or 2 1/2

5. $(2 + 5)/10 = 7/10$ or 0.7

6. $6 + 3^2 - 5 = 10$

7. $(3 + 2)^2 = 25$

8. $[12(5) - 2(3)] / (3 \times 2) = 9$

9. $-(3 - 5x) = -3 + 5x$

10. $6(4 + 2x) - x(5) = 24 + 12x - 5x = 24 + 7x$

11. $3(4/x) = 12/x$ or $12x^{-1}$

12. $5x(4 - 2) = 10x$

13. $(5x + 6)(3) = 15x + 18$

Exponents, Roots, and Logarithms

1. $2^0 = 1$

2. $(1/4)^2 = 1/16$ or 0.0625

3. $(-x)^4 = x^4$

4. $(x^3)^2 = x^6$

5. $2^2 (2^3) = 2^5 = 32$

6. $x^5(x^{-2}) = x^3$

7. $(4 \times 2)^2 = 8^2 = 64$

8. $2^{-1} = 1/2$ or 0.5

9. $x^2/x^4 = x^{-2}$ or $1/x^2$

10. $(2/3)^2 = 4/9$ or $0.444...$

11. $(7y^2)^1 = 7y^2$

12. $(5/9)^{-1} = 9/5$ or $1\ 4/5$ or 1.8

13. $x^5/x^{-2} = x^7$

14. $(27/8)^{-1/3} = 2/3$

15. $(4/9)^{1/2} = 2/3$

16. $\sqrt{x^4} = x^2$

17. $\sqrt[3]{27y^3} = 3y$

18. $\sqrt{4 \times 16} = 2 \times 4 = 8$

19. $\sqrt{\dfrac{25}{81}} = \dfrac{5}{9}$

20. $\sqrt[4]{\dfrac{x^4}{y^6}} = x/y^{3/2}$

21. $e^0 = 1$

22. $\ln 1 = 0$

23. $\log_{10} 100 = 2$

24. $\log_{10} (5 \times 2) = 1$

25. $\ln e^3 = 3$

Solving Equations for x

1. $3x + 7 = 20$: $x = 13/3$ or $4\ 1/3$
2. $(1/3)x = 6$: $x = 18$
3. $3(x + 2) = 2(x + 1)$: $x = -4$
4. $4x = 3(x - 2) + 7$: $x = 1$

Systems of Linear Equations

1. $3x - 2y = 6$ and $x + 2y = 14$: solution = $(5, 4.5)$
2. $x + 3y = -1$ and $2x + y = 3$: solution = $(2, -1)$

Linear Equations and Cartesian Coordinates

1. Given a line with the equation $y = 3x + 2$, fill in the following table.

x	y
0	2
−2/3	0
1	5
−1	−1

2. In the equation $y = -x + 5$, what is the slope and what is the y-intercept? Slope = -1, y-intercept = 5

3. Given the equation $y = 6 - 2x$, if x increases by 2 what happens to y? y decreases by 4

4. Find the slope for the following pair of points: $(5, 3)$ and $(2, -1)$: $4/3$

5. Given a line with slope -1 that runs through the point $(2, 4)$, find the y-coordinate for the line when it passes through $x = -3$: $y_1 = 9$

Linear Equalities

1. If $a < b$, what is the relationship of $3a$ to $3b$? $3a < 3b$
2. If $a < b$, what is the relationship of $-2a$ to $-2b$? $-2a > -2b$
3. Solve down to an inequality for x: $5(2x - 1) > 8$: $10x > 13$ or $x > 13/10$
4. Solve down to an equality for x: $3x(2)$ GE 4: x GE $4/6$ or x GE $2/3$

Fractions, Decimals, and Percents

1. $\dfrac{3x^2y}{1} = 3x^2y$

2. $\dfrac{5xy^3z^2}{6y^5} = \dfrac{5xz^2}{6y^2}$

3. $\dfrac{8}{10} + \dfrac{3}{15} = \dfrac{30}{30} = 1$

4. $\dfrac{8y^3}{2y} + \dfrac{9y^2}{3} = 7y^2$

5. $\dfrac{5}{4} \times \dfrac{7}{3} = \dfrac{35}{12} = 2\dfrac{11}{12}$

6. $\dfrac{3x}{7} \times \dfrac{2}{x} = \dfrac{6}{7}$

7. $\dfrac{7}{5} \div \dfrac{14}{10} = 1$

8. $\dfrac{x}{3} \div \dfrac{2}{3x} = \dfrac{x^2}{2}$

9. What is 20% of 75? 15
10. What is the decimal equivalent of 7/21? 0.333
11. If we sold 500 units last year, and sales increased by 10% this year, how many units did we sell this year? 550
12. If sales declined by 20% this year, how many units did we sell? 400

Factorials, Permutations, and Combinations

1. 7! = 5040
2. 6P4 = 360
3. 8C3 =56

4. $\dfrac{x!}{(x-1)!} = x$

5. In baseball, how many ways are there to choose a batting lineup (9 players) from 15 players total (order does count)? 15P9 = 1,816,214,400
6. How many unique combinations (order does not count) of 5 items can you select from 10 unique items? 10C5 = 252

B

Introduction to Statistical Packages

At some point in your statistics career, you may need to use a statistical package: theoretical understanding and a pocket calculator can take you only so far. Fortunately, we live in an age when there are many types of software available to make the task of doing statistics easier. Most statisticians work with one or more of the standard *statistical packages*, such as SAS or SPSS: a statistical package is basically a collection of software routines with a common interface that has been designed to simplify the job of performing statistical analysis and related tasks such as data management. The main thing to remember with regard to statistical packages is that, like any computer software, they are only a means to an end. Each package has its advantages and disadvantages, and at least at the beginning level you will probably need to use whatever is available at your workplace or at your school. If you then need to learn a new package (say, for a different job) it should not pose great difficulty. If you have a good theoretical understanding of statistics and at least minimal computer aptitude, you can figure out how to use just about any statistical package.

However, starting to work with a new statistical package may seem a daunting task, particularly if your boss or instructor assumes that you are already an expert in it! Printed manuals or online help files may or may not be useful at the very start: a surprising number assume you are already familiar with the software in question, when that familiarity is the very thing you lack. So the purpose of this appendix is to give you a brief overview of several of the most popular packages, with particular emphasis on matters that may be crucial to the new user and/or not clearly stated in most documentation.

Another thing I have tried to accomplish in this appendix is to provide a sense of the particular strengths and weakness of each package, and what are typical uses for each. Of course I can only speak from my experience, and my thoughts are certainly not the last word on the subject. If you are ever in the position of needing to choose a package to purchase for your department that will perform specific functions, you will want to consult several of the many online or published reviews that discuss the relative capabilities of the different statistical packages, such as *http://en.wikipedia.org/wiki/Comparison_of_statistical_packages*.

Minitab

Minitab is a statistical package developed at Pennsylvania State University in the 1980s and now sold by the privately owned company Minitab, Inc. It is commonly used as instructional software in beginning statistics classes and is also used for business and quality improvement applications because it includes routines to compute many statistics and create the charts most commonly used in those contexts. Although Minitab is a proprietary product, a free 30-day trial copy may be downloaded from the company web site at *http://www.Minitab.com*.

Minitab is favored in some beginning statistics classes because it is easy to use: according to the company web site, it is the most common statistical software used for instruction in colleges and universities worldwide. The standard installation includes an extensive system of help files and demonstrations, which also make it popular with beginners. However, the features that make it easy for beginners to learn, such as reliance on a menu interface while offering only limited syntax capability and the provision of only a limited number of analytical choices, may make it unsuitable for more advanced applications.

Minitab can import and export files in several formats, including its proprietary Minitab worksheet (identified with the extension *.mtw*) and Minitab project (*.mpj*) formats, and Excel (*.xls*) and text (*.txt*) files. Data is stored in rectangular files as shown in Figure B-1: rows are numbered and columns are identified as C1, C2, etc. Variable names may be added in the shaded row between the column label and data set. Both data and variable names may be typed directly into the Minitab worksheet.

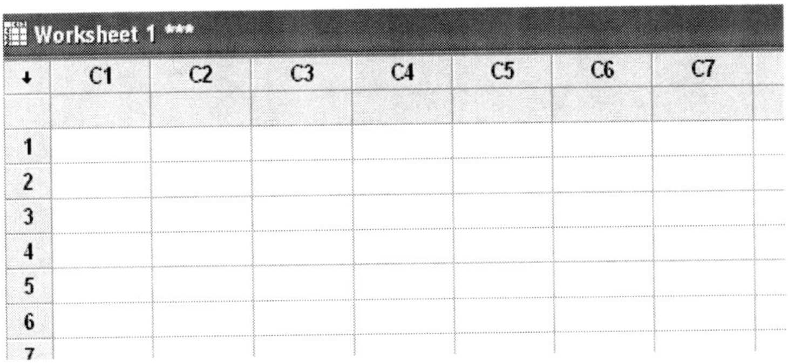

Figure B-1. Minitab worksheet

Commands in Minitab are usually generated through the menu interface; they are recoded in the *session window*, along with output that can be expressed as text. Graphical results are each written to a separate window (which can make for quite a proliferation of open windows during an analysis!). All results plus the data set for an analysis may be saved as a Minitab project, and data sets and graphs may also be saved as separate files in a number of different formats. An excerpt from the session window for a logistic regression analysis is shown in Figure B-2.

```
MTB > Blogistic 'CHD' = CHD CAT AGE CHL SMK ECG;
      T
```

Results for: evans

Binary Logistic Regression: CHD versus CAT, AGE, CHL, SMK, ECG

```
Link Function: Logit


Response Information

Variable  Value  Count
CHD       1         71   (Event)
          0        538
          Total    609


Logistic Regression Table

                                             Odds     95% CI
Predictor     Coef     SE Coef     Z      P  Ratio  Lower  Upper
Constant   -6.76472   1.13218   -5.97  0.000
CAT         0.776079  0.333091    2.33  0.020  2.17   1.13   4.17
AGE         0.0325374 0.0151541   2.15  0.032  1.03   1.00   1.06
CHL         0.0093670 0.0032332   2.90  0.004  1.01   1.00   1.02
SMK         0.828039  0.304211    2.72  0.006  2.29   1.26   4.15
ECG         0.416540  0.292459    1.42  0.154  1.52   0.85   2.69


Log-Likelihood = -201.337
Test that all slopes are zero: G = 35.884, DF = 5, P-Value = 0.000


Goodness-of-Fit Tests

Method            Chi-Square  DF     P
Pearson             588.700  586  0.461
Deviance            397.129  586  1.000
Hosmer-Lemeshow      16.062    8  0.041


Table of Observed and Expected Frequencies:
(See Hosmer-Lemeshow Test for the Pearson Chi-Square Statistic)

                              Group
Value   1     2     3     4     5     6     7     8     9    10   Total
1
  Obs    0     2     5     9     6     8     8     4     6    23      71
  Exp  1.8   2.8   3.7   4.4   5.2   6.3   7.3   8.8  11.5  19.2
0
  Obs   60    59    56    52    55    53    53    57    55    38     538
  Exp 58.2  58.2  57.3  56.6  55.8  54.7  53.7  52.2  49.5  41.8
Total   60    61    61    61    61    61    61    61    61    61     609


Measures of Association:
(Between the Response Variable and Predicted Probabilities)

Pairs        Number  Percent  Summary Measures
Concordant    25869    67.7   Somers' D              0.36
Discordant    11933    31.2   Goodman-Kruskal Gamma  0.37
Ties            396     1.0   Kendall's Tau-a        0.08
Total         38198   100.0
```

Figure B-2. Minitab session window

Running an analysis using Minitab means opening a data set and choosing analytical options from the menu system (often several layers of nesting are involved in
even simple analyses, which can be confusing). Minitab can do many basic
descriptive statistics, graphical displays, power and sample size calculations,
random number generation, and some more advanced statistical analyses such as
linear and logistic regression; however, the options available are often surprisingly limited compared to statistical packages such as SPSS or SAS. Therefore, if
Minitab is under consideration for purchase it is wise to run some proposed analyses using the trial copy to see if these limitations will be a problem for your
proposed uses.

The greatest strength of Minitab may be in quality control and related business
applications: it is the world leader in that context, according to the company web

site. Minitab is often the statistical package taught in conjunction with Six Sigma and similar types of quality improvement training. Specific business and quality control functions are easily produced in Minitab, including DOE (Design of Experiments) analyses, run charts, control charts (Minitab was used to create the control charts for Chapter 18 of this book), time series methods, fishbone diagrams (cause and effect diagrams), Pareto charts, and capability analyses.

Appendix C lists several useful references for Minitab. The Minitab home page (*http://www.minitab.com*) includes a number of tutorials and papers to assist Minitab users, including a downloadable basic textbook (*http://www.minitab.com/support/docs/rel15/MeetMinitab.pdf*). A web search will reveal many independent tutorials and help sites as well. Instructional books about Minitab or that teach it in conjunction with quality improvement include Ryan, Joiner, and Cryer's *Minitab Handbook* (Duxbury), Matthews's *Design of Experiments with Minitab* (American Society for Quality), and Henderson's *Six Sigma: Quality Improvement with Minitab* (Wiley).

SPSS

SPSS is a general-purpose statistical computing package sold by SPSS, Inc., which was first released in 1968. It is widely used by social scientists (the name originally meant Statistical Package for the Social Sciences) and is also used extensively in other areas including health research, business, and education. SPSS might be characterized as offering capabilities somewhere between Minitab and SAS: it is more complex and offers many more analytical possibilities than Minitab, but is more limited than SAS. On the other hand, many beginners find SPSS easier to learn than SAS, as SPSS offers both syntax and a menu interface, offers a spreadsheet interface as its default, and is superior in data formatting and documentation.

SPSS can import and export data in many formats and in nonrectangular configurations; however, the data set is always translated to an SPSS rectangular data file, known as a system file (which uses the extension **.sav*). *Metadata* (information about the data) such as variable formats, missing values, and variable and value labels are stored with the data set. Two views are offered of the data: the spreadsheet-like *Data View* (Figure B-3) and the *Variable View* (Figure B-4), which shows the metadata. You can type directly into either window, so data may be entered directly into the Data View, and variable names, labels, and so on into the Variable View window.

SPSS can be operated entirely through syntax (computer code), which may be typed directly into the syntax window, or written using any text or word processing program and pasted into the syntax window (Figure B-5). SPSS syntax files are stored with the extension **.sps*. This is a great advantage over programs that are entirely menu-driven because the syntax preserves a record of how an analysis was performed, can be shared with others (for instance, by emailing the text file to a collaborator), and can be reused (for instance, to produce daily or weekly reports). SPSS syntax is easy to write and interpret (at least relative to many other programs!), as should be evident in the code excerpt in Figure B-6. You can probably guess what this code is doing without ever having used SPSS.

	id	educ	jobtime	prevexp	minority	salary	salbegin
1	1	15	98	144	0	$57.00	$27.00
2	2	16	98	36	0	$40.20	$18.75
3	3	12	98	381	0	$21.45	$12.00
4	4	8	98	190	0	$21.90	$13.20
5	5	15	98	138	0	$45.00	$21.00
6	6	15	98	67	0	$32.10	$13.50
7	7	15	98	114	0	$36.00	$18.75
8	8	12	98	0	0	$21.90	$9.75
9	9	15	98	115	0	$27.90	$12.75

Data View / Variable View /

SPSS Processor is ready

Figure B-3. SPSS Data View

	Name	Type	Width	Decimal	Label	Values	Missing	
1	id	Numeric	4	0	Employee Code	None	None	E
2	educ	Numeric	2	0	Educational Level (years)	None	0	E
3	jobtime	Numeric	2	0	Months since Hire	None	0	E
4	prevexp	Numeric	6	0	Previous Experience (months)	None	None	E
5	minority	Numeric	1	0	Minority Classification	{0, No}...	9	E
6	salary	Dollar	7	2	Current salary (thousands)	None	None	E
7	salbegin	Dollar	7	2	Beginning salary (thousands)	None	None	E
8								
9								
10								

Data View / Variable View /

Figure B-4. SPSS Variable View

Here's a hint: lines beginning with * are comments, i.e., notes to the programmer rather than executable lines of code. The actual program recodes the continuous variable *exercise* into the dichotomous variable *exerc_cat*, adds labels to the new variable and its values, and creates a cross-tabulation table of the two variables to check that the coding was executed correctly.

However, some people prefer to use the menu interface, and almost any statistical analysis or data management function in SPSS can be accomplished by either means. I prefer to think of the menu system as an alternative way of generating code that can be saved in a syntax file. This lets me enjoy the best of both worlds: I can use the menus to write the syntax for an unfamiliar command, then save the syntax as a record of the analysis performed, which I can also reuse or alter if I desire. The second paragraph of syntax in Figure B-5 was created this way: the tell-tale sign of menu-generated syntax is the capitalized commands (RECODE, VARIABLE LABELS, etc.). To generate syntax using the menu system, make all relevant selections in the menu command interface, then select "Paste" rather than "OK" as the final step, as shown in Figure B-6. This results in the syntax being saved in a syntax file, or appended to an existing syntax file, if one is already open. On the other hand, if you simply want to run an analysis and don't care about saving the syntax, click "OK" instead and the analysis executes immediately. The statistical results are the same either way.

```
class demo week 3.sps - SPSS Syntax Editor
File  Edit  View  Data  Transform  Analyze  Graphs  Utilities  Run  Add-ons  Window  Help

* 2) Recoding.
* If not using predetermined categories, look at the distribution of your data first (but don't print huge frequency tables!).
freq exercise/histogram normal/stats = mean median.
* NB: no missing values in this data set (missing < 0).
recode exercise (0 thru 5= 1) (5.01 thru hi = 2) into exerc_cat.
val labels exerc_cat 1 "0-5 hrs/wk" 2  "> 5 hrs/wk".
var label exerc_cat "exercise categories".
* check the recoding.
crosst exercise by exerc_cat.
freq exerc_cat.




* Recoding through the menus: Transform/Recode into different variables.
RECODE   exercise
  (0 thru 4.8=1)  (4.801 thru Highest=2)  INTO  exc_cat2 .
VARIABLE LABELS exc_cat2 'categorized exercise'.
EXECUTE .
crosst exercise by exc_cat2.
freq exc_cat2.
```

Figure B-5. SPSS syntax window

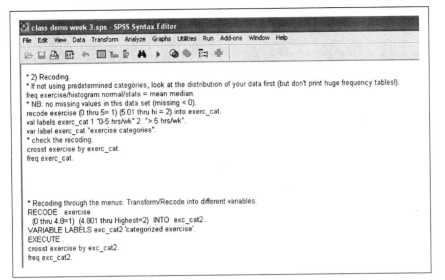

Figure B-6. Using the SPSS menu system to generate syntax

It would be impossible in this brief space to enumerate all the types of analyses available; an overview of SPSS capabilities can be found on the SPSS web page (*http://www.spss.com*). It is an expensive program beyond the range of most individuals, but educational prices are lower and often universities obtain a site license so they are able to provide students and employees with access to SPSS for free or at much lower cost.

Appendix C lists several sources of help for the SPSS user. For someone who wants to learn SPSS using the menu system, a useful series of books written by Marija Norusis have been published by SPSS, Inc.: these include the *SPSS 15.0 Guide to Data Analysis* (2007), the *SPSS 15.0 Statistical Procedures Companion* (2006), and the *SPSS 15.0 Advanced Statistical Procedures Companion* (2007). Note that slightly older versions of these books may also be useful, particularly at the beginning level, since many of the changes in new versions of SPSS involve adding new programs rather than changing the more basic procedures. For those wishing to learn SPSS syntax, books available include Boslaugh's *An Intermediate Guide to SPSS Programming: Using Syntax for Data Management* (Sage) and Levesque's *SPSS Programming and Data Management: A Guide for SPSS and SAS Users* (which may be downloaded in PDF format or ordered in book format from *http://www.spss.com/statistics/base/data_management_book.htm*). Tutorials, sample code, and guidance about using SPSS are available from a number of web sites: two of the best are located at UCLA (*http://www.ats.ucla.edu/STAT/spss/*) and Texas A & M University (*http://www.stat.tamu.edu/spss.php*).

SAS

SAS is a statistical software package that was developed at North Carolina State University in the 1960s, and since 1976 has been a commercial product sold by SAS Institute. It is another step up in complexity from SPSS: SAS is somewhat more difficult to use but offers much more in terms of types of analyses available and flexibility in specifying and executing those analyses. The major disadvantage for beginners is that SAS is a syntax-based system and there are so many choices to be made for even a simple analysis that it can overwhelm people who don't have a background in or aptitude for programming. SAS is also less friendly in terms of managing data files and metadata; for instance, it stores formats in files separate from the data file and requires that the format location be specified in the syntax every time the data file is opened (rather than attaching the format information to the data file, as SPSS does). However, SAS has become the standard language in many professional fields, and there is more assistance for learning and using SAS, both from the SAS web page (*http://www.sas.com*) and help desk and from many published books and web sites, than is available for SPSS.

SAS is similar to SPSS in many ways: it is a comprehensive statistical package that can conduct more types of analyses than can possibly be enumerated here; it can read and write data sets in many different formats; and it is prohibitively expensive for an individual to buy but may be affordable if your school or place of business has a site license. The major difference is that SAS is primarily a syntax-based system, with the exception of JMP, a menu-driven, interactive statistics and graphics package that is now sold by a division of SAS. Many statisticians prefer to work with syntax anyway, partly because they (like me!) are so old they learned to use computers before graphical interfaces were available and partly because (as mentioned in the SPSS section above) syntax may be shared and reused. In addition, writing syntax forces programmers to think through their analysis in a way that can be avoided while clicking on menus. To someone just starting out in statistics, however, the lack of a menu interface may seem more of a barrier than an advantage. This may be somewhat ameliorated by using the time-tested method of taking someone else's code and altering it to fit your needs, and there is so much annotated SAS code available on the Internet that you could probably teach yourself to write SAS programs just by using this method.

SAS has three main windows: the *syntax window*, where you can type your syntax or copy it in from another text or word processing program; the *log window*, which contains a record or log of everything done in a particular session, including warnings and other messages from the SAS system; and the *output window*, where output from statistical procedures is sent by default (it can be directed to other locations, such as an *html* or *.rtf* file, through use of the ODS system). To use SAS, you must open an SAS data set or import another type of data (such as a file stored in Excel or text format), submit commands through the syntax window, and check the output in the output window. The log and syntax windows are illustrated in Figure B-7.

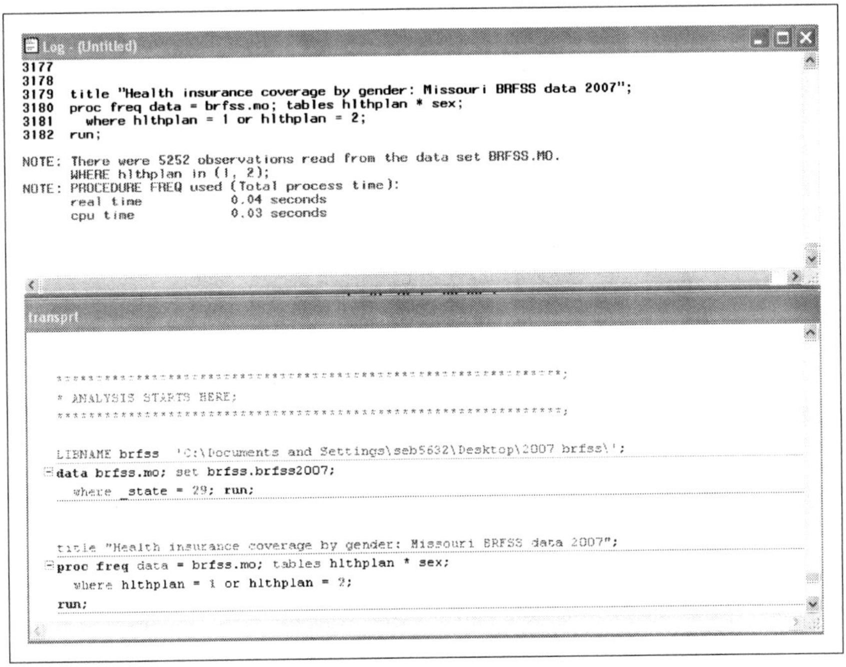

```
Log - (Untitled)
3177
3178
3179    title "Health insurance coverage by gender: Missouri BRFSS data 2007";
3180    proc freq data = brfss.mo; tables hlthplan * sex;
3181      where hlthplan = 1 or hlthplan = 2;
3182    run;

NOTE: There were 5252 observations read from the data set BRFSS.MO.
      WHERE hlthplan in (1, 2);
NOTE: PROCEDURE FREQ used (Total process time):
      real time          0.04 seconds
      cpu time           0.03 seconds
```

```
transprt

***************************************************************;
* ANALYSIS STARTS HERE;
***************************************************************;

LIBNAME brfss 'C:\Documents and Settings\seb5632\Desktop\2007 brfss\';
data brfss.mo; set brfss.brfss2007;
  where state = 29; run;

title "Health insurance coverage by gender: Missouri BRFSS data 2007";
proc freq data = brfss.mo; tables hlthplan * sex;
  where hlthplan = 1 or hlthplan = 2;
run;
```

Figure B-7. SAS log and syntax windows

The syntax window (*transprt*) illustrates three main features of SAS programming. The first is that the location of SAS data files are declared using the *libname* command and the data files themselves referenced with a two-part name: *library. datasetname*. In this case we declared the library *brfss* (the actual name is arbitrary) to exist at the physical location:

C:\Documents and Settings\seb5632\Desktop\2007brfss\

then referenced the data set *brfss.brfss2007* that is stored in that location. The second main point is that SAS programs consist primarily of two types of steps:

1. DATA steps, which open, manipulate, and save data files.
2. PROC steps, which perform statistical analyses on the files.

In this case, our DATA step opened the file *brfss.brfss2007*, selected cases for Missouri (state code = 29), and stored the selected cases in a new file called *brfss.mo*. We then created a crosstabulation table for the variables hlthplan and sex using this new data file, selecting only cases with a value of 1 or 2 on the variable hlthplan.

The log window (*Log – (Untitled)*) echoes the syntax submitted and also contains messages from the SAS system. Messages in the output window tell us that 5,252 cases were used to create the cross-tabulation table, and give us information about processing time and CPU usage.

An excerpt from the output created by this syntax is shown in the SAS output window in Figure B-8.

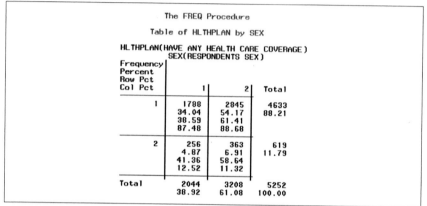

Figure B-8. SAS output

To the far left of the log, syntax, and output windows are two other windows that may be toggled between by use of the tabs in their lower corners. The *Results* window (Figure B-9) shows an outline of the results produced during a session: clicking on any folder causes the next greater level of detail to be displayed. The *Explorer* window (Figures B-10 and B-11) allows access to different SAS libraries (any libraries created by the user, such as *y* in this case, must have been declared by a *libname* command during the current SAS session): clicking on the folders moves the display to the next greater level of detail.

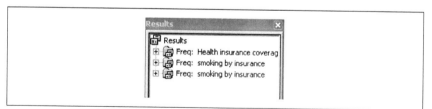

Figure B-9. SAS Results window: output from the Frequency procedure is displayed at one level greater of detail than that of the Means or Corr procedure

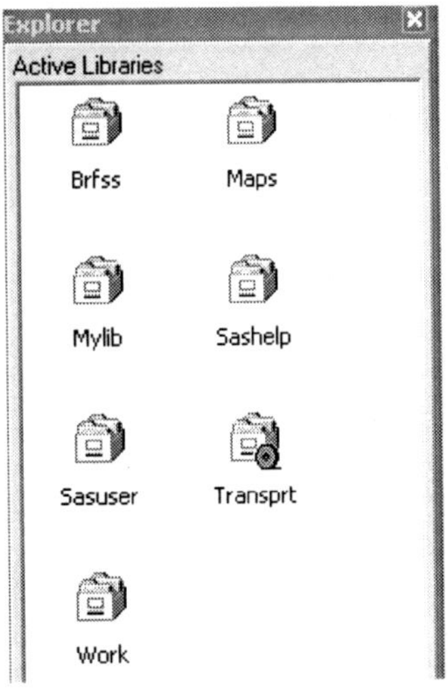

Figure B-10. SAS Explorer window

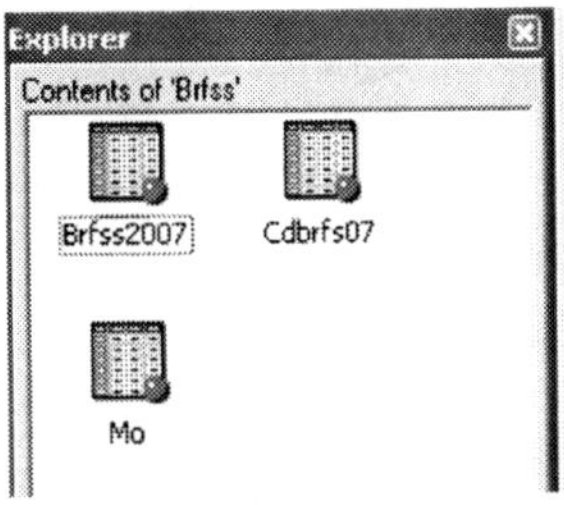

Figure B-11. Contents of a data library (three SAS data files) from SAS Explorer window

Note that it is possible to open an SAS data set in spreadsheet form (as in Figure B-12), which SAS calls *Viewtable* format, by clicking on it in the Explorer window, and that it is possible to enter or edit data directly by this method: normally, however, in SAS these procedures are accomplished using syntax.

Several resources to learn SAS are listed in Appendix C. This is only a beginning: there are many resources available on the Internet to help you learn SAS. Because it is a syntax-based language, examples of annotated code are particularly useful in learning SAS. Good sources of annotated code include the support section of the SAS web site (*http://support.sas.com*), the UCLA web site (*http://www.ats.ucla.edu/stat/sas/modules/*), and the Texas A & M web site (*http://techdocs.tamu.edu/Completed/SASUG*); many more can be found by searching the Internet. There are many more books published about SAS than about any of the other packages discussed in this appendix, so the trick is not finding a book about SAS but finding the book that meets your needs. Two good books for beginners that

VIEWTABLE: Brfss.Mo

	STATE FIPS CODE	GEOGRAPHIC STRATUM CODE	HOUSEHOLD DENSITY STRATUM CODE	PRE-CALL STATUS CODE	REPLICATE NUMBER	REPLICATE DEPTH	FILE MONTH	INTERVIEW DATE
1	29	7	1	1	1061	20	1	01082007
2	29	7	1	1	1062	18	1	01122007
3	29	7	2	1	1068	10	1	01042007
4	29	7	2	1	1069	24	1	01202007
5	29	7	1	1	2061	11	2	02052007
6	29	7	1	1	2062	17	2	02122007
7	29	7	1	1	2063	11	2	02062007
8	29	7	1	1	2064	6	2	02012007
9	29	7	1	1	2071	24	2	02092007
10	29	7	1	1	3061	1	3	03122007
11	29	7	1	1	3063	5	3	03172007
12	29	7	1	1	4061	13	4	04062007
13	29	7	1	1	4061	25	4	04242007
14	29	7	1	1	4062	28	4	04032007
15	29	7	1	1	4065	28	4	04042007

Figure B-12. SAS data set in Viewtable format

include discussions about file types, importing and exporting data, etc., are Delwiche and Slaughter's *The Little SAS Book: A Primer* (SAS), and Cody's *Learning SAS by Example: A Programmer's Guide* (SAS). Cody and Smith's *Applied Statistics and the SAS Programming Language* (Prentice Hall) focuses more on statistical procedures. There are also many discipline-specific SAS books, such as Walker's *Common Statistical Methods for Clinical Research with SAS Examples* (SAS).

R

R is a programming language that functions as a statistical package because of the many pre-written statistical routines (computer code written to perform a particular task) that are available. It differs from the other packages discussed in this appendix because, rather than being a proprietary product sold or licensed by a corporation, it is a product that is freely available for download. R is sometimes described as "GNU-S" because it is an implementation of the proprietary language S-Plus, which is sold by the Insighful Corporation. It is an extremely powerful system, and new routines are being written and made available on the Internet every day by statisticians and programmers all over the world. Graphics available in R are superior to those produced by almost any other system. Another advantage of using R is that every algorithm in R is available to be read and interpreted by anyone (so you can find out exactly what the computer is doing when it executes a command), in contradistinction to proprietary packages such as SAS and SPSS.

Free is a tough price point to beat, so you may wonder why everyone isn't already using R to do statistical work. The answer is that R is much harder to use than the other packages discussed in this appendix, particularly for someone who doesn't have a lot of aptitude or experience as a programmer. Even the R tutorials and help files may be baffling to the naive user. Using R also requires the programmer to think about what they are doing, to a greater extent than programming in SAS. While this is certainly an educational advantage, people who just want to produce a few simple statistics may not feel that the initial difficulty is worth the investment.

On the other hand, if you start out using R at the same time you learn statistics it may be no more difficult to learn than any other package. There are several GUI implementations available, and as R becomes increasingly common, even more user-friendly adaptations may be developed. A sort of natural experiment is currently taking place as R is increasingly being adopted as a teaching language for beginning statistics, so perhaps in 10 years we will be able to answer this question. One thing is certain: as R becomes more popular and is used more as the first language for introductory statistics classes, more instructional materials appropriate to absolute beginners are being written and distributed. And if you are serious about statistics as a career, you need to become familiar with R because it is the most powerful and flexible language available, and may become the *lingua franca* of statistical programming in the near future.

To use R, you first need to download it to your computer. The easiest way to do this is to go to the CRAN (Comprehensive R Archive Network) web page (*http://cran.r-project.org*) and follow the instructions. The next step, unless you are very stout of heart (or already an ace programmer), is to find a good instructional text for R; there are some on the market, and others that may be downloaded from the Internet. You may also want to check out the resources available at *http://www.r-project.org/*.

R is a command-oriented language: you type commands at a command prompt and the R-interpreter responds interactively, either executing the command or giving you an error message. The commands are quite compact compared to those used in SPSS and SAS, and can appear cryptic to the uninitiated; however, once you learn to use R, you will come to appreciate its efficiency. Even more so than with the other languages discussed in this appendix, the best way to get comfortable with R is to get some basic instructional materials and run through some very simple examples on your computer. The R language is really quite logical, but that logic is easier to recognize through use and practice than by reading someone else's explanation.

Another thing you should know about R is that it is an *object-oriented language* (as are Java, C++, and Smalltalk, among others, but in distinction to the other packages discussed in this appendix); this basically means that everything you create in R is an object that can be further manipulated by other commands. An object is also a member of a *class*, meaning that it has certain characteristics and internal organization that allow you to perform operations on it. Again, those are concepts that are easier to understand when you have some experience using an object-oriented language.

Several resources for learning R are listed in Appendix C, and there are more becoming available every day. An Internet search is one good way to turn up resources, since many instructors using R have made their instructional materials freely available. Instructional books for R include Dalgaard's *Introductory Statistics with R* (Springer), Maindonald and Brown's *Data Analysis and Graphics Using*

R: An Example-Based Approach (Cambridge), Braun and Murdloch's *A First Course in Statistical Programming with R* (Cambridge), and Crawley's *A Handbook of Statistical Analyses Using R* (Chapman & Hall). Instructional materials available from the Internet include *Using the R Statistical Computing Environment to Teach Social Statistics Courses* by Fox and Anderson (*http://www.unt.edu/rss/Teaching-with-R.pdf*), Verzani's *Using R for Introductory Statistics* (*http://cran.r-project.org/doc/contrib/Verzani-SimpleR.pdf*), and Baron and Li's *Notes on the use of R for psychology experiments and questionnaires* (*http://www.psych.upenn.edu/~baron/rpsych/rpsych.html*).

Microsoft Excel

Microsoft Excel is, properly speaking, not a statistical package at all, although it is sometimes used as one. Excel is a spreadsheet application produced by Microsoft Corporation that is frequently used for data management because of its ubiquity (it is preloaded on most new computers sold in the United States, for instance), ease of use, and the fact that the major statistical packages have prewritten routines to import and export data in Excel format. Excel also has the capability to produce graphs and charts and perform some statistical analyses, although you should know that Excel has some well-known flaws in statistical accuracy (discussed, for instance, at *http://www.daheiser.info/excel/frontpage.html*), so the advisability of using it for anything beyond the most basic displays and calculations is arguable. On the other hand, Excel may be entirely adequate for your needs, or it may be the software of choice in a class you are taking. Just remember that Excel is a spreadsheet application, not a statistical package, and proceed accordingly. If you have to justify a decision to use Excel versus some other package for analysis or teaching, this is one time I would highly recommend reading some of the (often heated) discussion on this issue, which you can easily find through an Internet search.

Excel stores data in individual *spreadsheets*, which it calls *worksheets*; multiple worksheets are collected into a *workbook*. Individual data points are stored in *cells* (the rectangular boxes in the worksheets) identified by column and row, e.g., cell A1. Both individual worksheets and workbooks use the extension **.xls*. A spreadsheet looks like a rectangular data set, but has many more capabilities, including built-in functions to perform computations on sets of cells such as rows or columns of data. Excel also offers many choices regarding how data is stored, how it appears on the screen, and how it is printed: a given cell, column, or row can be formatted for string or numeric data, to appear in a number of date formats, and so on.

In Figure B-13, you can see a worksheet (Sheet1) within a workbook that includes three worksheets: you maneuver between worksheets by clicking on the tabs at the bottom of the window (labeled Sheet1, Sheet2, and Sheet3 in this example). Rows are horizontal, as in the standard rectangular data set, so we have row 1, row 2, etc. Columns are vertical, so we have column A, column B, and so on. Individual cells are defined by row and column, so the cell in the upper-lefthand corner is A1, the next to its right is B1, and the next below is A2. A1, A2, etc. are called *cell references*.

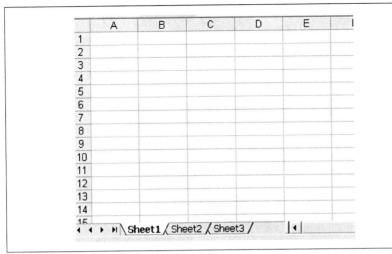

Figure B-13. Microsoft Excel worksheet

Data can be entered simply by typing in the worksheet, in which case Excel applies default formats based on its "best guess" as to the type of data entered. These formats can be changed using the menu commands Format/Format Cells; Figure B-14 shows some of the choices available for date-format data. If you are using Excel to collect data that will be transferred to a different program for analysis, you should be aware that formatting is often lost or garbled in the transfer process. For this reason, particularly when working with time and date variables (which because of their complexity and the different ways they are stored in the different programs, are frequently mistranslated between programs), some researchers prefer to use text format for all Excel data to be imported, and to format it after importation in the program where it will be analyzed.

Variable names can be added in the first row, and many packages have the option to retain those names when importing data (i.e., they will not be confused with data in the new program but will be attached to the data used as variable names). However, because the row containing the variable names is counted as a data row in Excel but not in programs such as SPSS and SAS, the imported file will have one fewer row than the Excel file. This may cause panic, as it appears a case has been lost, although the discrepancy is just due to differing ways of storing data.

Another trap for the unwary when transferring data between systems is the fact that each system has a different set of rules for variable names; it can be disheartening to spend a lot of time entering meaningful variable names in a spreadsheet, only to have them appear as "Var1", "Var2", and so on when the file is imported into a statistical package. If you are going to import variable names, you need to follow the rules of the target program, so if you are going to import the data into SPSS, follow the SPSS naming conventions when entering the names in your Excel spreadsheet. One solution is to use simple names (such as v1, v2, etc.) in Excel and then write code in the target program to add meaningful names to the variables.

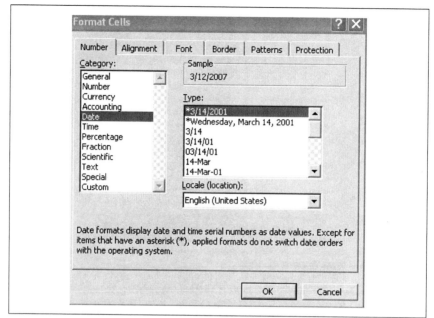

Figure B-14. Some examples of formatting available in Excel

Excel has the capability to create many types of charts and graphs. Somewhat confusingly, to create a chart or graph you insert it into a worksheet, using the menu commands Insert/Chart and making choices from a series of menus. It can then be saved as a separate object if desired and/or inserted into other programs such as Microsoft Word. Types of charts and graphs available include column, bar, line, pie, scatterplot, area, bubble, donut, radar, and stock (hi-low-close).

You can do quite a lot of basic arithmetic in Excel, and the spreadsheet capabilities are particularly useful if you need to do arithmetic on many rows or columns of numbers. Excel includes a number of built-in functions that allow you to compute basic statistics for any collection of cells, and you can also perform arithmetic operations by specifying the equation. In either case, the function or formula is entered into a cell, which will also be used to store the results of the calculation. Note that in the examples below, cells are referred to by their location (row and column). For instance, to add numbers you can specify the equation using the + sign or the function SUM; for large groups of numbers, using a function is more practical. Both methods are shown in Figure B-15, and both produce the identical result (30). Normally cells containing a formula display the result of the calculation, but you can cause the formulas to be displayed by selecting a range of cells and typing CTL ` (hold down the control key and type a backtick); the same sequence will reverse the process, i.e., hide the formulas and display the results. Excel has hundreds of built-in functions: you can find a list of them under "Function Reference" in the help menu or by searching the help files. Note that only the results of formulas, not the formulas themselves, are generally transferred when a data set is moved from Excel to some other program.

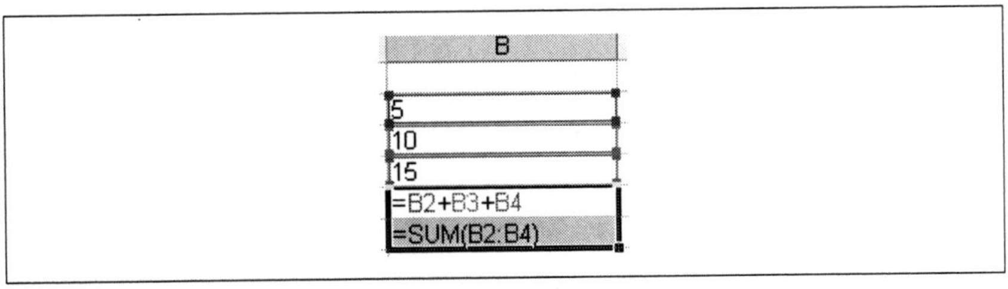

Figure B-15. Two ways to perform addition in Excel

Often, when working with spreadsheets, a formula may be entered once and then copied to other locations; for instance, you may sum one column of numbers, then wish to sum several more columns without respecifying the formula. This can be done quickly in Excel by dragging the formula from one cell to the next: the cell references will change automatically. This is demonstrated in Figure B-16.

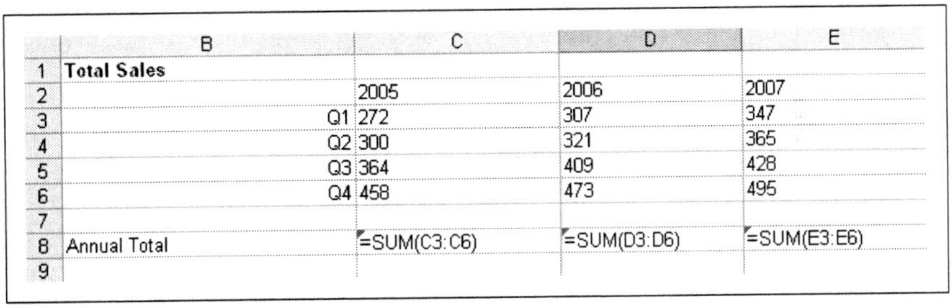

	B	C	D	E
1	Total Sales			
2		2005	2006	2007
3		Q1 272	307	347
4		Q2 300	321	365
5		Q3 364	409	428
6		Q4 458	473	495
7				
8	Annual Total	=SUM(C3:C6)	=SUM(D3:D6)	=SUM(E3:E6)
9				

Figure B-16. Repeating the same formula across columns

Note that cell references can be either *relative* or *constant*. The references in the formulas above are relative: the "Annual Total" line for each column is simply the sum of the four cells in rows 3–6 of that column (which represent quarterly sales totals for that year). When the formula was dragged from column C to column D, the references in the formula were also changed, so "=SUM(C3:C6)" became "=SUM(D3:D6)" and similarly for column E. In some cases you want a cell reference to be constant, meaning that it remains the same when the formula is dragged to a new column or row. To hold a cell constant, precede both the column and row designations by a dollar sign: C1 always refers to cell C1, whether the rest of the formula refers to column A or Z, or row 1 or 100. The reference $C1 would keep the column constant but allow the row to change, while C$1 would keep the row constant but allow the column to change.

Excel includes a set of tools called the Analysis ToolPak, which can be used to do more complex statistical analyses. The Analysis ToolPak is an *add-in* or supplemental program that can perform procedures such as random number generation and exponential smoothing, as well as statistical procedures such as *t*-tests and ANOVA. To access these options, click on Tools/Data Analysis in the Excel menu system: if you don't see the Data Analysis option, Analysis ToolPak has not been installed on your system. Bear in mind also that the accuracy of the results using some of these procedures has been severely criticized (as mentioned above), and the choice to use them should take those criticisms into account.

Appendix C lists several sources of help if you are trying to learn to use Excel for statistical purposes. An Internet search using terms such as "Excel AND statistics" should locate a number of tutorials offering more instruction (and more criticism!) regarding using Excel to do statistical calculations. In addition, there are many handbooks for Excel on the market, and several statistics books have been written that use Excel, including Triola's *Elementary Statistics Using Excel* (Addison Wesley) and Knight's *Analyzing Business Data Using Excel* (O'Reilly).

C

References

Preface

Abelson, Robert P. *Statistics as Principled Argument*. Hillsdale, NJ: Lawrence Erlbaum Associations (1995).

> Abelson, who taught at Yale University for 42 years, provides an excellent presentation of how to think through, and with, statistics.

Frey, Bruce. *Statistics Hacks: Tips and Tools for Measuring the World and Beating the Odds*. Sebastopol, CA: O'Reilly (2006).

> *Statistics Hacks* is a collection of entertaining short essays that use everyday examples to introduce statistical concepts, from testing the randomness or lack thereof in your iPod's "random" shuffle feature to using Benford's law to detect fabricated data.

Huff, Darryl. *How to Lie with Statistics*. New York: W.W. Norton (1954; reprint 1993).

> Originally published in 1954, Huff's work remains a classic introduction to how even the simplest statistical techniques can be used to mislead, confuse, or even outright lie. Readers who can look past the dated examples and (in particular) stereotypical illustrations will find this slim volume an excellent resource and a lot of fun as well.

Salsburg, David. *The Lady Tasting Tea: How Statistics Revolutionized Science in the Twentieth Century*. New York: W.H. Freeman (2001).

> This popular history examines the application of statistics and probability to scientific problems in the twentieth century, shaping the story around the lives and accomplishments of pioneers such as Ronald Fisher, Karl Pearson, and Jerzy Neyman.

Chapter 1

Carmines, Edward G., and Richard A. Zeller. *Reliability and Validity Assessment.* Thousand Oaks, CA: Sage (1979).

One of the earliest entries in the Sage "little green books" series, this volume provides a basic introduction to classical methods to evaluate reliability and validity assessment, plus a brief discussion of factor analytic methods.

Hand, D.J. *Measurement Theory and Practice: The World Through Quantification.* London: Arnold (2004).

Hand provides an excellent discussion of the theory and practice of measurement, including chapters devoted to special problems in the fields of psychology, medicine, the physical sciences, and economics and the social sciences.

Uebersax, John. Kappa coefficients, *http://ourworld.compuserve.com/homepages/jsuebersax/kappa.htm* (accessed April 4, 2008).

Uebersax provides a thorough discussion of the strengths and weaknesses of kappa as part of his discussion of agreement statistics in general (*http://ourworld.compuserve.com/homepages/jsuebersax/agree.htm*).

Chapter 2

Hacking, Ian. *An Introduction to Probability and Inductive Logic.* Cambridge: Cambridge University Press (2001).

This volume was written as an introductory text for philosophy students but will be appreciated by anyone who would like a primarily verbal, rather than mathematical, introduction to the basic ideas of statistics.

Mendenhall, William, et al. *Introduction to Probability and Statistics.* 13th ed. Pacific Grove, CA: Duxbury Press (2008).

This is a popular textbook for probability and statistics for students who have not taken calculus.

Packel, Edward W. *The Mathematics of Games and Gambling.* Washington, D.C.: Mathematical Association of America (2006).

Packel traces the connections between games and gambling (including backgammon, roulette, and poker) and mathematics and statistics, in a manner that assumes only standard high school preparation in statistics. Many illustrations and exercises are included.

Ross, Sheldon. *A First Course in Probability.* 7th ed. Prentice Hall (2005).

Ross provides a basic introduction to probability theory, illustrated with many examples, for students who have taken elementary calculus.

Chapter 3

Boslaugh, Sarah. *An Intermediate Guide to SPSS Programming: Using Syntax for Data Management*. Thousand Oaks, CA: Sage (2004).

Covers the basic aspects of data management, including SPSS code to perform many relevant tasks.

Cody, Ron. *Cody's Data Cleaning Techniques Using SAS Software*. Cary, NC: SAS Institute (1999).

Presents techniques for checking and cleaning data, illustrated with examples using SAS code.

Hernandez, M.J. *Database Design for Mere Mortals: A Hands-On Guide to Relational Database Design*. Upper Saddle River, NJ: Addison Wesley (2003).

A good guide to the theory and practice of setting up databases, discussed in terms of principles applicable to any database rather than instructions in using any particular software product.

Levesque, Raynald. *Programming and Data Management for SPSS 16.0: A Guide for SPSS and SAS Users*. Chicago: SPSS Institute. Available for download from *http://www.spss.com/spss/data_management_book.htm* (accessed April 4, 2008).

Discusses data management tasks in the context of SPSS programming. Assumes more facility with SPSS than the Boslaugh text.

Little, Roderick J.A. and Donald B. Rubin. *Statistical Analysis with Missing Data*. 2nd ed. Hoboken, NJ: Wiley (2002).

Little and Rubin wrote the book on missing data and this is the standard reference on the subject. However, it's not for the faint of heart, and assumes considerable mathematical sophistication on the part of the reader.

Chapter 4

Cleveland, William S. *Visualizing Data*. Summit, NJ: Hobart Press (1993).

Discusses effective graphical presentation of data, with many examples, including a discussion of the visual and psychological principles that lie behind effective graphical presentation of information.

Education Queensland. "The six characteristics of data sets." Available from *http://exploringdata.cqu.edu.au/six_char.htm* (accessed April 4, 2008).

An excellent discussion of histograms and the effect of bin width on their appearance, using the Old Faithful data set (duration of eruptions of the famous geyser in Yellowstone National Park, U.S.A.). The entire site is worth checking out as well: it's intended for secondary school teachers but is so well written that many college students and professionals may find they profit from the examples and presentations.

Robbins, Naomi. *Creating More Effective Graphs*. Hoboken, NJ: Wiley (2004).

An easy-to-use guide that shows good and bad examples of graphs presenting the same information, always with an eye to using graphical techniques to communicate statistical information more effectively.

Tufte, Edward R. *The Visual Display of Quantitative Information*. 2nd ed. Cheshire, CT: Graphics Press (2001).

A landmark that forever changed the way we use graphics to display information. Admirers of Tufte's sometimes contentious approach will want to check out his other works as well, including the most recent, *Beautiful Evidence* (2006).

Chapter 5

Christensen, Larry B. *Experimental Methodology*, 10th Ed. (Hardcover). Allyn & Bacon (2006).

A very readable and comprehensive introduction to research and experimental design with a focus on educational and psychological topics.

Fisher, R.A. *Statistical Methods, Experimental Design, and Scientific Inference: A Re-issue of Statistical Methods for Research Workers, the Design of Experiments, and Statistical Methods and Scientific Inference*. Oxford University Press (Paperback, 1990).

If you want to read the original rationale for many of the designs and issues described in this chapter, there is no better place than the original source.

Chapter 6

Good, Phillip I. and James W. Hardin. *Common Errors in Statistics (and How to Avoid Them)*. Wiley (Paperback, 2006).

An excellent book that shows you how to avoid mistakes in statistical methodology and reasoning.

Darryl Huff's *How to Lie with Statistics*, cited under Chapter 1, is also highly relevant to this chapter.

Chapter 7

Cohen, J. "The World Is Round." *American Psychologist*, 49: 997–1003 (1994).

Classic article by one of the most vocal critics of the enshrinement of alpha = 0.05 as absolute indicator of the statistical significance or lack thereof.

Dorofeev, Sergey and Peter Grant. *Statistics for Real-Life Sample Surveys: Non-Simple-Random Samples and Weighted Data*. Cambridge: Cambridge University Press (2006).

A well-written guide to sampling and the analysis of survey data when simple random sampling is not possible (which is most of the time).

Mosteller, Frederick, and John W. Tukey. *Data Analysis and Regression: A Second Course in Statistics*. Reading, MA: Addison Wesley (1977).

A classic textbook in inferential statistics, including a chapter on data transformation.

National Institute of Standards and Technology. Engineering Statistics Handbook: Gallery of Distributions, *http://www.itl.nist.gov/div898/handbook/eda/section3/eda366.htm* (accessed April 4, 2008).

A nice presentation of 19 common statistical distributions, including ample illustrations, formulas, and common uses for each.

Rice Virtual Lab in Statistics: Simulations/Demonstrations, *http://onlinestatbook.com/stat_sim/index.html* (accessed April 4, 2008).

An Internet site with links to many Java simulations demonstrating statistical concepts, including the Central Limit Theorem, confidence intervals, and data transformations.

Chapter 8

Fisher, R.A. "Applications of "Student's" Distribution." *Metron* 5: 90–104 (1925).

Discusses testing for differences between means using the characteristics of the *t* distribution.

Goset, William Sealy. "The probable error of a mean." *Biometrika* 6 (1): 1–25 (1908).

The original paper describing characteristics of the *t* distribution.

Chapter 9

Holland, Paul W. "Statistics and Causal Inference." *Journal of the American Statistical Association*, Vol. 81, No. 396. (Dec., 1986), 945–960.

Describes the problematic relationship between the need to determine causal inference and the statistical tools available to analyze certain types of data.

Spearman, C. "The Proof and Measurement of Association Between Two Things." *American Journal Psychology* 15 (1904) pp. 72–101.

Probably the most influential paper on measures of association in psychology.

Stanton, Jeffrey M. "Galton, Pearson, and the Peas. A Brief History of Linear Regression for Statistics Instructors." *Journal of Statistics Education*. Volume 9, Number 3 (2001).

An excellent and very readable introduction to the development of ideas underlying correlation and regression.

Chapter 10

Agresti, Alan. *Categorical Data Analysis.* 2nd ed. Hoboken, NJ: Wiley (2002).

> This is the standard textbook for advanced classes on categorical data analysis. It can be heavy going for the beginner, but is clearly written and covers everything from 2×2 tables to linear models.

Conover, W.J. *Practical Nonparametric Statistics.* Hoboken, NJ: Wiley (1999).

> This is one book that lives up to its title: it's a great reference for people who need to learn a particular technique and don't want the theoretical detail provided in Agresti's book.

Chapter 11

Wilcoxon, F. "Individual Comparisons by Ranking Methods." *Biometrics Bulletin* 1: 80–83 (1945).

> The original paper describing the Wilcoxon Mann Whitney-U test for equal sample sizes.

Mann, H. B., & Whitney, D.R. "On a test of whether one of two random variables is stochastically larger than the other." *Annals of Mathematical Statistics*, 18: 50-60 (1947).

> The subsequent paper extending the test to unequal sample sizes.

Chapter 12

Cohen, J., Cohen P., West, S.G., & Aiken, L.S. *Applied multiple regression/ correlation analysis for the behavioral sciences.* (2nd ed.) Hillsdale, NJ: Lawrence Erlbaum Associates (2003).

> An excellent textbook introduction to simple and multiple regression.

Galton, Francis. "Regression Towards Mediocrity in Hereditary Stature," *Journal of the Anthropological Institute*, 15: 246–263 (1886).

> The original paper, which can be found reprinted in Cohen et al (2003).

Chapter 13

Fisher, R.A. "Studies in Crop Variation. I. An examination of the yield of dressed grain from Broadbalk." *Journal of Agricultural Science*, 11, 107-135 (1931).

> The original experiments and formulation underlying ANOVA.

O'Brien, R. G., & Kaiser, M. K. "MANOVA method for analyzing repeated measures designs: An extensive primer." *Psychological Bulletin*, 97, 316-333 (1985).

> A very readable introduction to the key uses and approaches for MANOVA.

Chapter 14

Achen, Christopher H. *Interpreting and Using Regression*. Series: *Quantitative Applications in the Social Sciences*, No. 29. Thousand Oaks, CA: Sage Publications (1982).

An excellent introduction to the correct (and cautious) interpretation of multiple linear regression models.

Iverson, Gudmund R. *Contextual Analysis*. Thousand Oaks, CA: Sage Publications. Series: *Quantitative Applications in the Social Sciences*, No. 81 (1991).

Introduces contextual analysis and the role of variables and groups of variables in model development.

Chapter 15

Bates, Douglas M., and Donald G. Watts. *Nonlinear Regression Analysis and Its Applications*. Wiley Series in Probability and Statistics (1988).

Very practical textbook introduction to curve fitting and nonlinear modeling.

Hosmer, David W., and Stanley Lemeshow. *Applied Logistic Regression*, 2nd ed. New York; Chichester: Wiley (2000).

Specialist coverage of logistic regression and its applications.

Chapter 16

Gould, Stephen Jay. *The Mismeasure of Man*. W.W. Norton & Company (1996).

An excellent book that sets out the historical context of intelligence testing and the (mis)use of various multivariate techniques in the understanding of individual differences.

Hartigan, J.A. *Clustering Algorithms*. New York: John Wiley & Sons, Inc. (1975).

A modern classic, with complete coverage of foundation concepts in clustering, including distance measures. Sufficient detail to implement all of the algorithms.

Chapter 17

Clemen, Roger T. *Making Hard Decisions: An Introduction to Decision Analysis*. Pacific Grove, CA: Duxbury Press (2001).

A textbook that emphasizes the logical and philosophical problems behind decision-making while discussing different approaches to decision analysis.

The Economist Newspaper. *Numbers Guide: The Essentials of Business Numeracy.* Hoboken, NJ: Wiley (1997).

A handy pocket guide to numerical operations useful in business, including index numbers, interest and mortgage problems, forecasting, hypothesis testing, decision theory, and linear programming.

Gordon, Robert J. "The Boskin Commission Report and Its Aftermath." Paper presented at Conference on the Measurement of Inflation, Cardiff, Wales, September 1, 1999. Available online from *http://faculty-web.at.northwestern. edu/economics/gordon/346.pdf* (accessed April 4, 2008).

Summarizes criticisms regarding the U.S. Consumer Price Index, including those identified by the 1995 Boskin Commission report, which suggested that the CPI overstated inflation.

Levitt, Steven D., and Stephen J. Dubner. *Freakonomics: A Rogue Economist Explores the Hidden Side of Everything.* New York: HarperCollins (2005).

New York Times bestseller in which a University of Chicago economist uses economic theory and statistical analysis to examine questions from cheating in sumo wrestling to whether legalizing abortion lowered the crime rate. Although written for the general public, it has been adopted as a textbook at some universities: study guides and other materials, including a link to the author's blog, are available from *http://freakonomicsbook.com/studyguide/ index.html.*

Shumway, Robert, and David S. Stoffer. *Time Series Analysis and Its Applications: With R Examples.* New York: Springer (2006).

A popular time series textbook that includes code in R (a free computer language) to execute times series analyses.

Tague, Nancy. *The Quality Toolbox.* 2nd ed. Milwaukee, WI: American Society for Quality (2005).

A reference book that provides an overview and brief history of Quality Improvement, followed by an alphabetical guide to QI tools, including standard statistical and graphical procedures such as the box plot and hypothesis testing, and more specialized tools such as control charts and fishbone diagrams.

Chapter 18

Cohen, Jacob. (2002). "A Power Primer." *Psychological Bulletin* vol. 112 #1 (July 1, 2002). Available online from *http://www.math.unm.edu/~schrader/biostat/ bio2/Spr06/cohen.pdf* (accessed April 4, 2008).

Very readable introduction to power concepts, prefaced by research by Cohen and other into the neglect of power considerations in published studies.

Ahrens, Wolfgang, and Iris Pigeot, Eds. *Handbook of Epidemiology*. New York: Springer (2004).

> Guide to epidemiology consisting of chapters on specialized topics written by experts in each field. The chapter on sample size calculations and power analysis includes formulas and examples for the most common study designs used in medicine and epidemiology.

Hennekens, Charles H., and Julie E. Buring. *Epidemiology in Medicine*. Boston: Little, Brown (1987).

> An easy-to-read introduction to epidemiology, from basic concepts through study design and types of analysis.

Pagano, Marcello, and Kimberlee Gauvreau. *Principles of Biostatistics*. 2nd ed. Pacific Grove, CA: Duxbury Press (2000).

> An introduction to biostatistics suitable for an undergraduate course; less detailed and easier to use than Rosner's text.

Rosner, Bernard. *Fundamentals of Biostatistics*. 6th ed. Pacific Grove, CA: Duxbury Press (2005).

> An excellent introduction to biostatics for graduate students or those who are willing to grapple with more theoretical details than are provided in Pagano and Gauvreau's text.

Rothman, Kenneth J., et al. *Modern Epidemiology*. 3rd ed. Philadelphia: Lippincott, Wilkins, and Williams (2008).

> A very thorough discussion of epidemiology, including several chapters written by guest authors, for students willing and able to grapple with the subject.

Chapter 19

Crocker, Linda, and James Algina. *Introduction to Classical and Modern Test Theory*. Wadsworth (2006).

> An updated version of a standard textbook that is strongest in its descriptions of models based on classical test theory.

Ebel, R.L. *Measuring Educational Achievement*. Englewood Cliffs, NJ: Prentice Hall (1965).

> This text is the source of the rules to interpret item discrimination that are cited in Chapter 19.

Embretson, Susan, and Steven Reise. *Item Response Theory for Psychologists*. Mahwah, NJ: Erlbaum (2000).

> This is an introductory textbook that takes an intuitive approach to IRT, with many graphical displays and analogies with classic measurement approaches.

Hambleton, Ronald K., et al. *Fundamentals of Item Response Theory*. Thousand Oaks, CA: Sage (1991).

A very clear introduction to item response theory that explains how it overcomes some of the limitations of classical test theory.

Appendix A

"Ask Dr. Math." *http://mathforum.org/dr.math/* (accessed April 4, 2008).

Searchable archive of answers to mathematical and statistical questions, ranging in difficulty from elementary school through college.

Department of Mathematics and Statistics, McMaster University. (2005). *Mathematics Review Manual*. Hamilton, Ontario: McMaster University. Available for download from *http://www.math.mcmaster.ca/lovric/rm/MathReviewManual.pdf* (accessed April 4, 2008).

Provides a review of mathematical concepts, from basic algebra through calculus, along with advice about how to learn and understand mathematics. There are solved problems and a quiz for each topic covered.

Appendix B

Minitab

Henderson, Robin G. *Six Sigma: Quality Improvement with Minitab*. Hoboken, NJ: Wiley (2006).

An introduction to statistics using Minitab for people working in QI.

Matthews, Paul G. *Design of Experiments with Minitab*. Milwaukee, WI: American Society for Quality (2005).

An introduction to statistics using Minitab, with particular emphasis on design of experiments.

Minitab. (May 2007). "Meet Minitab for Windows." Available for download from *http://www.minitab.com/support/docs/rel15/MeetMinitab.pdf*.

Basic manual on using Minitab, created by the manufacturer.

Ryan, Barbara F., et al. *Minitab Handbook: Updated for Release 14*, (2004).

A basic statistics book including examples of analyses in Minitab.

SPSS

Norusis, Marija. *SPSS 15.0 Guide to Data Analysis*. Prentice Hall (2007).

An introduction to data analysis using SPSS, intended for people who will be using the graphical interface rather than syntax. This is the most basic book in a series by this author that uses a common approach to SPSS and statistics. More specialized volumes include the *Statistical Procedures Companion and Advanced Statistical Procedures Companion.* All volumes are regularly updated as new versions of SPSS are released.

Statistics Department, Texas A & M University. "SPSS Tutorials." *http://www.stat.tamu.edu/spss.php* (accessed April 4, 2008).

Web-based tutorials on a variety of SPSS topics.

UCLA Academic Technology Services. "Resources to help you learn SPSS." *http://www.stat.tamu.edu/spss.php* (accessed April 4, 2008).

Includes tutorials, sample code, and guidance to help SPSS users, from beginners to advanced programmers.

The volumes by Levesque and Boslaugh, cited under Chapter 3, are excellent guides to using SPSS syntax.

SAS

Cody, Ron. *Learning SAS by Example: A Programmer's Guide.* Cary, NC: SAS Institute (2007).

A thorough introduction to SAS programming, from data input through macros and SQL, with many examples of code and more coverage of statistical topics than Delwiche and Slaughter.

Cody, Ron, and Jeffrey K. Smith. *Applied Statistics and the SAS Programming Language.* Prentice Hall (2005).

Applied book on using SAS to solve common programming problems, illustrated with examples from business, medicine, education, and psychology.

Delwiche, Lora D., and Susan J. Slaughter. (2003). *The Little SAS Book: A Primer.* 3rd ed. Cary, NC: SAS Institute.

Straightforward introduction to SAS, particularly useful for absolute beginners; coverage of statistical topics is limited.

UCLA Academic Technology Services. "Resources to help you learn SAS." *http://www.ats.ucla.edu/stat/sas/* (accessed April 4, 2008).

Includes tutorials, sample code and guidance to help SAS users, from beginners to advanced programmers.

R and S-Plus

Baron, Jonathan, and Yuelin Li. "Notes on the use of R for psychology experiments and questionnaires." *http://www.psych.upenn.edu/~baron/rpsych/rpsych.html* (accessed April 8, 2008).

Basic introduction to R and code useful for psychology applications.

Dalgard, Peter. *Introductory Statistics with R.* New York: Springer (2002).

Introductory statistics textbook that also introduces the R language and is illustrated with examples using R code.

Everitt, Brian S., and Torsten Hothorn. *A Handbook of Statistical Analyses Using R*. Boca Raton, FL: Chapman & Hall (2006).

A more advanced introduction to R, assuming more statistical understanding than Dalgard or Maindonald and Braun. Topics covered include cluster analysis, multidimensional scaling, principal components analysis, meta-analysis, and recursive partitioning.

Fox, John, and Robert Anderson. "Using the R Statistical Computing Environment to Teach Social Statistics Courses." *http://www.unt.edu/rss/Teaching-with-R.pdf* (accessed April 4, 2008).

A 36-page guide to teaching statistics with R; much more basic than any of the published textbooks in this section.

Maindonald, John, and John Braun. *Data Analysis and Graphics Using R: An Example-Based Approach*. 2nd ed. Cambridge: Cambridge University Press (2007).

Introduction to R and to statistical analysis.

The R Project for Statistical Computing. *http://www.r-project.org/* (accessed April 4, 2008).

The first place to go for information about R, including instructions on how to download the programs, online manuals and FAQs, and an annotated bibliography of reference books and textbooks.

Verzani, John. "Using R for Introductory Statistics." *http://cran.r-project.org/doc/contrib/Verzani-SimpleR.pdf* (accessed April 4, 2008).

A 114-page introduction to R and some basic statistical applications; more elementary than the textbooks cited in this bibliography.

Excel

Heiser, David A. "Microsoft Excel 2000, 2003 and 2007 faults, problems, workarounds and fixes," *http://www.daheiser.info/excel/frontpage.html* (accessed April 4, 2008).

A regularly updated web site discussing, among other things, problems with using Excel as a statistical application.

Knight, Gerald. *Analyzing Business Data with Excel*. Sebastopol, CA: O'Reilly (2006).

A textbook illustrating how to use Excel for common business statistical applications, including modeling, quality control, queuing, and graphical presentation.

Triola, Mario. *Elementary Statistics Using Excel*. 3rd ed. Addison Wesley (2006).

An elementary statistics handbook illustrated with examples using Excel.

Index

We'd like to hear your suggestions for improving our indexes. Send email to *index@oreilly.com*.

prospective longitudinal studies, 86
Proxscal algorithm, 312
proxy measurement, 6
pseudo-chance-level parameter, 387
psychometrics, 366
    composite tests, reliability, 374
    IRT, 383–388
    item analysis, 379–383
    measures of internal
        consistency, 375
    percentiles, 367
    test construction, 370–373
    uses of, 366
p-values, 145

Q

QI (Quality Improvement), 328–335
    choosing a charting method, 334
    control charts, 329
    out of control processes, 334
    products, processes, and
        systems, 328
    statistical process control, 329
quadratic polynomial regression, 288
quality, 328
Quality Improvement (see QI)
quantitative variables, 171
quasi-experimental studies, 86
quota sampling, 134

R

R × C table, 189
r (Pearson's product-moment correlation
    coefficient), 176
R programming language, 424
r2 (coefficient of determination), 180
random error, 8
random measurement error, 373
random selection, 91
randomization, 354
randomized block design, 104
range, 59
rank correlation coefficient, 183
rank sum, 209
ranking, alphabetic versus numerical
    categories, 208
Rasch model, 386
rate, 340

ratio, 340
ratio data, 4
Raudenbush, Stephen, xiii
raw time series, 321
real numbers, 394
recall bias, 17, 86
records, 45
rectangular coordinates, 399
rectangular data file, 45–47
reference groups, 270
regression algorithms, 272
regression coefficients, 132
regression equations, 132
regression towards the mean, 98
regressors, 133
relational databases, 47
reliability, 8–11
reliability coefficient, 375
reliability index, 374
reliability of experimental results, 101
repeated measures ANOVA, 255–257
repeated measures t-tests, 160
replication, 99
research design, 85
    evaluating, 111, 111–113
        coincidence, 113
        controls, 113
        population, 112
        sampling, 112
        variation, 112
    experimental studies (see
        experimental studies)
    mixed designs (see mixed designs)
    observational studies, 86–88
    response variables, 85, 95
response variables, 95, 133
restriction, 354
retrospective longitudinal studies, 86
risk assessment, 326
risk ratio, 348–352
$r\varphi$ (phi correlation coefficient), 186
robustness, 208
rolling average, 321
roots, 395–396
rule of double negatives, 394
run charts, 330

About the Authors

Sarah Boslaugh holds a Ph.D. in Research and Evaluation from the City University of New York and has 15 years of experience as a statistical analyst in a variety of professional settings, including the New York City Board of Education, the Institutional Research Office of the City University of New York, Montefiore Medical Center, Magellan Health Services, and Washington University School of Medicine. She is currently a Performance Review Analyst for BJC HealthCare in St. Louis, Missouri, and teaches a two-semester sequence in Statistics for the Health Sciences at Washington University School of Medicine. Her previous books include *An Intermediate Guide to SPSS Programming: Using Syntax for Data Management* (SAGE) and *Secondary Data Sources for Public Health* (Cambridge University Press). She was also editor-in-chief of *The Encyclopedia of Epidemiology* (SAGE).

Paul Andrew Watters, Ph.D., CITP, is an associate professor in the School of Information and Mathematical Sciences and Centre for Informatics and Applied Optimization (CIAO) at the University of Ballarat. Until recently, he was head of data services at the Medical Research Council's National Survey of Health and Development, which is the oldest of the British birth cohort studies, and an honorary senior research fellow at University College London. He uses multivariate statistics to develop orthogonal and nonorthogonal methods for feature extraction in pattern recognition, especially in biometric applications.

Colophon

The animal on the cover of *Statistics in a Nutshell* is a thornback crab, also known as a spiny spider crab (*Maja squinado*, *Maja brachydactyla*). Found in the northeast Atlantic Ocean and the Mediterranean Sea, the thornback crab is the largest of the European crabs, with a carapace diameter of two to seven inches. It is easily identifiable by the two hornlike spikes between its eyes, and the six or so smaller spikes that extend from each side of its shell. The thornback's body is reddish, with pink, brown, or yellow markings, and its surface is also covered with small spikes, as the crab's name implies.

Thornback crabs are occasionally found on the shore, but they prefer depths of 90 to 600 feet. They are solitary animals except during mating season, when they form large breeding mounds. In years when their numbers are particularly abundant, they can be a source of frustration for lobster fisherman, as they infest the lobster pots. Thornbacks are themselves fished for their delicious claw meat.

Male thornbacks are effective predators; their delicate-looking claws are actually quite powerful and can open small mussels to feed on them. Their claws are also double-jointed, so although it is generally safe for a person to hold crustaceans by each side of their shells, thornbacks are able to reach over their backs to pinch the offender. Females have smaller, less flexible claws and are thus more vulnerable to attack. To defend against their predators—which include lobsters, wrasses, and cuttlefish—many species of spider crabs decorate their spiny shells with seaweed, sponges, or aquatic debris to better blend in against the seabed.

The cover image is from Lydekker's *Library of Natural History*. The cover font is Adobe ITC Garamond. The text font is Linotype Birka; the heading font is Adobe Myriad Condensed; and the code font is LucasFont's TheSansMonoCondensed.